AF838786

The Cuban Tourism Industry

Lexington Studies on Cuba

Series Editor: John M. Kirk, Dalhousie University
and Mervyn Bain, University of Aberdeen

This series will publish texts on all aspects of Cuba, focusing on the post-1959 period. It seeks to be truly interdisciplinary, with studies of all aspects of contemporary Cuba—from foreign policy to culture, sociology, to economics. The series is particularly interested in broad, comprehensive topics (such as women in Cuba, economic challenges, human rights, the role of the media, etc.). All ideological positions are welcomed, with solid academic quality being the defining criterion. In exceptional circumstances edited collections will be considered, but the main focus is on high quality, original, and provocative monographs and innovative scholarship.

Recent Titles in This Series

The Cuban Tourism Industry: Evolution, Challenges, and Prospects
By Paolo Spadoni and José Luis Perelló Cabrera

Afro-Cuban Religions and the Arts: A Dog Has Four Legs But Takes Only One Path
By Alan West-Durán

Citizen Engagement in Cuba: Neighbors and the State in Pogolotti
By James A. Baer

Social and Solidarity Economy in Cuba: Foundations and Practices for Socialist Development
Edited by Rafael J. Betancourt and Jusmary Gómez Arencibia

Cuba, Africa, and Apartheid's End: Africa's Children Return!
By Isaac Saney

Disaster Preparedness and Climate Change in Cuba: Management and Adaptation
Edited by Emily J. Kirk, Isabel Story, and Anna Clayfield

Cuban International Relations at 60; Reflections on Global Connections
Edited by Mervyn J. Bain and Chris Walker

Soviet Influence on Cuban Culture, 1961-1987: When the Soviets Came to Stay
By Isabel Story

Entangled Terrains: Empire, Identity, and Memory of Guantánamo
By Asa McKercher and Catherine Krull

God and the Nation: A Social History of Cuba's Protestants
By James A. Baer

The People's Professors: How Cuba Achieved Education for All
By Kathryn Moody

Youth and the Cuban Revolution: Youth Culture and Politics in 1960s Cuba
By Anne Luke

Cuba's Forgotten Decade: How the 1970s Shaped the Revolution
Edited by Emily J. Kirk, Anna Clayfield, Isabel Story

The Cuban Tourism Industry

Evolution, Challenges, and Prospects

Paolo Spadoni and José Luis Perelló Cabrera

BLOOMSBURY ACADEMIC
NEW YORK • LONDON • OXFORD • NEW DELHI • SYDNEY

BLOOMSBURY ACADEMIC
Bloomsbury Publishing Inc, 1359 Broadway, New York, NY 10018, USA
Bloomsbury Publishing Plc, 50 Bedford Square, London, WC1B 3DP, UK
Bloomsbury Publishing Ireland, 29 Earlsfort Terrace, Dublin 2, D02 AY28, Ireland

BLOOMSBURY, BLOOMSBURY ACADEMIC and the Diana logo are trademarks of Bloomsbury Publishing Plc

First published in the United States of America 2025

Copyright © Bloomsbury Publishing, 2025

Cover images:
© [Alxpin]/iStock
© [Eloi_Omella]/iStock
© [Travel_Motion]/iStock
© [Eloi_Omella]/iStock

All rights reserved. No part of this publication may be: i) reproduced or transmitted in any form, electronic or mechanical, including photocopying, recording or by means of any information storage or retrieval system without prior permission in writing from the publishers; or ii) used or reproduced in any way for the training, development or operation of artificial intelligence (AI) technologies, including generative AI technologies. The rights holders expressly reserve this publication from the text and data mining exception as per Article 4(3) of the Digital Single Market Directive (EU) 2019/790.

Bloomsbury Publishing Inc does not have any control over, or responsibility for, any third-party websites referred to or in this book. All internet addresses given in this book were correct at the time of going to press. The author and publisher regret any inconvenience caused if addresses have changed or sites have ceased to exist, but can accept no responsibility for any such changes.

Library of Congress Cataloging-in-Publication Data on File

ISBN: HB: 978-1-49859-043-3
ePDF: 979-8-76515-513-4
eBook: 978-1-49859-044-0

Typeset by Deanta Global Publishing Services, Chennai, India

For product safety related questions contact productsafety@bloomsbury.com.

To find out more about our authors and books visit www.bloomsbury.com and sign up for our newsletters.

To my wife Inés Avilés-Spadoni, my mother Maria Pia, my brother Mirco, my nephew Matteo, and my dearest friend Luca Gaudenzi.
—Paolo Spadoni

To my wife Teresa.
—José Luis Perelló Cabrera

Contents

Figures

Tables

Foreword

John M. Kirk, Dalhousie University

When Americans think of Cuba, they often conjure up images of vintage 1950s cars, cigars, rum—and perhaps pictures of bearded revolutionary icons such as Fidel Castro or Che Guevara. Older Americans might also remember scenes from "Godfather, Part II," and the designs of the mafia in turning Havana into a second Las Vegas. Most tourists in Cuba today (Canadians, West Europeans, and increasingly Russians) have different experiences—beach-and-sun tourism (forbidden to US tourists by Washington), salsa, and perhaps the ubiquitous Havana Club rum.

Understanding Cuba is never easy—the reality is complex, confusing, and occasionally contradictory. The study of tourism on the island over the past sixty years illustrates this complex reality well. Tourism is one of the most significant hard currency earners for the government—yet for many years was ignored, largely because of its tawdry image from pre-revolutionary times, when Havana was seen as a "sin city," replete with prostitution, gambling, narcotics, mob-run casinos, and pornographic shows. The moralistic revolutionary government that took power on New Year's Day 1959 wanted nothing to do with that shameful past, and so shelved all interest in its commercial potential.

For fully thirty years afterward, tourism was deliberately forgotten by the Castro government. It was only in the early 1990s when the revolutionary government, faced with the traumatic impact of the implosion of the Soviet Union (and the disappearance of billions of dollars of subsidies, as well as its principal source of petroleum and trade), was forced to come up with a rescue plan for the socialist system. Reluctantly, the revolutionary government decided that tourism offered the potential of a significant source of hard currency. Now that the Soviet Union had disappeared from the political scene, sun, sand, and salsa were to be sold to hard currency clients from the Global North.

Since then, we have seen tourism become one of the three principal anchors of the Cuban economy, along with remittances from family members abroad and the exportation of professional services—mainly in healthcare. It has not been an easy process, as this book attests. Fidel Castro was openly skeptical about this newfound development strategy and is alleged to have claimed that tourism would prove to be "a pact with the devil." As the authors of this book show, while tourism has indeed generated significant amounts of hard currency, the challenges associated with developing the tourism industry have been many.

Tourism, however, didn't start in the 1990s or indeed in the pre-revolutionary epoch of glamor, when Ava Gardner, Frank Sinatra, Errol Flynn, and Marlon Brando among many Hollywood stars visited. As the authors of this book point out, the roots of tourism can be traced back to the early days of the Spanish colonial period—when Cuba became a resting place for would-be *conquistadores* about to explore and take control of the Latin American mainland. The military Conquest of Mexico in the sixteenth century gave rise to annual fleets going to and from Spain, both bringing Spanish military and settlers to the mainland of Latin America and transporting them on their return to the "motherland," laden with gold and silver. Given its strategic geographical location, Cuba was the logical place for both sets of expeditions to rest and recuperate before major transportation challenges—and as a result became a crossroads for centuries. In many ways, tourists today are doing what many Spanish did centuries ago, seeking respite on the island.

In the late nineteenth and early twentieth centuries, Cuban tourism reached a new pinnacle, as Americans flooded into the island. The new generation of explorers mainly came in search of trade and investment opportunities. This wave followed the successful American military intervention after Teddy Roosevelt and his "Rough Riders" took advantage of the sinking of the USS *Maine* in 1898 and overthrew the Spanish military on the island. US interests soon controlled the island. It is no accident that the first two leaders of an "independent" Cuba were General John Brooke and General Leonard Wood. US investment flooded into Cuba, commerce flourished, American settlers arrived, bringing a distinctive culture to the island—and tourists started a new phase in the industry. In the early twentieth century, there were over twenty steamship companies taking Americans to Cuba, and air travel to the island started in 1927. Cuba became a wonderful place to invest in—but also to explore on a vacation.

Until the Cuban revolution in 1959 US commercial interests reigned supreme on the island. The sugar industry was controlled by American investors, as were banking and most public services (transportation, electricity, water). The major source of tourism was also from the US mainland—just 90 miles away. Dozens of flights and car ferries operated daily, major construction projects in luxurious hotels took place, and tourists took advantage of this tropical paradise so close—and so Americanized.

All of the major hotels in Havana (built with American financing) boasted a state-of-the-art casino, many of which were owned by US gangster syndicates. In all, twenty-eight new hotels were built from 1952 to 1958, mainly in Havana as tourists flocked to the island for pleasure in what was billed on gaudy posters as a tropical paradise. This was America's playground, the principal attraction in the Caribbean. There were some 270 brothels, with 11,500 sex workers, modern hotels, an attractive and affordable vacation lifestyle for tourists with dollars, and easy access from the East Coast—round-trip airfare was just $36 from Miami.

This all ended in 1959 as dictator Fulgencio Batista was overthrown and young bearded revolutionaries took power. Cuba now witnessed decades in the construction of a revolutionary socialist society, a state-controlled economy, and massive social change. Some 10 percent of the population rejected the new model and voted with their feet, most heading to Florida. The pre-revolutionary tourist model was studiously ignored and indeed vilified: casinos were now illegal, a program to close brothels was implemented, and a campaign against drugs was instituted. The number of foreign tourists understandably plummeted, especially after Washington broke relations with Cuba in January 1961. From 180,000 tourists in 1959, the numbers rapidly dropped—to just 4,180 in 1961.

Following the implosion of the Soviet Union in the 1990s, Cuba witnessed the reluctant embrace of tourism by a desperate government in Havana. Forgotten was the erstwhile scorn for anything that smacked of bourgeois morals. In its place, the government talked about promoting "*turismo sano*" or "healthy" tourism. To a large extent, this simple "beach and sun" philosophy was imposed and was successful. It was even termed the locomotive of the economy, dragging other industries to support its growing capabilities while providing employment. The number of tourist arrivals rose dramatically and brought badly needed funds to the state coffers.

However, while tourism has been one of the mainstays of the Cuban economy since the collapse of the Soviet Union, it has not been without major problems. Tourism numbers continued to increase steadily until 2019 but came crashing to an abrupt standstill in recent years. There were two major problems—the Covid-19 pandemic and US policy under the Trump administration.

Covid-19 was a disaster for the tourism industry in Cuba, as virtually the entire industry shut down. What made it worse was the policy of the Trump administration (2017–21), which adopted an openly aggressive approach toward the island. In many ways, Trump sought to undo completely the major planks of the previous Obama administration program in Cuba—which had rejected the traditional policy of isolating Cuba and reopened diplomatic relations. Under Trump, the tourism industry was targeted in several ways, and American tourism (in essence the logical market, given its geographical proximity) rapidly decreased.

Sadly, the Biden administration (2021–5) failed to follow through on a commitment to overturn many of the key Trump policies on Cuba, and bilateral relations remained in diplomatic limbo. This has had a clear knock-on effect in terms of US tourism to the island: under US law, Americans are not allowed to vacation on Cuba's beaches, and people-to-people travel is forbidden, as is cruise traffic. Nor can they stay in scores of hotels seen as providing financial benefits to the Cuban government, much less consider investment in managing Cuban hotels. European tourists to Cuba who also wanted to travel to the United States now faced increased bureaucratic hurdles from Washington—and as a result, European tourism started to decline. The re-election of Donald Trump in November 2024 understandably poses further unknown challenges for the tourism industry in Cuba.

As a result of almost a decade of recent US policy (and the potential of further challenges under a new Trump administration), Cuba has never returned to the glory years of American tourism it experienced under the Obama years, when hundreds of thousands of Americans came to explore this previously forbidden fruit. This understandably has had a major detrimental impact on the Cuban economy. The data in the book provide a stark reminder of just how the US role had plummeted and how misplaced was the optimistic expectation of Cuban officials about increasing waves of American tourists.

The authors also show, however, that the decline in Cuban tourism is not just the result of the pandemic and US policy. The Cuban government, which clutched onto the concept of beach-and-sun vacations in the 1990s, remains mired in a mindset that continues this strategy, combined also with the construction of luxury hotels in Havana. The trouble is that the hordes of American tourists who arrived during the Obama years were no longer able to come after Donald Trump was elected. The occupancy rates in these state-of-the-art hotels are remarkably low.

In short, Cuban officials reacted late to the radical change in the political climate. As a result, more hotels were built, with the expectation—not fulfilled—that they would be occupied by ever-increasing numbers of tourists. Perhaps more serious, the economic crisis in Cuba led to questions about why expansion in tourist facilities was occurring at a time of major social tensions: shortages of food and medicines, blackouts due to the erratic supply of fuel, an increase in crime, and professionals in the state economy leaving for the private sector. The departure of some 800,000 Cubans for the United States within two years illustrated the level of social frustration. Yet still, the emphasis on building luxury hotels continued, with little innovation in other areas of the tourist industry. Moreover, while other countries in the region have seen their tourism industry rebound (and in fact increase) after Covid, the situation in Cuba has had only moderate success.

A personal note might be of interest, since in some ways it illustrates some of the recent trends noted here. I first went to Cuba in 1976 as a graduate student, and in the almost fifty years since, have returned to Cuba on scores of occasions—for purposes of research, trade and investment consulting, university liaison, diplomacy, and tourism. In the early years of my experiences on the island, the only groups of tourists encountered were "heroes of work" from the Soviet Union and East European socialist countries. They were exemplary workers in their factories and fields who were now rewarded with a vacation on the beaches of their COMECON partner. We drank vodka and talked about legendary soccer and hockey figures.

From the early 1990s (when the first modern hotel was built in Varadero and Spanish investors came to manage the new hotels) until Covid-19 times, my family took vacations at beaches overrun with Canadian vacationers (about 40 percent of tourists to the island), and to a lesser extent Europeans. For a few years (the second Obama administration), while the beaches remained off-limits to Americans, the cities (and particularly Havana) were packed with American tourists, and finding a taxi became a challenge. Most recently (April 2024), I vacationed on the beach—as did over 900,000 Canadians that year—enjoying two weeks of respite from colder weather in the "Great White North." My personal experiences confirm many of the claims made by the authors about tourist trends on the island today.

This book, written by two eminent scholars, has been many years in the making. It offers an extraordinary overview of the history and role of tourism on this island. Perhaps one of its most important features is the extensive use of data to support arguments made—and the number of detailed tables offers a treasure trove of material for any researcher. The comparison of the industry in Cuba and other countries in the Caribbean (Chapters 4 and 5) is also particularly helpful in assessing the recent evolution of different models in several countries in the region.

Just as pertinent is the call for "rebirth rather than recovery" in Chapter 6—since it is clear that the tourism industry needs to strike out in fresh, innovative directions. This will hopefully happen, given the enormous potential of human talent and natural beauty on the island. Cuban tourism reinvented itself in the 1990s after deliberately ignoring it for thirty years. It will be interesting to see what the next decade brings—for Cuba and its tourist industry.

Introduction

Full of natural beauty and recreational attractions, Cuba is one of the most popular tourist destinations in the Caribbean, receiving millions of visitors each year from all over the world. Despite a number of shortcomings, above all critical efficiency problems, the tourism industry is an essential part of the Cuban economy in terms of contribution to the country's gross domestic product (GDP), generation of hard currency revenues, job creation, and other benefits.

International tourism in Cuba began in the early part of the twentieth century, particularly in the 1920s, when Havana, due to its permissive rules toward alcohol consumption, gambling, and other indulgent pastimes, became a favorite destination for wealthy Americans who flocked to Cuba to engage in forbidden pleasures and enjoy winter bouts of horse racing, golfing, and various country club activities. Several hotels were built or renovated in Cuba around this time. Aside from two periods of decline, one during the Great Depression in the first half of the 1930s, and the other during the Second World War, tourism remained an important sector of the island's economy until Fidel Castro came to power in 1959. After the Second World War, Havana experienced a surge in tourist activity with a new cycle of investments in hotels and a growth in tourist flows, with the opening of casinos adding to the attraction. Inbound tourists, especially in the 1950s, were mostly middle-class Americans and celebrities looking for an exciting Caribbean getaway. A tour of Havana at night in the late 1950s would show the great casinos, luxurious hotels, and splendid advertisements up in lights that were symbols of a city of great attractions. Its magnificent tourist facilities gave the appearance of development in full swing. In reality, the comeback of tourism during that decade had a dark side since it occurred with the mafia, gambling, drugs, and prostitution as its backdrop (Villalba 1993, 106–15).

The development of tourism in Cuba virtually ended after the Cuban revolution in 1959. The casinos and illegal businesses were closed. International tourist arrivals on the island plunged dramatically as a consequence of worsening relations with the United States. The latter imposed a comprehensive economic embargo against Cuba in the early 1960s, including travel restrictions that effectively banned US tourism to one of the favorite destinations of US travelers. As Cuba rapidly disappeared from the regional tourist map, other countries pushed to fill the void, namely the Bahamas, Puerto Rico, Jamaica, the US Virgin Islands, and even smaller islands like Martinique, Guadeloupe, and Aruba. During the 1970s, inbound tourist arrivals, primarily from the United States, started to grow tremendously also in the Dominican Republic and in the tourist center of Cancún in the Mexican Caribbean. Meanwhile, Cuba began to receive small groups of foreign tourists, but by and large, they were visitors from Eastern European countries whose trips were motivated by political interests and solidarity. At that time, the Castro government continued to view tourism, above all from Western source markets, in a negative light because of its perceived adverse social, cultural, and ideological effects on Cuban society (Salinas, Mundet, and Salinas 2018, 221). In the 1980s, Canadian tourists, who would soon become the largest group of overseas visitors to Cuba, found it favorable to travel to the island. Various tour operators introduced some Cuban destinations in their catalogues with attractive vacation programs, chartering flights for the trips. Moreover, in 1977, during the administration of Jimmy Carter, the US government essentially lifted the travel ban with respect to Cuba. Travel agencies quickly created tour packages to Cuba, new US-based charter flights launched their operations, and small cruises brought American tourists to Havana (Perelló Cabrera and Betancourt 2018, 63–4). However, the administration of Ronald Reagan reimposed severe restrictions on travel to Cuba in 1982. For the rest of the 1980s, international visitor arrivals in Cuba grew steadily, but their level at the end of that decade was roughly the same as in the late 1950s (Quintana et al. 2004, 100; Villalba 1993, 116).

The early 1990s saw the collapse of the Soviet Union and of socialism in Eastern Europe, which triggered a deep economic crisis in Cuba. The United States tried to capitalize on this situation by intensifying sanctions against the island. In response, as part of a broad program of economic reforms, Cuba decided to target tourism as a vital sector and aggressively promote it in capitalist countries. In order to obtain the necessary capital for investment and to ensure a constant influx of tourists, Cuba invited prominent European hotel chains to form joint ventures with national hotel groups and administer hotels and beach resorts on the island. International tourism in Cuba expanded markedly during the 1990s, and by the mid-2000s, more than two million tourists traveled to Cuba every year, generating over $2 billion in foreign exchange earnings (ONEI 2006). It was around this time that the government of Fidel Castro launched a process of recentralization of the Cuban economy that reinforced state control over hard currency operations and the tourism sector. And after Fidel Castro fell ill in late July 2006 and was replaced as head of state by his brother and former defense minister Raúl Castro, the Cuban military, through its conglomerate GAESA and its sub-entity Gaviota, essentially took over tourism development projects in Cuba.

Annual visitor arrivals continued to grow over the next decade or so to exceed the three million mark, but it was the historic rapprochement between Cuba and the United States during the Obama administration, which was announced in late 2014 and entailed the loosening of restrictions on US-based travel to the island, among other measures, that could have been a true game-changer for the Cuban tourism industry. Commercial flights and cruise ship travel between the two countries resumed. Fueled by US overnight and cruise visitors, and by tourists from Europe and other key source markets, annual visitor arrivals in Cuba rose noticeably over the next few years. However, the Trump administration, beginning in November 2017, moved decisively to reverse Obama's policy with a series of measures that significantly tightened travel restrictions and other US economic sanctions with respect to Cuba. It is worth noting that the harshest Trump-era sanctions against Cuba are still in effect today despite a modest easing during the Biden administration.

In all fairness, tourist arrivals in Cuba from European countries and other markets were already declining when Trump's moves caused a drastic plunge in US trips to the island. As it was coping with the adverse effects of US sanctions, Cuba's tourism sector was hit by the Covid-19 pandemic. The latter wreaked havoc on the tourism industry of Cuba and all Caribbean countries, forcing border closures and the introduction of strict entry rules that practically shut down inbound tourism. Yet, Cuba has struggled in its tourism recovery from the Covid-induced shock far more than its principal Caribbean competitors. It is clear that the Cuban tourism industry has several critical problems that require major changes to be tackled effectively, and that the Cuban economy also needs to be reformed to strengthen the ability of tourism to serve as a driving force for economic growth and development.

This book begins with a historical review of the development of tourism as an important industry in pre-1959 Cuba and the subsequent decline in the first fifteen years of the revolution before a modest recovery in the second half of the 1970s and during the 1980s. It continues with a thorough analysis of the remarkable expansion of international tourism in Cuba in the post-Cold War era and in the era of improved relations with the United States that officially began on December 17, 2014, along with an assessment of the sector's economic impact and its main challenges. An examination of the damaging effects on the Cuban tourism sector of stiffened US sanctions imposed by Donald Trump is also provided. Furthermore, a comparative analysis of tourism in Cuba and in the Caribbean, with a focus on Cuba's main competitors in the region, is carried out to highlight the strengths and weaknesses of the Cuban tourism industry. Finally, this book examines tourism trends in Cuba and the Caribbean in the post-pandemic era and sheds additional light on Cuba's recent tourism performance to underscore key flaws of the Cuban tourism industry and suggest potential adjustments.

1

Historical Development and Trends of Tourism in Cuba

Cuba is an attractive tourist destination not only because of its sandy white beaches and impressive landscapes but also because of its important historical and cultural inheritance. Cuba was Spain's last colony in the Western Hemisphere. A seductive natural beauty, the island offered no immediate wealth to Spanish conquerors as it was poor in gold, silver, and precious gemstones. Except for a few villas (Baracoa, Bayamo, Trinidad, Sancti Spiritus, Puerto Príncipe, and Santiago de Cuba) generally established on the coast,[1] economic and social activities as well as linkages with the outside world were concentrated in Havana, which received the vast majority of visitors from abroad. Foreign visitors to the remaining villas mainly consisted of filibusters, smugglers, and buccaneers.

The last of seven villas founded by Don Diego Velázquez, San Cristóbal de La Habana (Havana) was established in 1519. By 1532, its population was the largest on the island, and on December 20, 1592, King Don Felipe II conferred it the title of city. Havana owes its prominence to the fact that during the colonial period it became a crucial stopover and the hub of trade and shipping for Spanish galleons and traders carrying the wealth of the American continent back to their motherland. The daily presence of numerous ships in the port of Havana fueled the development of a service economy specifically aimed to provide lodging, food, and amusements to visitors of different nationalities and habits. The importance of the city was officially recognized on May 24, 1634, when a *Real Cédula* (Spanish royal decree) declared Havana "Key to the New World and Bulwark of the West Indies" (Kapcia 2005, 32) and practically sealed its role as a vital travel destination in the trade between the Americas and Europe. Havana's population grew from about 25,000 at the end of the sixteenth century to some 90,000 at the end of the eighteenth century, and by 1760 it was a larger city than New York (De las Cuevas Toraya 2001, 41; Simoni 2016, 34).

Antecedents of Tourism in Cuba

Tourism, in its modern conception, started to develop in Cuba only in the second half of the nineteenth century and became a significant sector within the Cuban economy only in the twentieth century. But Cuba has been a major travel destination in the Caribbean region for nearly four hundred years. As a matter of fact, even if the words "tourism" and "tourist" had yet to be invented, the history of tourism in Cuba can be traced back to the seventeenth century when Havana's lodging houses and taverns began to be filled with many transient foreign travelers. Overseas visitors continued to grow in the eighteenth century and became so important for the city that in 1764 then Captain General of Cuba, Conde de Ricla, authorized a small printing shop in Havana run by the Sevillian printer Blas de los Olivos to create a guidebook for foreign travelers to the island (Sánchez Baena 2009, 41). While it took this project a few years to come to fruition, the first *Calendario Manual y Guía de Forasteros en la Isla de Cuba* (Almanac and Visitors' Guide to the Island of Cuba), compiled by Diego de la Barrera, was finally published in 1781 (González-Ripoll Navarro, 1999, 74). Published annually from 1794, the guidebook represented the first tourist guide printed in Cuba and in the New World. Figure 1.1 shows the cover page of the issue of 1800.[2]

Besides a calendar of saints, feast days, and astronomical events, the guide for foreign visitors to Cuba included a comprehensive list of names of all Cuban authorities since the first villa of Baracoa was established in 1511, among them captain generals, governors, bishops and various religious authorities, military officers, members of honorary orders, and local officials. It also provided the names of courts and lawyers, hospitals, schools and teachers, and other professionals. Finally, it included some statistical information on marriages, births, and deaths on the island, on hospital patients (admitted, cured, and deceased) in Havana, on government revenues, and on the number of ships entering and leaving the port of Havana and the number of slaves (Capitanía General 1800). Around the mid-1830s, the guide's name was changed to *Guía de Forasteros en la Siempre Fiel Isla de Cuba* (Visitors' Guide to the Always Loyal Island of Cuba), which was published uninterruptedly until 1884 (Trelles 1907, 100).

Cuba experienced considerable development in the nineteenth century as its economic prosperity no longer depended almost exclusively on the role of the port of Havana as a mandatory stopover point for Spain's returning treasure fleets, but also on the island's own production of sugar and tobacco.

CALENDÁRIO
MANUAL,
Y
GUIA DE FORASTEROS
EN LA ISLA DE CUBA,
Para el Año de 1800.

HAVANA:
En la Imprenta de la Capitanía General.

Figure 1.1 Cover page of Foreign Visitors' Guide, 1800. Unknown. 1799. "Calendário Manual, y Guia de Forasteros en la Isla de Cuba, para el àño de 1800." Havana: En la Imprenta de la Capitania General.

A thriving sugar industry, in particular, fostered progress in Cuban society and supported technological advancements that enhanced communication and travel activities across Cuba's regions. The first steamboat, Neptuno, serving the Havana-Matanzas route, began its operations in 1819 (Leal Spengler 1988, 66), which coincidentally was the year when the first travel agency was formed in Cuba. Travel activities received further stimulus with the inauguration of the Havana-Bejucal railway line in 1837 and with the creation in 1840 of the first bus company covering the route between Havana and Cerro. One of the busiest commercial ports in Latin America, Havana by then was a flourishing and fashionable city with a vibrant life that attracted a considerable number of foreign travelers. Along with its pleasant insular weather, overseas visitors enjoyed the Cuban capital's numerous theaters, cafes, shops, parks with music bands, and boulevards (Rodríguez Domínguez 2009, 3). Showing the intensity of commercial and diplomatic relations, Havana's directory of the early 1800s included specific information on eight foreign consulates (Torres-Cuevas and Loyola Vega 2001, 177).

Relations with the United States, in particular, were assuming a growing importance. In 1818, a Spanish royal decree opened Cuban ports to international trade. As a result, Cuba's trade ties with the United States deepened dramatically in the next few decades, while commercial connections with Spain witnessed a steady decline. By the 1880s, more than 60 percent of Cuba's total merchandise exports (primarily sugar) were destined for the US market (Pierre-Charles 1978, 22).[3] Furthermore, although nowhere near the levels experienced in the twentieth century, the flow of travelers to Cuba from the United States began to increase in the second half of the nineteenth century. Americans visited the island for business purposes, health reasons, or pure pleasure. In the mid-1860s, the New York and Cuba Mail and Steamship Company, commonly known as the Ward Line, started to offer maritime services between Cuban and North American ports using newly built side-wheel wooden steamers.[4] Later, propeller-driven ships replaced the side-wheel vessels for ocean voyages (Blume 2012, 509–10). The Ward Line soon emerged as the most reliable means of transportation for passengers and freight between New York and Havana, and between New York and the Cuban ports of Santiago and Cienfuegos via Nassau, Bahamas. The shipping company offered its services until 1954, when soaring fuel prices and competition from airlines forced it to cease operation.

Additional steamship lines maintained regular passenger services between various US cities and Havana at the end of the nineteenth century. For direct trips to the Cuban capital from New York, which generally took four days aboard big ships with excellent accommodations, Atlantic Mail Co. charged $60 gold. For trips from Baltimore and New Orleans, Baltimore Line and Bremen Line charged, respectively, $50 gold and $30 gold. Moreover, Atlantic Coast M.S.L. offered voyages from New York with stops in Sisal and Veracruz (Mexico),

Alliance Line U.S.M. connected New Orleans to Havana with stops in Cedar Keys, Tampa, Apalachicola, and Pensacola (Florida), and Philadelphia & Southern Mail offered voyages from Philadelphia and New Orleans. To promote US-based travel to Cuba, illustrated stories with pictures of the island appeared in top US magazines of that time, such as *McClure*, *Cosmopolitan*, *Munsey*, and *Century*. Major US newspapers ran articles on Cuba, and several books in English about Cuba were published in the United States (Villalba Garrido 1993, 7).

Meanwhile, Cuba's need to build new accommodation facilities for the rising number of overseas visitors had grown urgent. Initially used for military lodging and later to accommodate international visitors and travelers from other Cuban provinces, the first hotel to appear in Havana (in a new area of the city called *extramuros*[5] and today known as Centro Habana) was the Hotel Perla de Cuba in 1835. Another one, the Hotel La Unión, was inaugurated in 1846 in the area within the city's old walls known as Old Havana. But it was only in the second half of the nineteenth century that several hotels, some of them with luxury comforts and amenities, sprung up in the Cuban capital. Reportedly built in 1856, the Hotel Inglaterra was recognized for its excellent quality. It had comfortable rooms, an exclusive a la carte restaurant, and balconies where guests could enjoy a magnificent view of the entrance to Havana's bay and of the entire Paseo del Prado, the city's most beautiful promenade that today officially separates Old Havana from Centro Habana (De las Cuevas Toraya 2001, 90). Built in 1860, not too far from the Hotel Perla de Cuba and moved in 1888 to Paseo del Prado near the best cafés and the Tacón theater, the Hotel Telégrafo distinguished itself with superior services, in particular bathrooms inside rooms, a renowned restaurant called La Vizcaína, and personnel who spoke multiple languages.

Among the new facilities, the Hotel Santa Isabel was the first American-style top-class hotel to become available in Cuba. Unlike the European system of administrative organization that envisaged hotels as single units without departments, the American system established hotels with separate management of private rooms and food services (MICONS 1980). Founded by Colonel Luis Lay of New Orleans in 1867 and widely regarded as Cuba's best hotel,[6] the Santa Isabel had a privileged location on the eastern side of Plaza de Armas, Havana's oldest colonial square where a military band played music every night. The hotel had spacious rooms, restaurant service, and employees who spoke English. An additional advantage was the fact that ladies who stayed there were attended by persons of their same sex, a service still unknown in Cuba (Rodríguez Domínguez 2009, 6).

As for other hotels, one of the most luxurious was the Hotel Pasaje that opened in 1877. It was the first hotel in Cuba equipped with a hydraulic elevator and it had a covered gallery with a structure of iron and glass and two monumental arcades that connected Zulueta street with Paseo del Prado. The Hotel Roma was constructed in 1881 on Teniente Rey street between Zulueta and Monserrate streets. Considered a resort hotel, it had bathrooms with built-in showers, a rather uncommon feature at that time (Lobo Montalvo 2000, 156). Taking the name from its Spanish owner Ventura Trotcha, the Hotel Trotcha was inaugurated in 1886 in the Vedado area, an aristocratic suburb which had developed on the seashore west of the Havana proper. The facility had elegantly furnished rooms with private bathrooms and small balconies overlooking the ocean. It was surrounded with large gardens of tropical flowers and fruits and had exclusive dining arrangements with refreshments and meals served at all hours. Curiously, it was at the Hotel Trotcha that US military leaders had their lodging during the negotiations ending the Spanish-American War of 1895–8 (Wright 1910, 62).[7]

Finally, it is worth emphasizing that by the end of the nineteenth century, almost all of Cuba's most important cities had hotels. Matanzas boasted the hotels León de Oro, Ciervo de Oro, and Ferrocarril. There were also the hotels Cabarrouy and Bardino in San Diego de los Baños, the hotels Fargas and Santa Rita on the Island of Youth, the Hotel Unión in Cienfuegos, the Hotel Niágara in Trinidad, and the Hotel Carrera y Co. in Guanajay. Santiago de Cuba had the hotels Comercio, Suss, and Madame Adela Lescalles (De las Cuevas Toraya 2001, 216–17). By then, however, lodging possibilities for international visitors were still very limited in Havana and across Cuba, as guest houses remained the most popular accommodation option.

The 1900–1930 Period

After the colonial period under Spanish rule came to a close, Cuba became formally an independent republic but de facto a US protectorate.[8] In 1901, the Platt Amendment to the Cuban Constitution gave the United States the right to intervene in Cuban internal affairs by either warlike or pacific measures when deemed necessary. A permanent US naval base at Guantanamo Bay was established. The Reciprocity Treaty of 1903 granted preferential treatment to Cuban agricultural products in the US market while lowering tariffs on US exports to Cuba. Needless to say, Cuba's political and economic ties with the United States and the flow of US visitors to the island intensified considerably. Albeit experiencing an uneven development with sharp differences between urban and rural areas, the Cuban economy witnessed a huge expansion (an annual average growth rate of 8 percent) between 1900 and 1925, mainly centered on the performance of the sugar industry that was recovering from the devastation brought about by the war against Spain. Fueled by booming exports of sugar and, to a lower extent, tobacco products, Cuba's total exports of goods grew sevenfold during this period while imports grew less than fivefold. As a result, the country's annual merchandise trade balance began to record surpluses.[9] This produced an increase in hard currency reserves and stimulated the process of capital formation with foreign resources, above all US investments, which in turn contributed to the economic expansion. (García Molina 2005a, 16–17).

The majority of US citizens who traveled to Cuba in the early twentieth century went there to scout for investment opportunities and to set up businesses. Within a couple of decades of the new Cuban republic, the United States had become a pervasive presence in Cuba, fully controlling the island's political process and dominating vital sectors of the Cuban economy. In 1920, US-owned sugar mills already produced almost 50 percent of Cuba's sugar, and by 1927 they accounted for 70–75 percent of total output. By then, US companies had invested some $1.5 billion in Cuba, about $600 million in the sugar industry and the rest in railroads, public utilities (gas, electricity, water, telephones), real estate, tobacco, commerce, mining, agriculture, banking, and various industries (Pérez Jr. 1986, 258–9). American firms also built resorts, casinos, racetracks, country clubs, golf courses, and marinas in Cuba as important changes aimed to lure foreign tourists, in particular US tourists, took place on the island. In 1909, Cuba reinstated the National Lottery and legalized cockfighting. The United States had prohibited any form of lottery sales and promotion in 1893, and cockfighting remained legal only in a few US states (Goodman 1995, 8; Davis 2016, 129).[10] In 1919, in a clear sign that tourism was beginning to be perceived by Cuban authorities as a key development tool, the government of Mario García Menocal formed the National Commission for Tourism Promotion (*Comisión Nacional para el Fomento del Turismo*) and enacted the country's first tourist law that legalized casino gaming and other gambling activities in Cuba.[11] At that time, gambling was almost uniformly outlawed throughout the United States, and state legislatures had banned horse track betting that was instead legal in Cuba.

But a crucial stimulus for US tourism in Cuba was the passage by the US Congress of the Volstead Law or Prohibition Act in late 1919 (it went into effect in January 1920), which ended the legal sale, distribution, and consumption of alcoholic beverages in the United States. Cuba's first tourism boom indeed occurred during the 1920s when the enforcement of Prohibition encouraged many US citizens to travel abroad to seek unrestricted access to alcohol. Cuba was then the most interesting nearby place because tourism had yet to fully develop in the Bahamas, Bermuda, and in most islands of the Caribbean except in Puerto Rico. Between 1920 and 1930, Havana alone came to receive one in every two Americans who traveled to the Caribbean (Figueras Pérez and Pérez Cárdenas 2015, 179). Americans though were not drawn to Cuba solely by the legality of alcohol. They also flocked there to try their luck at the lottery, gamble at casinos, bet on horse races and jai alai, buy the services of prostitutes, and pursue other illicit thrills. In essence, Cuba, and especially Havana, became a playground for affluent Americans looking for pleasures they could not enjoy at home. Of course, additional magnets were Cuba's pristine beaches and its tropical climate, winter bouts of golfing and other recreational sports, and a wide range of amenities at Cuban seaside resorts. It must be pointed out that Cuban authorities did not focus only on wooing upscale tourists. New marketing strategies specifically designed to attract middle-class vacationers gathered momentum in the second half of the 1920s. One such strategies was to make Cuba's carnival season a less distinctively Cuban holiday by replacing traditional local themes with American ones and by selling grandstand seats at carnival functions as part of tourist package deals (Schwartz 1997, 83–4).

The lack of legal restrictions on alcohol drinking, along with the success of Cuba's tourism promotion strategies, translated into growing arrivals of tourists from the United States. As exhibited in Figure 1.2, tourist arrivals in Cuba rose from 31,566 in the 1924–5 season to 86,270 in the 1929–30 season.[12] During this period, Cuba's share of total tourist arrivals to the Caribbean region skyrocketed from little more than one-third to about 80 percent.[13] There were also sizable annual numbers of excursionists whose visits to the island did not include an overnight stay (Quintana et al. 2004, 74). By the late 1920s, international tourism in Cuba generated between $10 million and $15 million a year, representing the country's third largest source of hard currency revenues after sugar and tobacco exports (Villalba Garrido 1993, 44).

Americans mainly relied on maritime transportation services to travel to Cuba during the 1920s even though air travel emerged as an alternative option toward

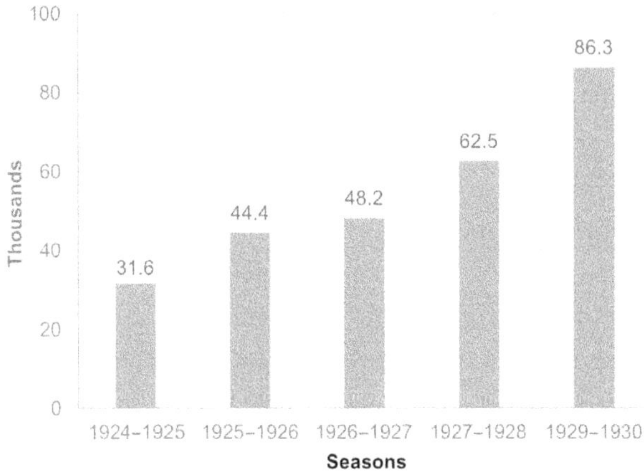

Figure 1.2 Tourist arrivals, 1924–30. Generated by the authors, based on Villalba Garrido 1993; Quintana et al. 2004.

the end of the decade. During the tourist season from December to March, more than twenty companies operated steamers plying between US ports and Havana. There were regular steamship services to Havana from New York (Ward Line, Cunard Line, Munson Line), Key West and Jacksonville (Peninsular and Occidental Steamship Company), and New Orleans (United Fruit Company, Vaccaro Line). Scheduled rail services connected a number of US cities to these locations, where travelers could embark on ships for Cuba. Ferries loaded with passenger cars also made regular trips between Key West and Havana. As for air travel, in October 1927 Pan American Airways inaugurated flights from Key West to Havana using small 10-passenger tri-motor Fokker planes.[14] The following year, Pan American launched Miami-Havana passenger flights and initiated a flight connecting Havana to Puerto Rico with stopovers in the Cuban cities of Cienfuegos and Santiago de Cuba (Villalba Garrido 1993, 49–50; Sáenz Rovner 2008, 20–1).

During the first three decades of the twentieth century, Havana witnessed strong urban growth as its population rose from 250,000 to more than 500,000. While the Cuban government devoted substantial resources to improving and upgrading highways, bridges, ports, city streets, and other basic infrastructures throughout Cuba, it was Havana that experienced the highest degree of modernization and development. Rising world sugar prices during the First World War that boosted the circulation of hard currency on the island, along with the return of Cuban capital that had been deposited temporarily in US banks for safekeeping during the war against Spain, helped ignite a construction boom in Havana (Scarpaci, Segre, and Coyula 2002, 57). This building boom created residential areas with magnificent mansions and resulted in the construction of several national icons and urban landmarks that significantly altered the city's architectural content. After a few years of virtual paralysis in the wake of the world sugar crisis of 1920 that pushed many Spanish and Cuban companies into bankruptcy, construction activity picked up again. The Cuban government continued to invest its own money, but from the time Gerardo Machado came to power in 1925, it began to rely heavily on loans from US banks and on US investments to modernize Havana.[15] In the tourism sector, where pressure had mounted to add new hotels and recreational facilities, different types of business practices, often involving US entrepreneurs, were employed. In some cases, Cuban companies built the property and then sold it to a Cuban or foreign tourism firm. In other instances, foreign tourism chains negotiated the construction of the property directly with Cuban authorities (Colantonio and Potter 2006a, 96–7). And in yet other instances, Cuban owners hired foreigners to run their facilities.

Many hotels were built or upgraded in Havana between 1900 and 1930 to accommodate a growing number of foreign (mainly US) tourists, businessmen, and investment-seekers. In the early 1900s, the hotels Inglaterra, Telégrafo, Isla de Cuba,[16] and Trotcha underwent major renovations to meet modern expectations of comfort and luxury. Around that time, the hotels Gran Hotel Habana (built in 1900 and remodeled in 1906), Plaza (1902), Miramar (1906), and Sevilla (1908) opened just outside the districts of Old Havana. The Gran Hotel Habana was owned by the Cuban entrepreneurs Guillermo del Toro and Pilar Samoano, a married couple that also owned the Hotel Telegrafo. Founded by US Captain Walter Fletcher, the Hotel Plaza (renovated in 1919) was actually built by the US firm Purdy and Henderson, which also erected the Bank of Cuba building, the Cuban Telephone Company, the Centro Gallego, and the Havana Yacht Club. William Thomas Burbridge of New York opened the Hotel Miramar, a small luxury facility overlooking the Morro Castle. Constructed by the Cuban company El Guardián, the Hotel Sevilla was sold in 1919 to the American hoteliers John McEntee Bowman and Charles Flynn who renamed it Sevilla-Biltmore. In 1924, they added to the facility a separate ten-story tower with 300 rooms (Schwartz 1997, 44–5).

Numerous hotels opened during the construction boom of the late 1910s and in the following decade, among them the hotels Ohio, Lafayette, New York,

Lincoln, Albany, Vanderbilt, Packard, Royal Palm, Cecil, St. Louis, Biscuit, Bristol, Ritz, Astor, Parkview, and Ambos Mundos.[17] Several new facilities were owned or managed by Americans and were designed for the comfort and convenience of US guests. Even Cuban- and Spanish-owned hotels catered largely to American tastes. Regardless of their ownership, many hotels in Havana had English-language names, English-speaking waiters, and all kinds of American-style services. As Louis A. Pérez Jr. (2008, 173) wrote, "It was precisely the juxtaposition of the foreign with the familiar, the odd with the new, to be abroad without being away that was at the heart of the Cuban appeal to North American sensibilities."

During the 1920s, however, the construction of new hotels began to move further westward into the emerging residential district of Vedado as the completion of Havana's famous Malecón (seaside promenade) in 1901 by the US General Leonard Wood stimulated the development of new upscale neighborhoods along the city's coastline. Situated on the gardened Avenida de los Presidentes, one of the most splendid streets in Vedado, the Hotel Presidente was built with US capital in 1928. It was erected around the same time as the nearby Hotel Palace and Hotel Victoria. Also in 1928, Purdy and Henderson started the construction of the iconic Hotel Nacional, which stands on a hill in Vedado's heart at the junction of La Rampa (a seven-block strip of 23rd street) and the Malecón. This project was the result of a leasing agreement between the Machado government and a group of US entrepreneurs backed financially by the National City Bank of New York. Under the terms of the agreement, the US investors allocated $4 million for the construction of the hotel, and Cuban authorities granted the use of land in exchange for a presidential suite to accommodate government guests. After sixty years, the hotel's property would be transferred to the Cuban state at no cost (De las Cuevas Toraya 2001, 220). Designed by the New York architectural firm McKim, Mead and White, and operated by the American managers of the Plaza, Savoy-Plaza, and Copley-Plaza hotels, the Hotel Nacional had a glamorous opening on December 30, 1930, with the presence of business executives of the National City Bank and other US companies that had invested heavily in Cuba.

As previously noted, the involvement of American investors in the Cuban tourism sector went well beyond hotel projects to include a wide array of recreational businesses. In 1915, the US entrepreneur H. T. Brown founded the Oriental Park Hippodrome (operated during the winter by the Cuban American Jockey Club) in Marianao, a newly developed resort area immediately west of

Vedado and the Almendares river. Built by a Cuban family and managed by the US hotelier Alfred Gamard, the upscale Hotel Almendares opened a few years later in the same area. In 1925, the owners of the Hotel Sevilla-Biltmore, John Bowman and his associate Charles Flynn, formed the Cuban National Syndicate. Composed of powerful Cuban officials and US financiers, the Bowman-controlled syndicate leased the Oriental Park Hippodrome and Marianao's Gran Casino Nacional (Schwartz 1997, 59). The casino, which together with the Havana Yacht Club earned Marianao the title of "the Montecarlo of America" (Villalba Garrido 1993, 47), reopened fully refurbished in 1928 and quickly became a gambling hotspot for US tourists. Cuba's tourism season indeed began in early December with the opening of the horse racing season and ended in March when the Gran Casino Nacional closed its own season.

In any case, there were plenty of recreational facilities and venues, amusements, and forms of entertainment for US tourists all over Havana during the 1920s. The city was filled with bars,[18] cafés, and restaurants that catered to US travelers. Several of these establishments were owned or administered by US entrepreneurs. Retail stores selling a wide assortment of American products proliferated. Theaters, cabarets, and nightclubs contracted US companies and performers to entertain their customers with North American productions. There were also different kinds of clubs that centered around North American sports and offered memberships to affluent people from the United States and Cuban aristocrats. Referred to as the "Big Five," the most exclusive organizations were the Havana Yacht Club, the Havana Country Club, the Miramar Yacht Club, the Casino Español, and the Havana Biltmore Yacht and Country Club (Scarpaci, Segre, and Coyula 2002, 111). The Bowman-controlled syndicate inaugurated the oceanfront Biltmore Club in 1928 while the US company Frederick Snare Corporation built the Havana Country Club in Marianao that contained Cuba's first golf course (a spectacular eighteen-hole facility) to lure US businessmen and tourists. Other organizations like the Almendares Yacht Club, the Havana Sports Club, the Vedado Tennis Club, and the Riverside Yacht Club sponsored various sports activities such as tennis, baseball, basketball, football, swimming, handball, and ice skating (Pérez Jr. 2008, 256).

American investors also became involved in tourism-related projects outside of Havana. The Cuba Company, a New Jersey-based corporation that was the single largest foreign investor in Cuba during the first two decades of the twentieth century, operated the Hotel Antilla in the Nipe Bay (Holguín province), the Hotel Camagüey in Camagüey, and the Hotel Casa Granda in

Santiago de Cuba.[19] Among various properties, US entrepreneurs owned the Hotel Burnside in Nueva Gerona, the Hotel New York in Camagüey, and the Castle-Pullman Hotel and Hotel Happiness in Varadero in the province of Matanzas. The DuPont family bought a sizable portion of premier beach on the Varadero peninsula and established a lavish three-story mansion named "Xanadú" and the DuPont Country Club (Ward 2007, 214–15). Other exclusive clubs seeking American memberships appeared in tourist areas across Cuba. In 1929, a US firm completed an eighteen-hole golf course at the Country Club de Santiago de Cuba. Put simply, as the leading provider of tourists, transportation services, and products to Cuba and as a vital source of investments in hotels and other recreational facilities on the island, the United States dominated virtually every major area of the Cuban tourism sector.

The 1930–45 Period

Cuba's first tourism boom ended abruptly in the early 1930s as the devastating global economic crisis (Great Depression) unleashed by the collapse of US stock market prices in October 1929 hit hard on tourism activities in Cuba and the island's overall economy. Arrivals of international tourists, above all Americans, decreased sharply, and construction projects in the Cuban tourism sector virtually stopped, with the notable exception of the Hotel Saratoga (previously located on the Monte street property that became the Hotel Isla de Cuba) that was built in 1933 on Havana's Paseo del Prado across from the National Capitol. Other Caribbean countries, including Puerto Rico and emerging destinations like the Bahamas, also saw a drop in tourist arrivals. Although alcohol prohibition in the United States provided some respite from the tourist slowdown, the repeal of the Volstead Act in 1933 delivered another serious blow to the Cuban tourism industry.

In mid-1934, following a coup led by Fulgencio Batista that resulted in the appointment of Carlos Mendieta as president,[20] the Cuban government invited a group of US experts from the Foreign Policy Association to visit Cuba to study the country's economic situation and offer proposals on how to tackle the most urgent problems. With respect to tourism, these experts recognized its potential but recommended to deemphasize gambling attractions and prioritize the promotion of sun-and-beach tourism, nature-related experiences, and the development of tourist areas outside of Havana (Foreign Policy Association 1935). Also in 1934, the National Tourism Corporation (*Corporación Nacional de Turismo*, CNT) was formed through Decree Law 599. The latter laid out the legal statutes regulating tourism operations in Cuba and called for increased coordination of tourism policies between state, provincial, and municipal institutions. Nevertheless, the CNT had little impact because it lacked financial resources and had very limited influence on governmental entities at all levels (Maribona 1959).

Figure 1.3 highlights the tremendous negative impact of the Great Depression and the end of Prohibition in the United States on the Cuban tourism industry. Annual tourist arrivals in Cuba plummeted from nearly 90,000 in 1930 to just 20,677 in 1934 and 29,113 in 1935. The island's gross revenues from tourism activities dropped as well, to $6.6 million in 1934 and $7.9 million in 1935 (Santamarina 1948). Even if they never surpassed the level of 1930, arrivals of tourists from abroad recovered substantially in the second half of the 1930s to reach 77,814 in 1940 and 79,895 in 1941. The annual number of cruise excursionists to Cuba, which had been less affected by the post-1929 global economic downturn than the flow of overnight tourists, also increased from 70,620 in 1934 to a peak of 116,152 in 1937 before suffering a sharp decline (like overnight tourists) during the years of the Second World War. As a result, gross tourism revenues expanded considerably to $14.4 million in 1940 (Villalba Garrido 1993, 50), practically the same level as a decade earlier.

Overall, despite the severe troubles of the early 1930s, the Cuban tourism sector retained a major role within the island's economy. As described in Table 1.1, Cuba obtained a cumulative $98.7 million in hard currency revenues[21] from

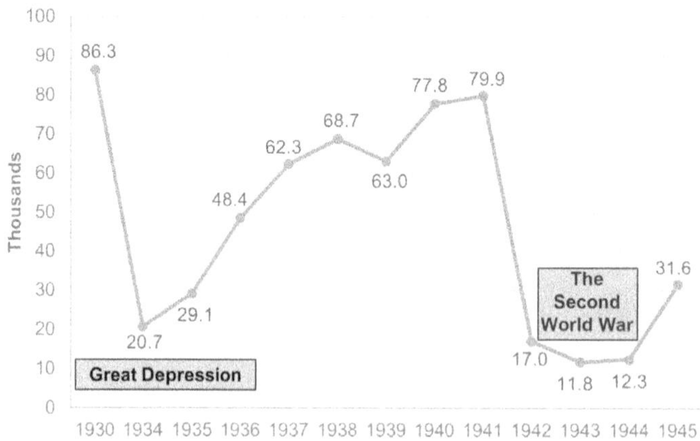

Figure 1.3 Tourist arrivals, 1930–45. Generated by the authors, based on Villalba Garrido 1993; Quintana et al. 2004.

Table 1.1 Exports of Goods and Gross Tourism Revenues, 1934–41 ($US millions)

	Cumulative Value (1934–41)	Annual Average
Sugar and sugar products	931.0	116.4
Tobacco	112.3	14.0
Food products	49.4	6.2
Mineral products	43.4	5.4
Animals and offals	21.1	2.6
Forest products	12.2	1.5
Metals and metal products	8.1	1.0
Chemical and pharmaceutical products	1.2	0.1
Other goods	7.7	1.0
Total goods	1,186.4	148.3
Tourism	98.7	12.3

Source: Santamarina 1944.

tourism activities between 1934 and 1941, or an annual average of $12.3 million. Albeit significantly lower than sugar exports that generated an annual average of $116.4 million and a cumulative $931 million during this period, international tourism receipts were only slightly lower than earnings from sales abroad of tobacco products ($14 million a year on average) and much higher than exports of any other Cuban goods. After sugar and tobacco, the two largest categories of merchandise exports in terms of revenues were food and mineral products, which brought in, respectively and on average, only $6.2 million and $5.4 million a year between 1934 and 1941 (Santamarina 1944, 110). Furthermore, while in absolute terms the flow of tourists dropped more conspicuously in Cuba than in many other countries of the Caribbean in the first half of the 1930s, Cuba's share of total tourist arrivals to the Caribbean increased from around 52 percent in 1935 to nearly 70 percent in 1938,[22] which means that the island recovered at a faster pace than the rest of the region.

Cuba actually never really lost its appeal as a tourist destination even in the aftermath of the Great Depression. On December 31, 1931, about thirty of the world's largest and most luxurious vessels in the world flocked to Havana's bay to celebrate the arrival of the new year. Among them was the British ocean liner RMS Mauretania of the Cunard Line, famous for its speed and elegant interiors. The newly launched Italian liner SS Rex could not find space and had to anchor in the bay of Matanzas. At that time, Cuban regulations exempted cruise ships whose first stop was Havana from paying port taxes and fees. This measure benefited businesses on the ground because excursionists disembarked with their travel budget intact (Villalba Garrido 1993, 50). Regular passenger steamship services between the United States and Cuba, which accounted for the vast majority of US trips to the island, continued during the 1930s notwithstanding a sizable reduction in traffic. Steamship passenger traffic from New Orleans to Cuba plunged nearly 80 percent between 1930 and 1937, but traffic from New York and other US cities only fell by about 12 percent during the same period (US Civil Aeronautics Board 1945).

Regarding air travel, the US carrier Pan American Airways continued to operate flights on its Miami-Havana route despite financial difficulties and also became a dominant factor in Cuban air transportation when it acquired 100 percent of the shares of the Cuban national airline Cubana de Aviación in May 1932. One of the earliest airlines to emerge in Latin America, the small Cuban carrier had started service from Havana to Santiago de Cuba in October 1929. By the late 1930s, as a wholly owned subsidiary of Pan American Airways, the renamed Compañía Nacional Cubana de Aviación S. A. had a fleet of three Lockheed Electras and four Ford Trimotors serving on the domestic Havana-Camagüey, Havana-Guantanamo-Baracoa, and Santiago-Baracoa routes.[23] For its part, Pan American Airways was then the only airline with scheduled international passenger services to the Caribbean region, offering flights from Miami to selected Caribbean destinations like Nassau (Bahamas), Kingston (Jamaica), Port-au-Prince (Haiti), Port of Spain (Trinidad), San Juan (Puerto Rico), and, of course, Havana.[24]

A major transportation project that instead never saw completion was the Sunshine Circle Route (*Ruta del Golfo*), a network of highways and car ferry services aimed at connecting Florida, Cuba, and the Yucatan Peninsula. The impulse for this plan came from a confluence of factors such as the prevailing sentiment for hemispheric unity and cooperation, the emergence of international tourism as a tool for economic development, and the growing use of cars for tourism purposes. In 1923, at the Fifth International Conference of American States in Santiago, Chile, representatives of the US automotive and construction industries proposed the creation of a road route known as the Pan American Highway, stretching from the extreme north to the extreme south of the Americas. Originally conceived as a single route, the project soon evolved into a system of roads with long pre-established routes linking almost all of the Pacific coastal countries of the Western Hemisphere. Isolated from continental land, Cuba nonetheless asked to be part of the system. At the convention of the American Road and Transportation Builders Association in Cleveland in

January 1929, the Cuban Secretary of Public Works, Carlos Miguel de Céspedes, raised for the first time the idea of the Sunshine Circle Route that would enable car travelers to make a circular trip on the Pan American Highway via two ferries and three countries, and to reach Cuba from both coasts of the United States (Freeman 2011). This required ferry services between Key West and Havana (a distance of 90 miles) and between the small port of La Fé in southwestern Cuba and Puerto Juárez in Yucatan (a distance of 130 miles), along with the construction of highways from Miami to Key West, from Havana to La Fé, and from Puerto Juárez to Mexico City (Planas 1943).[25]

Cuban authorities formally proposed the creation of the Sunshine Circle Route in 1936 at the Inter-American Conference for the Maintenance of the Peace in Buenos Aires. In 1931, to facilitate transportation across the island and stimulate tourism, the Boston-based Warren Brothers Company and the Compañía Cubana de Contratistas had finished building Cuba's own Central Highway (*Carretera Central*) that ran from Pinal del Rio, via Havana, all the way to Santiago de Cuba.[26] Mexico completed the section of the Pan American Highway from Nuevo Laredo to Mexico City in 1936, and two years later the Overseas Highway from the US mainland to Key West was officially opened for traffic. By the early 1940s, the only missing portions of highway for the circular trip were a 54-mile stretch to connect Pinar del Rio with La Fé, a few miles across the flat limestone plain west of Puerto Juárez, and a much longer segment of more than 450 miles through the jungle and mountains west of Campeche in Yucatan (Planas 1943). In 1941, the Fourth Pan American Highway Congress in Mexico City approved the expansion of the system to include Cuba (OAS 1963, 8).[27] After the end of the Second World War, amid booming car sales in the United States, the Sunshine Circle Route was the subject of several international conferences and captured the imagination of US tourists, journalists, and writers.[28] Cuba inaugurated the extension of its Central Highway to La Fé in 1957 while a group of US and British experts were studying the port's nearby area to assess the prospects for car ferry services to the Yucatan Peninsula that would join those already operating between Key West and Havana (Perelló Cabrera 2011a). Mexico's highway segments were completed in 1960, but by then a revolution had taken place in Cuba. In the next couple of years, the island's relations with the United States collapsed, and so did the project of the Sunshine Circle Route.

The outbreak of the Second World War deserves a special mention because of its significant negative effects on the Cuban tourism industry. Unlike the

First World War, which had favored tourism activities in Cuba by diverting there many US tourists who previously vacationed in Europe, the Second World War almost completely stopped the flow of tourists to the island. The military operations of German U-boats and Italian submarines off the Atlantic coast of the United States and in the Caribbean Sea represented a threat to passenger ships, whose trips to Cuba decreased considerably. Several ships, especially during the 1942–5 period, were used to transport troops, weapons, and provisions to Europe. This situation benefited Florida, Mexico, and Canada as alternative tourist destinations (Quintana et al. 2004, 77). Annual tourist arrivals in Cuba instead fell to just 12,365 in 1944. That year, tourism activities generated a meager $2.2 million in hard currency revenues (Villalba Garrido 1993, 54).

At any rate, notwithstanding certain periods of tourism expansion in the 1920s and 1930s and the fact that the sharp contraction in tourist arrivals during the Second World War years was primarily the result of external events, the Cuban tourism industry remained far below its true potential. Key limiting factors were a general lack of knowledge of tourism issues among government officials and local entrepreneurs, insufficient state support for tourism development, and the strong dedication of various Cuban administrations to other sectors (like sugar) that were considered more vital to the country's economic prosperity (Rodríguez Domínguez 2009, 14). Furthermore, tourism activities continued to be heavily concentrated in Havana since other areas on the island lacked the necessary infrastructure to attract large numbers of foreign vacationers. As Victor Santamarina (1944, 78), who served on the executive committee of Cuba's National Tourism Corporation, observed, "In studying the natural resources of our country with the aim of developing the tourism industry, we have insisted that the latter should not be developed only in Havana but throughout the entire national territory."

The 1945–58 Period

Three major developments of great relevance to the Cuban tourism industry occurred after the end of the Second World War. First, growing average annual earnings in the United States, which had gone up 30 percent in real terms between 1940 and 1950 (Schwartz 1997, 220), boosted the ability of US citizens to travel within the country and abroad (to Cuba and other foreign destinations) for tourism purposes. A shortened workweek and paid vacations for many US workers were additional stimulating factors. A 1950 survey by the American Automobile Association estimated that 72 million Americans, or about half of the US population, participated in vacation travel that year and that 63 million of these vacationers made their trips by car.[29] A national travel survey conducted by the research department of the Curtis Publishing Company in 1951 also estimated that US tourists spent some $10 billion annually on vacation trips, especially trips of three or more days away from home. While cars were by far the preferred mode of transportation, American vacationers spent approximately $240 million for airline transportation, of which $195 million went for domestic air trips and $45 million went for overseas air travel.[30]

Second, the notable expansion of air passenger services in the second half of the 1940s and particularly in the 1950s further stimulated US tourism in Cuba. Used by the vast majority of Americans for their trips to Cuba during the first half of the twentieth century, ships continued to ply the Florida Straits and connect various US cities to Cuba following the Second World War, but air travel soon emerged as an effective means of reaching the island (Gustavsen 2009). Air transportation actually began to increase during the war years when passenger ships were being sent to Europe to help with the war effort. In 1943, Pan American Airways intensified services on its Miami-Havana route and started to fly from Miami to Cienfuegos and Camagüey, and from those cities to other islands of the Caribbean and to additional Latin American destinations. The following year, a majority share in Compañía Nacional Cubana de Aviación, which served a number of Cuban domestic routes, was sold to Cuban investors, with Pan American Airways retaining a 42 percent stake. In May 1945, Cubana de Aviación inaugurated its own international flights on the Miami-Havana route using Douglas DC-3 airplanes.[31]

Air service to Cuba from the United States reached a whole new level in the post-Second World War period. To meet rising demand, Pan American Airways doubled its one-and-a-half-hour flights between Miami and Havana to six a day for the winter of 1945–6, reinforcing connections with air and train arrivals to Miami.[32] In the next few years, several US air carriers entered the Cuban market. By the early 1950s, Pan American Airways, Braniff International Airways, National Airlines, and Chicago and Southern Air Lines offered hundreds of flights weekly to Cuba from New York, New Orleans, Houston, Chicago, Tampa, and Miami. Pan American Airways alone operated up to eight flights a week between Florida and Cuba (Honey 1999). As air traffic volumes to Cuba and the rest of the Caribbean increased, airline tickets became less expensive. The combination of postwar economic prosperity in the United States and declining airfares significantly altered the social base of international tourism in Cuba. Whereas the island's first tourism boom in the 1920s was driven predominantly by growing inflows of upper-class US tourists, Cuba's second (and much more intense) tourism boom in the 1950s was largely due to a surge in arrivals of a new wave of middle-class and working Americans who had more disposable income and more leisure time for taking vacations abroad (Pérez Jr. 2008, 167).

Third, beginning in the 1950s, Cuba faced increasing competition from other Caribbean countries that were trying to develop their own tourism industry by building new hotels and recreational facilities with the help of foreign investors and by attracting international travelers from Europe, Canada, Latin America, and above all from the United States. Between 1,500 and 2,000 new hotel rooms per year were built in the Caribbean as a whole during the 1950s. Modern conventional mass tourism in the Caribbean, and arguably in Cuba too, emerged with the introduction of jet aircraft that stimulated the propensity for travel to the region from key source markets such as those in North America and Europe, and with the considerable increase of disposable income for wage laborers in developed countries (Grandoit 2005, 89; Duval 2004, 10). While in the early 1940s there was only one airline (Pan American Airways) that offered passenger services to selected Caribbean locations from the United States, by 1956 there were forty-three international and local air carriers operating in the region.[33] Albeit still important, by the mid-1950s passenger ships were no longer the most popular mode of transportation for foreign tourists to the Caribbean since air travel already accounted for 60 percent of total traffic volume (Quintana et al. 2004, 79–80). In short, the growth of international travel and tourism activities in the postwar years created both significant opportunities and critical challenges for the Cuban tourism industry.

Fueled by a growing influx of US vacationers, international tourism in Cuba increased rapidly after the Second World War, and this upward trend continued virtually uninterrupted until the late 1950s. As shown in Figure 1.4, tourist arrivals on the island more than tripled from 31,605 in 1945 to 114,885 in 1946. That year, Cuba received more foreign tourists than ever before, and by the end of the decade, the annual number of tourist arrivals stood at 150,048. In 1950, Cuban president Carlos Prío Socarrás invited the International Bank for Reconstruction and Development (IBRD), an institution of the World Bank, to send a mission to Cuba to conduct a study of the Cuban economy and make practical recommendations on how to spur the country's economic

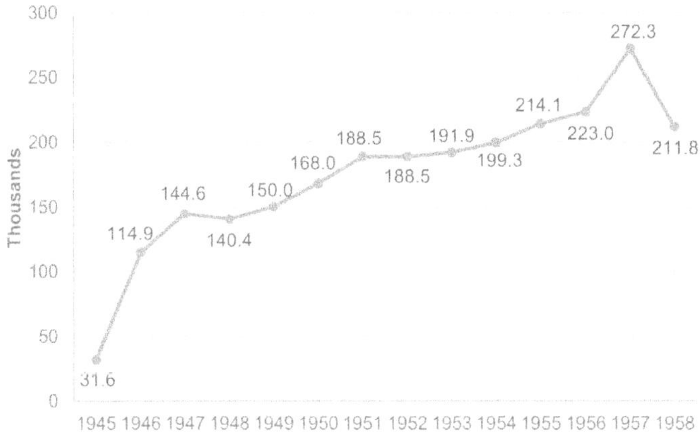

Figure 1.4 Tourist arrivals, 1945–58. Generated by the authors, based on Villalba Garrido 1993; Quintana et al. 2004.

development. Seventeen US experts from the IRBD spent three months in Cuba in the summer of 1950 and the next year they prepared a detailed report on virtually every aspect of the Cuban economy.[34] Regarding international tourism, the report proposed a strong development plan with the support of foreign investment but warned that a number of critical issues stifled the growth of the Cuban tourism industry. These issues included a rising competition from other Caribbean countries, overconcentration of hotel rooms (66 percent of the total) in Havana, poor hygiene and food handling practices in tourist facilities, a general lack of training and low levels of professionalism among hotel employees, and an insufficient diversification of tourism products despite the existence of beautiful natural areas in various regions of the island. The report, though, described Cuba's biggest problem this way: "Cuba as a nation is not at present equipped to develop a much larger tourist industry because she lacks such essentials as adequate and suitable accommodations, roads and sufficient things for the tourist to do or see" (IBRD 1951, 775).

According to the IBRD experts, a crucial reason for more foreign investment in the Cuban tourism industry was to tackle a chronic deficit in Cuba's tourism balance, which they blamed on the industry's inability to convince many Cubans to travel domestically and enjoy the resources of their own country (IBRD 1951, 767). Official statistics indicated that, since 1946, the annual amount of money that Cubans spent on trips abroad had always been higher than the expenditures of foreign tourists in Cuba. In 1949, for instance, Cubans spent $27.3 million on foreign travel while international tourists on the island spent $15.5 million.

What the US experts failed to acknowledge is that the Cuban tourism industry was by and large in American hands and that the real benefits of their plan would not be reflected in Cuba's tourism balance since the latter simply recorded the interplay between export and import travel spending. As Cuba and the United States maintained close economic ties, the IBRD essentially made the case for increased US investments in the island's tourism sector that would boost tourism activities and the hard currency earnings of US companies (Villalba Garrido 1993, 51).

Until the mid-twentieth century, Cuba generally did not have a well-defined policy for tourism development despite some measures during the previous three decades to regulate and foment international tourism on the island as well as timid but largely unfulfilled attempts to establish some sort of institutional coordination at various levels of government for dealing with tourism (Salinas, Mundet, and Eduardo Salinas 2018, 218). In June 1952, after Fulgencio Batista returned to power through a military coup, Decree Law 137 created the Cuban Tourism Institute (*Instituto Cubano del Turismo*, ICT), an autonomous entity under the Ministry of Commerce whose stated mission was to stimulate tourist arrivals and foreign exchange earnings from tourism in Cuba (González Núñez 2022, 84). Staffed with reformers, the ITC advocated planning activities that would benefit, first of all, Cuba and its own citizens. It also supported measures to protect foreign visitors from abuses and annoyances of every kind, increase advertising abroad, and promote natural and cultural attractions to move beyond the traditional focus on nightclubs, cabarets, horse tracks, roulette tables, and similar forms of entertainment. Moreover, between 1953 and 1955, the Batista government implemented a series of laws that, among other things, allowed travelers from North America and several Latin American and European countries to visit Cuba without a visa and granted generous tax concessions to all newly constructed hotels and motels on the island.[35]

Overall, the Cuban economy during Batista's second presidency from 1952 to 1958 was characterized by an increasing presence of US investors spreading over several sectors and a variety of industries like agriculture, public utilities, manufacturing, petroleum refining, and nickel mining, among others.[36] Regarding tourism, Batista selectively incorporated the ideas of the ITC and pursued a strategy that reinforced Cuba's economic dependence on the United States and was aimed primarily to enrich him and a small group of Cuban and US business associates.

To lure US visitors, Cuban authorities launched a major upgrade of the island's tourism infrastructure, improved urban sanitation, stepped up efforts to train English-speaking tour guides, set up programs to educate Cuban citizens on the benefits of tourism and make them more congenial hosts, and intensified advertising of Cuba's attractions in the United States via radio, movies, and the new medium of television. But instead of promoting Cuba's hundreds of miles of beaches, its mineral springs, colonial sites, and other natural and cultural beauties, Batista's planners were more interested in resurrecting the Cuban paradise of the 1920s and adapting it to the new era of mass tourism (Merrill 2009, 111–14).

Havana, in particular, became famous again in the mid-twentieth century for its nightlife, gambling halls, and brothels.[37] Along with a myriad of restaurants, bars, and cafeterias, the Cuban capital boasted the renowned Tropicana, San Souci, and Montmartre cabarets, numerous casinos inside hotels and nightclubs, big-name music acts, and even the all-nude Shanghai pornographic theater and burlesque house in the city's Barrio Chino (Chinatown). Havana played host to celebrities like Ava Gardner, Frank Sinatra, Walt Disney, Gary Cooper, and Marlon Brando. This time, though, Cuba also attracted US mafia kingpins such as Meyer Lansky, Santo Trafficante Jr., and others who set up gambling, drugs, and prostitution operations on the island. With the support of Batista and members of his inner circle who were eager to profit from these operations, the US-organized crime assumed control of major hotels and casinos in Cuba and created a tourist destination that rivaled Las Vegas (Schwartz 1997, 137).[38]

The Cuban tourism industry expanded dramatically in the 1950s. Annual tourist arrivals in Cuba rose from 168,024 in 1950 to 272,265 in 1957 (cruise passenger arrivals grew as well, from 27,137 to 75,242) before starting to decline in 1958 on the eve of the Cuban revolution (Table 1.2). During these years, the United States was the main source market, accounting for over 85 percent of total travelers to Cuba. Far behind were other emitting markets such as Canada, Mexico, Venezuela, Argentina, and some European countries like the UK, Spain, and France. Attracted in large part by Havana's fame as a center for gambling and prostitution, the annual number of US tourists in Cuba increased from 143,327 in 1950 to 236,217 in 1957 and dropped to 182,038 in 1958. By mid-decade, US air carriers offered regular flights to Cuba at affordable prices for most middle-class Americans. Round-trip flights to Havana cost $36 from Miami, $114 from New Orleans, and about $120 from New York (Merrill 2009, 122). Cuban rules facilitated US-based travel to the island in various ways. Americans

Table 1.2 Tourist Arrivals by Country of Origin, 1950–8

	1950	%	1954	%	1957	%	1958	%
United States	143,327	85.3	169,839	85.2	236,217	86.8	182,038	85.9
Canada	2,819	1.7	3,320	1.6	8,521	3.1	6,252	3.0
Mexico	5,139	3.0	6,142	3.1	4,391	1.6	3,851	1.8
Venezuela	2,484	1.5	2,971	1.5	2,858	1.1	2,555	1.2
UK	1,376	0.8	914	0.5	2,629	1.0	2,482	1.2
Spain	3,061	1.8	3,655	1.8	1,968	0.7	1,945	0.9
Argentina	763	0.5	914	0.5	1,550	0.5	1,330	0.6
France	768	0.5	919	0.5	1,624	0.6	1,177	0.6
Others	8,287	4.9	10,641	5.3	12,507	4.6	10.177	4.8
Total	168,024	100.0	199,315	100.0	272,265	100.0	211,807	100.0

Source: Villalba Garrido 1993.

were permitted to enter Cuba without a passport or a visa (only with a tourist card) and remain there for up to six months. They could also use ferry services from Key West to bring their cars into the country duty-free and without a special driver's license (Freeman 2011). They simply needed to present their own license and registration in order to drive freely across the island whose highways were intentionally marked similarly to those of the United States.[39] And even though the dollar-pegged Cuban peso became the only legal tender in Cuba after the US dollar ceased to circulate there on July 1, 1951, the two currencies remained on par with one another and Cuban banks did not charge for dollar-peso conversions.[40]

Cuba's second tourism boom in the 1950s unleashed another construction boom. Very few hotels were built on the island during the 1930s and 1940s because the existing capacity was sufficient to cope with the influx of foreign tourists. But many hotels had grown shabby after years of neglect and inadequate investments and needed to be renovated. Several new facilities were also required to accommodate much higher tourist numbers than those of previous decades. In effect, twenty-eight new hotels with a total of 2,867 rooms were completed between 1952 and 1958 (Table 1.3). Eighteen of these establishments (with nearly 2,400 rooms) were located in the province of Havana, almost all of them in the Cuban capital. Five were located in the province of Las Villas, and the rest in the provinces of Matanzas, Camagüey, and Oriente.[41] By the end of 1958, there were 128 hotels with 7,728 rooms in Cuba. The province of Havana had the biggest stock with 5,119 rooms, followed by Matanzas (where Varadero is located) with approximately 1,000 rooms. The city of Havana, which consolidated its

Table 1.3 Hotel Construction by Province, 1952–8

Province	Hotels	Rooms	Hotel Rooms in 1958	%
Pinar del Rio	0	0	237	3.1
La Habana	18	2,384	5,119	66.2
Matanzas	1	54	999	12.9
Las Villas	5	251	536	7.0
Camagüey	1	93	458	5.9
Oriente	3	85	379	4.9
Total	28	2,867	7,728	100.0

Source: Villalba Garrido 1993.

primacy as Cuba's main tourist site, had forty-seven hotels with 4,892 rooms, about 63 percent of the country's total (Villalba Garrido 1993, 65). Furthermore, large-scale government investment considerably improved Havana's basic infrastructure during the seven years of Batista's second administration that began in 1952.[42] Notable upgrades included the construction of a new water system, the expansion of the José Martí International Airport, also known as Rancho Boyeros, the completion of two underwater highway tunnels under the Almendares river and of another one under the Havana harbor that eased traffic congestion in the city, and the construction of the *Vía Blanca* ("White Way") highway that linked Havana to the beaches of Varadero (De las Cuevas Toraya 2001, 166–71).

The liberalization of the Cuban tourism sector under Batista's rule during the 1952–8 period resulted in a proliferation of new hotels in Havana that significantly increased the city's tourist accommodation capacity. The grandiose and ultramodern hotels Capri, Havana Riviera, and Havana Hilton, as well as the more affordable hotels Vedado, Colina, St. John's, and Flamingo, all occupied the area around the hilly strip of La Rampa in the Vedado district that was the site of many of the new nightclubs and restaurants. The hotels Lido and Caribbean were built near Paseo del Prado, while the Hotel Havana Deauville opened along the seawall at Galiano street. Another important trend was the expansion of construction activities across the Almendares river into the exclusive seaside residential district of Miramar where the hotels Copacabana, Chateau Miramar, Rosita de Hornedo,[43] and Comodoro were built. Among various tourist attractions, Miramar was situated fairly close to the newly constructed Marina Hemingway, had a greyhound racetrack (opened in 1951) and a modern amusement park called "Coney Island," and offered a casino inside the Hotel

Comodoro. Other casinos were installed at the hotels Riviera, Capri, Deauville, Nacional, Plaza, and Sevilla-Biltmore, and at the most popular nightclubs. It should be noted that the Hotel Nacional's lease was acquired by the Chicago hotelier Arnold Kirkeby in the 1940s, and later by the New York investor William Zeckendorf. In 1955, amid a surge in tourist arrivals to Cuba, Zeckendorf sold the management contract of Nacional for a profit to the Intercontinental Hotels Corporation (IHC), a wholly owned subsidiary of Pan American Airways (Schwartz 1997, 153).

Havana clearly monopolized the construction of tourist accommodation facilities since it accounted for almost 80 percent of all hotel rooms that were built in Cuba between 1952 and 1958. New facilities in other locations were typically very small and modest, except for the Gran Hotel in Santa Clara, the Hotel Santiago-Habana in Ciego de Ávila, and the Motel Oasis in Varadero that was owned by the IHC like the Hotel Nacional. In Varadero, there was also the lavish Hotel Varadero Internacional that the US hotelier William Liebow had opened in December 1950. The Internacional cost $3 million to build and featured a casino, a cabaret, and luxury stores. But even Varadero, which would become Cuba's prime tourism spot in the post-1990 period, was still largely undeveloped for tourism purposes in the 1950s despite its pristine beaches and other natural attractions (Quintana et al. 2004, 81–2). The striking disparity between Havana and the rest of the country was epitomized by the fact that more than 90 percent of the total investment (nearly $60 million) for the construction of new hotels in Cuba between 1952 and 1958 was destined for hotel projects in the Cuban capital. (Villalba Garrido 1993, 82–3).

Except for the Riviera and Deauville that received significant US investments, the leading hotels in Havana were built or renovated largely with funds provided by the Cuban Bank for Agricultural and Industrial Promotion (*Banco de Fomento Agrícola e Industrial de Cuba*, (BANFAIC), and by the Economic and Social Development Bank (*Banco de Desarrollo Económico y Social*, BANDES). Batista redirected BANFAIC's lending away from agricultural and industrial development and toward big hotel projects that would justify the issuing of casino licenses (Díaz-Briquets and Pérez-López 2006, 84). The remodeling of the Hotel Nacional after IHC took over its management was carried out with a BANFAIC loan. BANDES granted a sizable loan for the construction of the Riviera even though the hotel was financed primarily by a group of US investors led by Meyer Lansky and associated with gambling interests. The Havana Hilton, which cost an astounding $24 million to build, was financed with $18 million from the restaurant workers union pension fund and $8 million from BANDES. However, the exploitation of the most important hotels in Havana and of lucrative casinos was in the hands of the American mafia. These establishments included the Riviera, Hilton, Capri, Seville-Biltmore and Comodoro, the Tropicana, Internacional, San Souci and Montmartre casinos, and the Parisién casino of the Hotel Nacional (Pérez Jr. 2008, 196). And more hotels were under construction in 1958, involving investments worth over $90 million and a projected stock of 6,066 rooms (Baklanoff 2009, 329). Far from being a tool for economic development, tourism in Cuba served above all the financial interests of the American mafia and those of Batista and his cronics who profited handsomely from bribes and a cut of the revenues of the mob-run casinos.

Not surprisingly, foreign exchange earnings from tourism activities in Cuba skyrocketed during the 1950s, especially after 1955 when major hotel and casino projects in Havana were finalized. Tourism also generated substantial employment even though wages remained low and labor conditions, in particular for restaurant workers and women, were extremely poor (Villalba Garrido 1993, 107–9). As shown in Table 1.4, gross tourism revenues increased from $19.6 million in 1953 to $62.1 million in 1957. By that year, tourism was Cuba's second-largest earner of hard currency, far behind sugar ($656.3 million worth of exports) but ahead of tobacco ($47.9 million) and other goods (García Jiménez 1999). Despite a slight drop to $56.9 million in 1958, much higher levels of tourism earnings than in the past helped offset the deficit in Cuba's tourism balance. After at least a decade of negative signs,[44] the balance recorded surpluses of $4.3 million in 1956, $23.6 million in 1957, and $19.4 million in 1958. Batista's ITC advisers had been outspoken advocates of a planned tourism industry that could have a strong positive impact on the Cuban economy and serve as a backup for the sugar industry, which was prone to fluctuating performances caused

Table 1.4 Exports of Goods and Gross Tourism Revenues, 1953–8 ($US millions)

	1953	1954	1955	1956	1957	1958
Sugar and sugar products	530.1	432.9	474.4	525.1	656.3	593.1
Tobacco	41.9	41.2	43.5	43.8	47.9	49.6
Other goods	68.4	64.9	76.3	97.3	103.0	90.8
Total goods	640.4	539.0	594.2	666.2	807.2	733.5
Tourism	19.6	24.0	27.8	38.1	62.1	56.9

Source: García Jiménez 1999.

by volatile sugar market prices (Merrill 2009, 111). With his corrupt financial arrangements linked to gambling and illicit activities, Batista clearly did not envision such a useful role for tourism. Revenues from gambling ($16 million in 1956 and about $35 million a year in 1957–8), without which Cuba's tourism balance would have continued to record deficits, went to American mobsters and the overall tourism sector was essentially controlled by US citizens and companies. As Evaristo Villalba Garrido (1993, 106) put it, "Casinos with hotels became the [tourism] industry's highest expression of development, sustained by a degrading platform of sinful entertainments and amusements."

Moreover, Cuba remained the most popular tourist destination in the Caribbean until 1957, the year when the heyday of US tourism on the island (in terms of arrivals) reached its peak (Table 1.5). Yet throughout the 1950s, Cuba lost ground to its closest neighbors that were competing for a slice of the growing tourist pie (Colantonio and Potter 2006a, 93). Cuba's share of total tourist arrivals to the Caribbean region decreased from 26.4 percent in 1951 to 21.4 percent in 1957. During the same period, the Bahamas, Puerto Rico, US Virgin Islands, and Haiti, all countries that received the vast majority of overseas tourists from the United States, increased their relative share of regional tourist inflows. In 1958, amid mounting political instability and economic problems caused by the conflict between Fidel Castro's rebel army and the Batista government, tourism in Cuba started to decline again. That year, Cuba accounted for 16.1 percent of total tourist arrivals to the Caribbean and relinquished its number one spot among the region's leading tourist destinations to the Bahamas and Puerto Rico (Villalba Garrido 1993, 25–6). This declining trend intensified rapidly after 1958 as a socialist revolution undertook a radical transformation of the country's economic system and virtually put an end to international tourism in Cuba.

The Revolutionary Period, 1959–89

On January 1, 1959, Fidel Castro's rebel forces entered Havana and brought down the government of Batista, who departed into exile. The demise of international tourism in Cuba after the Cuban revolution can be attributed to two main reasons, in many ways connected to each other. First, economic and political relations between the United States and Cuba deteriorated quickly. Between May 1959 and October 1960, the Castro government expropriated seventy thousand acres of land owned by US sugar firms and took over US properties in key sectors such

Table 1.5 Tourist Arrivals to Caribbean Countries, 1951–8

	1951	%	1954	%	1957	%	1958	%
Bahamas	68,502	9.6	109,605	11.5	194,618	15.3	234,233	17.8
Puerto Rico	78,367	11.0	125,710	13.3	187,321	14.7	218,840	16.6
Cuba	188,519	26.4	199,315	21.0	272,266	21.4	211,807	16.1
US Virgin Islands	44,773	6.3	78,794	8.3	120,200	9.5	140,837	10.7
Netherlands Antilles	76,372	10.7	102,298	10.8	124,970	9.8	131,942	10.1
Bermuda	88,000	12.4	101,500	10.7	116,400	9.2	130,800	9.9
Jamaica	93,626	13.1	112,808	11.9	80,704	6.3	65,700	5.0
Trinidad and Tobago	31,900	4.5	34,300	3.6	51,200	4.0	60,600	4.6
Haiti	17,708	2.5	48,071	5.1	60,283	4.7	58,000	4.4
Dominican Republic	13,531	1.9	22,282	2.3	40,321	3.2	38,300	2.9
Barbados	11,239	1.6	14,074	1.5	23,899	1.9	24,980	1.9
Total	712,537	100.0	948,757	100.0	1,272,182	100.0	1,316,039	100.0

Source: Villalba Garrido 1993.

as telephone, oil, mining, banking, electricity, and tourism (Jatar-Hausmann 1999, 15). In response, the United States canceled Cuba's portion of the annual US sugar import quota in July 1960 and announced a ban on US exports to Cuba (except for some food and medicines) later in October. On January 3, 1961, the Eisenhower administration officially broke diplomatic relations with the Cuban government. Initially a reaction to Cuba's domestic reforms, US economic sanctions were notably expanded after Fidel Castro declared himself a "Marxist-Leninist," strengthened ties with the Soviet Union to reduce Cuba's economic dependency on the United States, and began to actively support revolutionary forces in Latin America.[45] President Kennedy announced a total ban on US trade with Cuba in February 1962 and, the following year, promulgated the Cuban Assets Control Regulations (CACR). Besides freezing all Cuban assets in the United States, the CACR imposed a comprehensive US economic embargo against Cuba that prohibited US-based travel to the island and made most financial and commercial transactions with the Cuban government illegal for US citizens (Spadoni 2010, 28).

Cuba's ideological concerns also played a key role in the virtual disappearance of foreign tourism on the island. The revolution of 1959 profoundly transformed Cuba's political landscape, economic organization, class structure, and foreign relations. Through a major process of agrarian reform, nationalization of private property, elimination of foreign investment, expansion of employment, and income redistribution, Cuba practically eradicated capitalism in favor of a socialist economic system emphasizing inclusive development (Pérez-Stable 1999, 84).[46] With respect to tourism, Cuban authorities undertook a major reorganization of the sector that was designed to wipe out the capitalist evils of gambling, drugs, and prostitution associated with tourist activities and was focused primarily on socioeconomic equality and the promotion of national tourism. Law 100 of February 1959 created the Department of Beaches for the People (*Departamento de Playas para el Pueblo*) and, two months later, Law 270 established free access to all beaches and coastal areas to every Cuban. In November of the same year, Fidel Castro formed the National Institute of the Tourism Industry (*Instituto Nacional de la Industria Turística*, INIT), whose principal task was to develop tourist facilities for use by Cubans at low prices. Meanwhile, the Cuban government closed all mob-run casinos and nationalized the Oriental Park Hippodrome in Marianao and all major US-owned hotels, among them the Hilton, Nacional, Riviera, Capri, Presidente, and Deauville in Havana, the Internacional in Varadero, and the Colony on the Isle of Pines (Miller and Henthorne 1997, 6). In

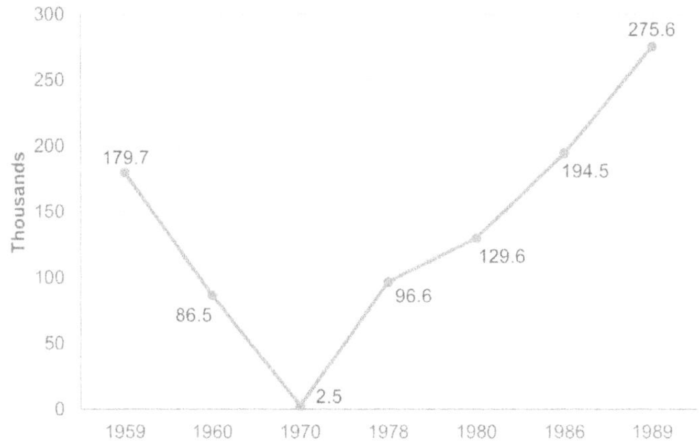

Figure 1.5 Tourist arrivals, 1959–89. Note: 1970–89 arrivals include excursionists. Generated by the authors, based on Quintana et al. 2004; CEE 1991, 1987.

sum, international tourist flows to Cuba all but stopped after 1959 as a result of both the US embargo and Cuba's socialist strategy that discounted tourism as a vehicle for economic growth and development, in part because of an aversion to the societal ills linked to tourism in the 1950s.

Approximately 180,000 foreign tourists, of which 80 percent were Americans, visited Cuba in 1959 (Figure 1.5), still a fairly high number considering the changes that were taking place in the country. In October of that year, Cuba hosted the annual convention of the American Society of Travel Agents (ASTA). Fidel Castro personally addressed the ASTA delegates by saying: "Everywhere you go in Cuba, you will be welcomed and received with open arms because our ambition, which is a noble ambition, is to convert our island into the best place for vacations, and the best and most important tourist center in the world."[47] But Castro's apparent enthusiasm for continuing the development of international tourism in Cuba proved to be short-lived. Already less than 90,000 in 1960, tourist arrivals plummeted to just 4,180 in 1961 and remained depressed during the rest of the decade to bottom out at 2,500 in 1970 as Cuba's favorable trade agreements with the Soviet Union, under which Cuban sugar was exchanged for Soviet oil, machinery, and credits, made the revenue-generating tourism business unneeded (Sharpley and Knight 2009, 242; Cerviño and Cubillo 2005, 227).[48] No major investment in tourism was carried out during the 1960s and early 1970s while existing tourist accommodation facilities were used by Cubans and international guests from socialist and other friendly countries, or simply closed down (Espino 1993, 50).

In the second half of the 1970s, international travelers to Cuba started to increase again. Mainly due to a surge of tourists from capitalist countries and, to a lower extent, growing arrivals from socialist nations through exchange programs and other special arrangements, Cuba attracted nearly 97,000 foreign visitors in 1978 (Figueras Pérez 2013, 219). Two years earlier, showing a renewed interest in tapping the potential economic benefits of international tourism, Cuban authorities had created the National Institute of Tourism (*Instituto Nacional de Turismo*, INTUR). The main responsibilities of INTUR were to plan tourism strategies, collect data on tourist arrivals and expenditures,[49] and coordinate the activities of new state companies such as Publicitur and Cubatur that were formed to publicize Cuba's vacation attractions abroad and organize trips to the island from Canada and several European countries (Colantonio and Potter 2006a, 106). Around the same time, another wave of hotel construction (and renovation of some existing facilities) began. From 1975 to 1981, twenty-nine hotels with a total of 4,097 rooms were built (Table 1.6). New hotels appeared in the principal tourist areas and in almost every Cuban province, from Havana and Matanzas to Cienfuegos, Ciego de Ávila, Camagüey, Holguín, and Santiago de Cuba, among others (De las Cuevas Toraya 2001, 347). However, most of these facilities were at best two-star or three-star hotels that could hardly compete with the luxury hotels and beach resorts that were being built, generally by foreign investors, in other countries of the Caribbean region. Furthermore, Cuba remained off-limits to

Table 1.6 Hotel Construction by Province, 1975–81

Province	Hotels	Rooms	%
Pinar del Rio	1	136	3.3
La Habana	3	664	16.2
Matanzas	2	272	6.6
Villa Clara	2	200	4.9
Cienfuegos	2	405	9.9
Sancti Spiritus	3	383	9.4
Ciego de Ávila	2	272	6.6
Camagüey	5	545	13.3
Holguín	3	528	12.9
Granma	3	400	9.8
Santiago de Cuba	2	164	4.0
Guantanamo	1	128	3.1
Total	29	4,097	100.0

Source: De las Cuevas Toraya 2001.

American tourists, even though restrictions on US travel to the island were essentially removed from 1977 until 1982.[50]

During the thirty years that followed the revolutionary takeover of January 1959, Cuba strongly rejected foreign direct investment (FDI) as a positive factor in economic development. Relations with the capitalist world were modest, and Cuba perceived them to be neither fair nor advantageous when compared with the preferential terms of trade and subsidized financing of economic projects offered by the Soviet Union. Even so, an important law concerning foreign investment was enacted in Cuba well before the Soviet Union and the socialist bloc in Eastern Europe ceased to exist. Cuba's Decree Law 50 of 1982 admitted and codified the juridical form of an economic association with foreign capital and allowed overseas companies to hold equity positions in Cuba. This legislation was basically designed for joint ventures and aimed primarily at stimulating the tourism sector as a recipient of FDI inflows (Spadoni 2014, 83). But despite its existence, concrete efforts to attract foreign investment were made only after the creation of the state corporation Cubanacán in 1987. By the end of 1989, only one joint venture with foreign participation had been authorized in Cuba. It was formed by Cubanacán with the Spanish hotel chain Sol Meliá in 1988 to build two hotels in Varadero (CEPAL 1995).

At any rate, foreign tourism in Cuba continued its upward trend during the 1980s. Exhibited in Table 1.7, data from the extinct State Statistics Committee (*Comité Estatal de Estadísticas*, CEE), which refer to visitors through travel agencies, show that international visitors to Cuba jumped from 129,591 in 1980 to 275,618 in 1989. During this period, arrivals from socialist countries remained stagnant at about 30,000 per year, while arrivals from capitalist countries more than doubled from 98,678 to 237,277 (CEE 1991, 1987). Canada emerged as the largest tourism source market for Cuba. In 1989, Canada sent 61,358 tourists to the island, followed by West Germany (46,556), Spain (27,327), Italy (17,493), and France (7,491) (Rodríguez 1993, 441). There are two additional series available, one from the Immigration and Foreigners Office (*Dirección de Inmigración y Extranjería*, DIE), which records all foreign visitors arriving at the border, and another from the Office of National Statistics and Information (*Oficina Nacional de Estadísticas e Información*, ONEI). Albeit with slightly higher figures than those reported by the State Statistics Committee, both series point to a rapid expansion of international tourism in Cuba in the 1980s mainly driven by visitors from capitalist nations.[51] And yet Cuba's tourism numbers paled when compared with those of the leading Caribbean destinations. In 1989,

Table 1.7 International Visitors by Area of Origin, 1980–9

	Socialist Countries	%	Capitalist Countries	%	Other[a]	%	Total[b]
1980	29,696	22.9	98,678	76.1	1,217	1.0	129,591
1981	34,634	32.7	69,543	65.6	1,859	1.7	106,036
1982	28,963	27.2	75,662	71.1	1,814	1.7	106,439
1983	20,007	16.9	95,000	79.5	4,208	3.6	118,308
1984	23,909	15.1	126,796	80.2	7,330	4.7	158,035
1985	28,322	16.4	139,856	81.0	4,398	2.6	172,576
1986	34,083	17.5	152,005	78.1	8,443	4.4	194,531
1987	31,446	15.1	166,606	80.0	10,189	4.9	208,241
1988	31,393	12.7	205,326	83.1	10,361	4.2	247,080
1989	28,647	10.4	237,277	86.1	9,694	3.5	275,618

Sources: Espino 1993; CEE 1991, 1987.
Note:
[a] Visitors not classified
[b] Visitors through travel agencies

Puerto Rico attracted about 2.4 million tourists, followed by the Bahamas (1.6 million), the Dominican Republic (1.4 million), Cancún, Mexico (1.15 million), and Jamaica (715,000).[52] The level of tourist arrivals to Cuba was practically the same as in the late 1950s.

Finally, in addition to its extensive trade dealings with the Soviet Union and other socialist countries, Cuba raised substantial hard currency revenues by exporting products to market economies. Mainly consisting of sugar, re-exported Soviet oil, and a few other goods, Cuba's hard currency merchandise exports averaged over $1 billion a year between 1981 and 1988 (Table 1.8) despite a sharp drop in sugar market prices in the first half of the 1980s (Ritter 1988, 126). Throughout the decade, international tourism also gained importance in the Cuban economy. Annual tourist expenditures in Cuba rose from $43.6 million in 1981 to $146.9 million in 1988. Gross tourism revenues, which include earnings from international transportation services, increased from $61.1 million to $189 million in the 1982–8 period. By 1987, excluding oil re-exports, tourist expenditures were the third largest generator of hard currency for Cuba after sugar and fish products exports, and they grew further in 1989 to $166 million. Gross tourism revenues grew as well that year to $260 million (Espino 1993, 58). It is safe to argue that tourists from capitalist countries made a crucial contribution. Highlighting a key structural change with respect to tourism earnings, Cuban statistics reveal that the share of total accommodation revenues generated by visitors from capitalist countries expanded from 46 percent in 1980 (visitors from socialist nations accounted for the rest) to 77 percent in 1989 (Durán Cárdenas 2000, 35). As the Cuban economy deteriorated toward the end of the 1980s and a major crisis would soon begin due to the collapse of the Soviet Union, the Cuban tourism industry was about to witness huge development. For now, though, it lagged far behind its Caribbean competitors. In 1989, tourist expenditures were $1.3 billion in the Bahamas, $1.2 billion in Puerto Rico, $818 million in the Dominican Republic, and $593 million in Jamaica (Thomas 2015).

Conclusion

Although it has attracted sizable numbers of international visitors since the port of Havana emerged as a key stopover in the trade between the Americas and Europe in the late 1500s, it was only in the twentieth century that Cuba became a popular tourist destination, above all for American travelers. Cuba

Table 1.8 Hard Currency Merchandise Exports and Tourism Receipts, 1981–8 ($US millions)

	1981	1982	1983	1984	1985	1986	1987	1988
Sugar	866.3	648.3	263.2	250.0	171.2	209.9	222.5	393.4
Oil re-exports	151.4	262.1	497.7	484.4	526.9	248.5	286.3	184.9
Other goods	388.1	445.5	473.0	401.3	546.3	448.9	455.9	535.0
Total goods	1,405.8	1,355.9	1,233.9	1,135.7	1,244.4	907.3	964.7	1,113.3
Tourist expenditures	43.6	50.9	59.4	75.7	87.3	97.7	111.7	146.9
Gross tourism revenues[a]	—	61.1	74.3	95.5	102.6	124.6	145.0	189.0

Sources: Espino 1993, Ritter 1988, BNC 1988.

Note:

[a] Include earnings from international transportation services.

experienced its first tourism boom in the 1920s when Prohibition and anti-gambling legislation in the United States helped turn Havana into a playground for wealthy Americans, who flocked there, by and large via steamers, to drink, gamble, and pursue other forbidden pleasures they could not enjoy in their own country. American developers built mansions, upper-scale hotels, casinos, race tracks, and exclusive country clubs on the island. A much bigger tourism boom occurred during Batista's second presidency from 1952 to 1958 when increased prosperity in the United States, as well as the expansion of air passenger service and package tours, fueled a major growth of middle- and working-class US vacationers in Cuba. During the 1950s, Cuba was the leading tourist destination in the Caribbean, with large casinos and luxurious hotels in Havana under the control of American mobsters who profited from gambling, drugs, and prostitution operations. Batista and his cronies also benefited from bribes and a cut of casino revenues. By the late 1950s, tourism had become a key sector for the Cuban economy, second only to sugar in terms of hard currency earnings.

Following the 1959 revolution, international tourism in Cuba practically disappeared, in part because the US market dried up as a result of the US embargo but also because the government of Fidel Castro nationalized most hotels and related businesses, rejected gambling and other capitalist vices linked to tourism, and prioritized egalitarian policies and domestic tourism. As socialist Cuba strengthened political relations and signed favorable economic agreements with the Soviet Union, international tourism was no longer promoted. It trickled back in the second half of the 1970s and gathered steam in the 1980s when a growing number of tourists from capitalist countries, especially Canadians and Western Europeans, visited the island. At the end of the 1980s, however, Cuba was practically the only Caribbean country without a strong tourism sector and without an aggressive tourism promotion policy. But as we will see in the next chapter, things changed dramatically in the post-Cold War era.

Notes

1 Of the original seven villas, only Sancti Spiritus was located inland. Puerto Príncipe was moved in 1528 to the current location of Camagüey.

2 The original copy of this document comes from the John Carter Brown Library at Brown University, https://archive.org/details/calendriomanualy00hava/page/n3/mode/2up.

3 Data for the 1880–6 period indicate that 62 percent of Cuba's total exports of goods during that period went to the United States. The rest went to Great Britain (22 percent), Spain (only 3 percent), and other countries (13 percent). As for imports, 30 percent of total goods still came from Spain, 20 percent from the United States, 20 percent from Great Britain, and 30 percent from other countries (Le Riverend 1967, 186).

4 The Ward Line began its operations in 1840 with a few small schooners trading between the United States and the main islands in the Antilles, among them Cuba.

5 Due to the burgeoning population, the city of Havana was transformed in the nineteenth century. With the three-quarter square mile of the original city increasingly crowded, many people moved westward into new residential zones beyond the old walls (*extramuros*) covering an area of about two square miles (Pérez Jr. 2001, 50).

6 Colonel Lay acquired the building that had been the residence of the Count of Santovenia and transformed it into the Hotel Santa Isabel.

7 In the early 1930s, after the Great Depression, the Hotel Trotcha was converted into a guest house with about sixty rooms. A fire completely destroyed it in 1986 (De las Cuevas Toraya 2001, 91).

8 The Cuban republic from 1902 (when US troops left the island) until 1959 (when the Cuban revolution began) is also called a "pseudo republic" due to the direct influence of the United States on Cuba's affairs (Gott 2005, 113).

9 Whereas in 1900 Cuba's annual merchandise trade balance posted a deficit of $18 million, by 1925 it had reached a surplus of $57 million (García Molina 2005a, 17).

10 The first anti-cockfighting laws in the United States were introduced in the first half of the nineteenth century. By 1915, thirty-six out of forty-eight states, along with the District of Columbia, had banned cockfighting.

11 Also in 1919, the cities of Havana and Miami signed an agreement to attract tourists from northern US states through joint promotional activities and combined trips (Villalba Garrido 1993, 44).

12 Data in Figure 1.2 refer to tourist arrivals during the winter season from December 1 to March 31 since summer tourism still did not exist.

13 Calculations of the authors from Villalba Garrido 1993. In 1925, the Caribbean region received 88,200 international tourists. By 1930, that number had reached 107,900.

14 Havana was one of the first cities in the world to operate an international airport. Pan American Airways initially flew to Havana's Camp Columbia Aviation Field, a military-civilian airport next to the Miramar district (Pérez Jr. 2008, 166–7). In

1930, the Rancho Boyeros Airport (later renamed Jose Marti Airport) replaced Camp Columbia.

15 For instance, Cuba's National Capitol (*Capitolio Nacional*) was built between 1925 and 1930 and was financed with about $20 million lent by the Chase Manhattan Bank (Colantonio and Potter 2006a, 97).

16 The Hotel Isla de Cuba was built in 1888 on Monte street across from Parque de la Fraternidad (Park of Fraternity) near Old Havana.

17 Opened in 1925, the Hotel Ambos Mundos in Old Havana became famous because it was home to the popular writer Ernest Hemingway for most of the 1930s.

18 By the 1920s, about seven thousand bars reportedly operated in Havana (Pérez Jr. 2008, 169). The favorite haunts of US vacationers were Sloppy Joe's, El Floridita, Plaza Bar, Café Sazerac, Inglaterra Bar and Patio, Donovan's, Paris Bar, Winter Garden, Ambos Mundos, and Jigg's Uptown Bar.

19 Incorporated in 1900 by some of the most affluent and powerful businessmen in the United States, the Cuba Company built the railroad system in eastern Cuba and invested substantially in the island's sugar industry, tourism industry, and other sectors. For a comprehensive review of the Cuba Company's business activities in Cuba, see Santamarina 2000, 2011.

20 Carlos Mendieta was the first of a succession of Cuban presidents (José Agripino Barnet, Miguel Mariano Gómez, and Federico Laredo Brú) until 1944, all of whom were controlled by Fulgencio Batista as commander-in-chief of Cuba's armed forces.

21 Tourism revenues reported in this chapter and in the rest of this book are gross revenues (total tourist expenditures) unless otherwise indicated.

22 Calculations of the authors from Villalba Garrido 1993.

23 "Airline Companies of the World," *Flight*, April 27, 1939 (433).

24 For a detailed analysis of Pan American Airways' operations in the Caribbean region and other parts of the world until the early 1940s, see Robert Winston Hawes. "The Empire of Pan American Airways," Master's Thesis, Boston University, 1943, https://archive.org/details/empireofpanameri00hawe.

25 In its edition of 1929, the popular Terry's Guide to Cuba (printed in the United States) included the map of the Americas with the routes of the Pan American Highway and the land-water routes connecting Cuba, Florida, and Mexico (Terry 1929).

26 Signed in February 1927, the total contract for Cuba's Central Highway was for approximately 692 miles of road, 481 of which was awarded to the little-known Warren Brothers Company. Required to give a portion of the work to Cubans, Warren Brothers subcontracted Compañía Cubana de Contratistas, in which then Cuban president Gerardo Machado held a stake (Skwiot 2010, 102).

27 The Fourth Pan American Highway Congress of 1941 also approved the expansion of the system to include Haiti and the Dominican Republic via ferry services connecting Cuba with Haiti (OAS 1963, 8).

28 In 1954, the *Popular Science* magazine featured an article that described a circular trip on the Pan American Highway, for instance, from Florida to Cuba and Mexico and then back to Florida, as a journey through five civilizations cutting a cross-section of the ages, namely the Anglo-Saxon, Spanish, Mayan, Indian, and Mexican civilizations (Griswold 1954).

29 American Automobile Association (AAA), "Americans on the Highway: A Survey of Recent Trends of Tourist Travel," Washington, D.C., AAA, 1950.

30 Curtis Publishing Company, "The Vacation Travel Market of the United States," Nationwide Survey No. 3, 1951.

31 Compañía Nacional Cubana de Aviación was the first Latin American carrier to establish scheduled air passenger services to Miami.

32 "Air Schedule Doubled," *New York Times*, September 24, 1945.

33 For a list of the most important airlines operating in the Caribbean and air traffic statistics for selected Caribbean countries in 1956, see Villalba Garrido 1993, 34–5.

34 The *Report on Cuba* (IBRD 1951) by the International Bank for Reconstruction and Development is also known as the Truslow Report from the name of the Chief of Mission, Francis Adams Truslow.

35 For an analysis of institutional and legislative changes in Cuba during the 1950s to foster the development of the Cuban tourism industry, see Rodríguez Domínguez 2009, 28–42.

36 By the end of the 1950s, US direct investments (worth about $1 billion) accounted for the following shares of Cuba's economic activity: 90 percent of electric power generation and telephone service; 37 percent of raw sugar production; 30 percent of commercial banking; 50 percent of public service railways; 66 percent of petroleum refining; 20 percent of insurance business; and 100 percent of nickel mining (Baklanoff 2009, 332).

37 A 1956 article in a now-defunct US tourism magazine provided a vivid account of the many pleasures that foreign visitors could enjoy in Havana at night, from casino gaming to music and dance acts, risqué plays, nude shows, pornographic movies, and plenty of rum. It described the Cuban capital as "the mistress of pleasure, the lush and opulent goddess of delights." Jay Mallin, "Havana," *Cabaret Quarterly* 5: 43–56, September 1956.

38 For an interesting examination of the American mafia's operations in Cuba in the 1950s, see English 2008.

39 R. Hart Phillips, "Cuba's New Deal for Tourists," *New York Times*, October 26, 1952.

40 "Cuba Substitutes Pesos for Dollars," *New York Times*, June 30, 1951.

41 The twenty-eight hotels with 2,867 rooms that were built in Cuba in 1952–8 had an accommodation capacity for 5,464 guests. Taking also into account the 285 rooms that were added to existing facilities during that period, the country's new room stock had a total accommodation capacity for 6,042 guests (Salinas 1998, 154–5).

42 For more information on Batista's economic policies in 1952–8, see McGuigan 2012.

43 Inaugurated in 1955 as a luxury facility, the Hotel Rosita de Hornedo became the Hotel Sierra Maestra after the Cuban revolution when it hosted foreign technicians working in Cuba (De las Cuevas Toraya 2001, 221).

44 Statistics on Cuba's tourism balance are available only from 1946 onwards.

45 Tougher US sanctions against Cuba at the height of the Cold War mainly aimed to raise the cost of Cuban adventurism in Latin America, raise the cost to the Soviet Union of maintaining its new relationship with the island, and reduce the latter's ability to find other capitalist trade and financial partners (Morley 1984, 28; Smith 1988, 47–9) It is worth noting that in 1964 the United States persuaded all members of the Organization of American States (OAS), with the sole exception of Mexico, to break diplomatic and trade relations with Cuba (Brenner 1988, 13).

46 For a concise review of Cuba's economic and social policies from 1959 until 1970, which began with strong measures to eliminate capitalism and the market and later pursued improbable idealistic goals with strengthened centralized decision-making, see Mesa-Lago and Pérez-López 2013, 5–9.

47 Fidel Castro's speech at the opening ceremony of the twenty-ninth ASTA Convention held in the Blanquita Theater on October 19, 1959, http://www.cuba.cu/gobierno/discursos/1959/esp/f191059e.html.

48 Cuban economists from the island claimed that US pressures on Canadian and European tour operators to halt their business operations in the Cuban market played a role in the steep decline of international tourism in Cuba after 1960 (Quintana et al. 2004, 91).

49 Apart from some sparse data and estimates, there are practically no international tourism statistics on Cuba for the 1960s and 1970s (Jayawardena 2003, 54).

50 In March 1977, Jimmy Carter did not renew the ban on travel to Cuba by US citizens that was supposed to be renewed every six months. Ronald Reagan reestablished the travel ban in April 1982.

51 The DIE figures show that international visitors to Cuba (including excursionists) rose from 150,700 in 1982 to 326,300 in 1989 (Espino 1993, 52). The ONEI figures indicate that tourist arrivals to Cuba grew from 238,000 in 1985 to 315,000 in 1989. See Oficina Nacional de Estadísticas e Información, "Series Estadísticas 1985–2015," http://www.onei.cu/series2015.htm.

52 Statistics of the Caribbean Tourism Organization (CTO), https://www.onecaribbean.org/statisticsarchive/.

2

International Tourism in Cuba in the Post-Cold War Era, 1990–2014

This chapter examines the evolution and results of the Cuban tourism industry from 1990, when the Soviet Union was about to dissolve, until the end of 2014 when Cuba and the United States launched a historic process of rapprochement with key implications for US-based travel to the island. Since the early 1990s, Cuba has targeted tourism as a priority sector because of its ability to generate foreign exchange and stimulate economic growth. Before the recent upsurge of exports of professional services, international tourism was, at least in gross terms, the single most important source of hard currency for the Cuban government. The leisure industry has also attracted foreign investment and know-how, created substantial employment, and established linkages with other economic sectors.

Mainly built around attractions such as beach resorts, historical sites and colonial architecture, and ecotourism sites, tourism development has concentrated in eight regions or poles around the island. These are Varadero and Havana, which account for the vast majority of revenues and lodging capacity, as well as the nontraditional regions of Jardines del Rey (Cayo Guillermo, Cayo Coco, and Cayo Santa Maria), Northern Camagüey (Playa Santa Lucia), Northern Holguín (Guardalavaca), the Canarreos Archipelago (Cayo Largo), Santiago de Cuba, and the cities of Trinidad and Cienfuegos along the central southern coast. Additional tourist development areas feature places of interest like the historic town of Baracoa near the eastern tip of Cuba,[1] the mountains, beaches, and caves of the provinces of Guantanamo and Granma, the cities of Holguín, Camagüey, Ciego de Ávila, and Santa Clara, the natural reserve park of Topes de Collantes in the Escambray mountains, and the Viñales Valley and other natural attractions of the Pinar del Rio province (Rutty and Richardson 2019, 2; Cerviño and Cubillo 2005, 230–1).

Policy and Institutional Framework

When its Soviet benefactor collapsed in the early 1990s, the Cuban economy took a nosedive.[2] With the loss of the external support that had sustained its economy, Cuba's real gross domestic product (GDP) shrank by an annual average rate of around 10 percent between 1989 and 1993 (ONEI 1998; CEE 1991). Cuban exports and imports also fell dramatically during this period. Thus, owing to the disappearance of the economic and financial system in which it had been situated during the Cold War era, Cuba suffered a debilitating blow and was forced to devise new and effective strategies to reinsert itself into the global market economy. The launch in September 1990 of an austerity program called the Special Period in Time of Peace stimulated a more pragmatic stance toward economic policy and resulted in the enactment of limited capitalist-style reforms. The program consisted of a series of measures intended to conserve energy and raw materials, stimulate food production, and expand markets for exports and imports. Between 1993 and 1994, several other measures were adopted, among them the legalization of the US dollar and of dollar-denominated remittances, the authorization of self-employment, the breakup of the state monopoly on land to set up agricultural cooperatives, the creation of free agricultural markets, the reorganization of the central administration of the state to reduce bureaucracy, and the promotion of foreign direct investment (FDI) (Jatar-Hausmann 1999, 61–2). Another important move, however, was the active promotion of international tourism as a tool for economic revival.

In explaining the measures that were being adopted to cope with an unprecedented economic crisis at the fourth congress of the Cuban Communist Party (*Partido Comunista de Cuba*, PCC) in October 1991, Fidel Castro touted the benefits of international tourism but made no secret that his decision to rely on foreign tourists for hard currency earnings was taken out of necessity and mostly against will (Jiménez 2008, 146). Fidel Castro specified, "We are building thousands of rooms every year for international tourism. . . .Earnings are growing at a remarkable pace and it is very important to understand that the country needs tourism even if that implies some sacrifices for us. We would like to enjoy all hotels but we need to save the homeland, the Revolution, and socialism." He then added,

> When we open a hotel for foreign tourists, there are people who think that something has been taken away from them, but that's not the case, we are

> collecting there to solve other problems . . . this is no act of disregard toward our fellow countrymen. I wish we had Venezuela's or Kuwait's oil for not having to think about international tourism and being able to build thousands of hotels [for Cubans].[3]

Regarding Cuba's appeal as a tourist destination and the country's efforts to further develop its tourism industry, the resolution on the economy of the 1991 PCC Congress stated,

> Cuba has numerous poles or places of interest for tourists across the entire national territory, among them large areas of beaches and keys, which together with socially positive factors like social stability, health security and coverage, hospitality, cultural richness, and the high level of education of its people, make the country an attractive destination for overseas vacationers and make tourism an important source of revenues for economic development, a source of employment for our workers, and a stimulating factor to develop various economic sectors associated with tourist activities. The efficient exploitation of existing installations and of those that will be built in the future is a permanent goal of the Cuban tourism industry, and thus the country will keep working toward a comprehensive development of its tourist poles. (Figueras Pérez 2002, 106–7)[4]

New laws, new institutions, and new work concepts were created to improve the quality of Cuba's tourism services and their ability to generate foreign exchange, to link tourism with the rest of the economy, and to increase the domestic supply of goods and services to the Cuban tourism industry. In 1992, the Cuban Constitution was amended, among other things, to recognize and protect property ownership by mixed enterprises and to allow the transfer of state property to joint ventures with foreign capital.[5] These changes established a less restrictive foreign investment regime within the framework of Decree Law 50 of 1982 and stimulated foreign investors from Canada, Mexico, and especially Spain and other European countries to enter the Cuban tourism market. Interestingly, while overseas partners in joint ventures with the Cuban government could generally control only up to 49 percent of a company's shares, the foreign investment regime permitted foreign ownership of 51 percent or more of an enterprise's shares in the tourism sector and in other special areas like mining and Latin America-based investments (Travieso-Diaz 1997, 106).[6] Moreover, two tourism circuits were created, one for international tourism that operated exclusively in foreign currency (mostly in US dollars even though the Cuban convertible peso or CUC, at par with the US dollar, was introduced in December 1994), and one for national tourism that involved lower quality lodging and catering facilities and generated revenues in Cuban pesos (CEPAL 2000). Although it has begun to complement international tourism in recent years, national tourism was not a priority in Cuba's development plans during the 1990s and most of the 2000s.

Also in 1992, the Cuban financial institution FINATUR (*Financiera para el Turismo*) was set up to provide funding for tourism-related production activities in Cuba and thus boost national supplies of goods used in the tourism sector (Figueras Pérez 2013, 237). Functioning as a development bank, FINATUR offered credits in hard currency for capital investment and projects in the agricultural sector and in various types of industries involving the manufacturing and processing of food, beverages, textiles, clothing, furniture, plastic, glass, and paper, among other products. The institution, which recovered the credits granted to domestic suppliers when the latter completed sales of goods to tourist enterprises, steadily increased its volume of operations throughout the 1990s.[7] In essence, FINATUR opened a market for depressed Cuban entities that were forced to innovate, introduce new technologies, and offer competitive products in order to meet the quality standards required by the tourism industry. Vendors that repeatedly failed to comply with their business obligations had their contracts suspended and were removed from the list of available suppliers. In the end, however, many of these suppliers encountered unsurmountable difficulties in adapting to the rules and requirements of commercial dealings with tourist enterprises. Thus, FINATUR disappeared in 2002, and its main functions were assigned to another bank, reportedly to Cuba's Commerce and Credit Bank (*Banco de Crédito y Comercio*, BANDEC) that had been established in 1997 (Figueras Pérez and Pérez Cárdenas 2015, 180; García Molina 2005b, 67).

A major restructuring of the Cuban tourism industry with significant decentralization of its business activities began in 1994. The National Institute of Tourism (INTUR) was replaced that year by the new Ministry of Tourism (*Ministerio de Turismo*, MINTUR), which was created to provide planning and policy direction, coordinate all tourist agencies and companies, regulate their commercial operations, and supervise the modalities in which foreign investors could enter the sector.[8] In addition, several state-run yet semi-independent tourist corporations were formed, in particular hotel chains such as Gran Caribe, Horizontes, Islazul, and Habaguanex, as well as non-hotel entities (travel agencies, hard currency chain stores, and various types of service companies) such as

Rumbos, Palmares, Transtur, Cubarte, Cubamar, Caracol, Puertosol, ITH, and Fintur, among others. While in 1991 there were only a small handful of entities responsible for developing the tourism sector in Cuba—besides INTUR (1976), the hotel chains Campismo Popular (1981), Cubanacán (1987), and Gaviota (1988, with ties to the Cuban military) and the travel agency Havanatur (1988)—these had grown to around fifty by 2002 (Colantonio and Potter 2006a, 40). These companies operated in hard currency, enjoyed some autonomy in decision-making, and to a certain extent competed against each other for business even if they focused on different segments of the tourist market (Ross and Fernández Mayo 2002, 279–80; Espino 2000, 363).[9] Unlike Islazul and Campismo Popular (with its five agencies) whose activities concentrated on domestic tourism, all the other tourist companies focused on international tourism.

Although no foreign investment has been introduced into the ownership of Cuba's tourist corporations, the latter have formed joint ventures with overseas companies in projects involving the building of new hotels in Cuba, extensive and costly refurbishments of existing installations, and the production of goods and services for the Cuban tourism industry. The number of these mixed enterprises rose notably after 1993 as Havana's authorities intensified their promotion of foreign investment. The Ministry of Foreign Investment and Economic Collaboration (*Ministerio para la Inversión Extranjera y la Colaboración Económica*, MINVEC) was created in April 1994. Through visits to foreign countries, participation in international investment events, and meetings with potential investors, Cuban officials actively promoted the advantages of business activities on the island (Pérez-López 1999). The opening of foreign investment and international tourism, matched by increasing interest but also growing complaints from foreign companies, led the Cuban government to design an updated and more attractive legislation in 1995. While repeating some of the basic aspects of Decree Law 50, Law 77 of September 1995 set out specific guarantees for foreign firms by establishing full protection and security against expropriation and it opened all sectors of the Cuban economy (except public health, education, and the armed forces) to foreign investment.[10] It also abolished the limit of 49 percent of foreign shares for joint ventures and authorized, for the first time, the possibility of 100 percent wholly foreign-owned investments. Finally, in an attempt to speed up and streamline the authorization process of new agreements, the law required that approval or denial of an investment must be given within sixty days of the presentation of the formal request (Travieso-Diaz and Trumbull 2002, 183).

As expected, Law 77 convinced a sizable number of foreign companies to enter the Cuban market with investments in tourism and other key sectors of the island's economy. Many investment projects not specifically associated with tourism were actually carried out to develop tourism-related infrastructures and activities such as transportation, telecommunications, energy, water, sanitation, agricultural production, and food processing (Henthorne 2018, 66). Yet, as the tourism industry expanded and consolidated, Cuba's strategy of fostering joint ventures between foreign firms and Cuban corporations for the development of hotels was replaced by a strategy of relying mostly on domestic resources to build tourism facilities and turning over the administration of hotels to foreign chains through management contracts (Pérez-López and Díaz-Bríquets 2011, 315). Sun-and-beach tourism was the main focus of Cuba's tourism growth policy even though other modalities like ecotourism, heritage or cultural tourism, and health tourism were promoted.

Moreover, a major policy shift on FDI occurred in 1998 when Havana's authorities declared their preference for economic partnerships that involved higher amounts of capital inflows and loan financing (Morris 2008, 780). Owing to banking reforms and economic recovery, Cuban vice president Carlos Lage Dávila announced that year the intention of the Cuban government to prioritize foreign investment in large development projects while putting a brake on smaller projects unless they included the introduction of new technologies or new export markets. He added that local state-run banks were now in a position to provide small amounts of capital (USCTEC 1998). As a result, the overall number of FDI projects in Cuba practically stalled for a few years and then experienced a steep decline.

It is worth mentioning that the beginning of this downward trend coincided with a process of recentralization of the Cuban economy launched by the government of Fidel Castro in 2003, essentially to reverse some of the liberalizing reforms it had implemented a decade earlier to secure the survival of a system then on the verge of collapse (Pérez-Stable 2007, 17). Besides reducing the number of Cuban agencies responsible for imports and ordering state enterprises to deposit all the hard currency they obtained through business activities into a single account at the Cuban Central Bank (and then request bank permission to use the money), Cuban authorities began to reassert central control over the tourism industry. Unhappy about the loose spending and corruption that had limited profits, in late 2003 Cuban authorities fired several top executives from the island's largest tourism group, Cubanacán. Early in 2004, they replaced the tourism minister, Ibrahim Ferradaz García, with Manuel Marrero Cruz, who, at the time of his designation, was heading the army-controlled Gaviota tourism group.[11] Foreign media revealed that these moves were part of a plan to merge most, if not all, activities of four major state corporations (Cubanacán, Gran Caribe, Horizontes Hoteles, and Isla Azul) of Cuba's Ministry of Tourism that controlled 75 percent of the hotel rooms on the island. The merger plan reportedly was supposed to be completed in the second half of 2004 along with the centralization of retail stores, transportation, and car rental services for tourists.[12]

In reality, the restructuring of MINTUR carried out by the government of Fidel Castro was somehow narrower in scope than the one envisaged in the aforementioned plan, at least when it comes to hotel activities. Focused primarily on the lower-end tourist market (two- or three-star properties), Horizontes was the only Cuban hotel chain whose operations were absorbed by another state corporation, namely Cubanacán. Nevertheless, various non-hotel activities were merged, some enterprises underwent significant structural changes, and new companies were formed as part of a recentralization of Cuba's entire tourism sector. For instance, services provided by the travel agencies Rumbos and Viajes Gran Caribe were taken over by Cubatur and Viajes Cubanacán. Transtur (transportation services including buses, taxis, and car rentals) and Emprestur (construction services) saw a major reorganization of their business operations. The fusion of the chain stores Caracol and Universo into Grupo Empresarial y Comercial Caracol to establish a new hard currency shopping network for tourists and the creation of the entertainment and restaurant chain Palmares also represented efforts to strengthen the non-hotel segment of the Cuban tourism industry (Silveira, Martins, and Castellanos 2010).[13]

Moreover, in November 2004, Cuban authorities put an end to the circulation of the US dollar in Cuba in favor of the CUC and slapped a 10 percent fee on dollar/CUC exchanges. Then, in April 2005, they revaluated the CUC by 8 percent against all international currencies. It goes without saying that Cuba's shift toward a recentralization of its tourism sector and its overall economy caused serious concerns among existing and potential foreign investors and that Fidel Castro's currency moves made Cuba a more expensive and thus less competitive tourist destination in the Caribbean region (Sharpley and Knight 2009, 249). It would take another six years (until March 2011) for the CUC to be devalued back to the level of parity with the US dollar, while the 10 percent surcharge on dollar/CUC exchanges remained in effect until the CUC was finally removed from circulation in January 2021.[14]

Most notably, the nomination of Marrero Cruz as tourism minister in February 2004 signaled the beginning of a process of militarization in Cuba's tourism sector that resulted in Gaviota progressively breaking away from MINTUR and assuming a leading role in the execution of hotel projects on the island, a role that it continues to play today (Figueras Pérez 2020; Henthorne 2018, 75; Espino 2008, 131). This process gathered pace after longtime defense minister Raúl Castro temporarily took over power from his ailing older brother Fidel in late July 2006, and especially after he officially became Cuba's president in February 2008. Whereas in 2000 Gaviota controlled only about 10 percent of all hotel rooms in Cuba, its share of total rooms had reached 28.3 percent by 2010 and nearly 37 percent by the end of 2014. Gaviota incorporated over 19,000 new rooms into its hotel portfolio between 2000 and 2014, approximately three-fourths of all rooms that were added to Cuba's total hotel capacity during this period.[15]

After taking the reins of Cuba, Raúl Castro introduced several economic reforms (including significant structural changes) primarily intended to improve living standards on the island, boost domestic production and efficiency, and overcome chronic shortcomings typical of highly centralized economic systems like that of Cuba.[16] Some of these reforms had important implications for the Cuban tourism industry. First, in line with his stated commitment to reduce bureaucratic bottlenecks and make life easier for Cuban citizens, Cuba's new president did away with a number of restrictions and limitations that angered people on the island. Among various measures that included the legalization of personal cellphones, the removal of long-standing bans on car and private home sales, and an immigration reform giving Cubans greater freedom to leave and enter the country, Raúl Castro repealed the prohibition on Cubans staying at luxury beach resorts and hotels formerly reserved exclusively for international visitors. Cubans were also allowed to rent cars at state-run agencies. The end of this type of segregation, which had fueled the spread of the expression "tourist apartheid" to describe Cuba's tourism policies (Cabezas 2009, 106; De la Fuente 2001, 329; Hagedorn 2001, 25; Schwartz 1997, 210), led to a sizable increase in hotel stays by Cubans and in hard currency (CUC) revenues generated by national tourism (Perelló Cabrera 2015b).

Second, Havana's government implemented new rules in October 2010 allowing more Cubans to work for themselves and set up small private businesses in various areas as a way to create jobs for 1 million excess state workers (20 percent of the country's total labor force) that were supposed to be dismissed over the next five years (Vidal Alejandro and Pérez Villanueva 2012, 44).[17] Although the layoff plan soon fell behind schedule and was put on hold indefinitely, the expansion of self-employment proceeded strongly. As a result, Cuba witnessed a surge of private entrepreneurs or *cuentapropistas* in tourism as many Cuban residents began to offer new products and services to tourists or expanding traditional businesses such as *paladares* (private restaurants), *casas particulares* (private bed and breakfasts), and taxi services. In October 2013, Resolution No. 145 of Cuba's Ministry of Tourism authorized state companies in the tourism sector to contract Cuban private entrepreneurs for lodging, meals, transportation, excursions, entertainment, and other support services. State-run travel agencies also started to promote private offers in their tourism packages and programs.[18]

Third, a noteworthy change was Raúl Castro's decision to ease rules on land rights for foreign investors in Cuba with an eye toward developing golf courses (and marinas) surrounded by luxury housing for tourists.[19] Decree Law 273 of August 2010 allowed overseas investors to lease state land on the island for up to ninety-nine years instead of the previous limit of fifty years. Its preamble said that the law aimed at "facilitating the process of participation of foreign investment in international tourism" by giving "greater security and guarantee to the foreign investor in the real estate business."[20] Finally, along with declarations by Havana's officials that foreign investment will now transcend its current complementary role to domestic investment efforts to soon occupy a major role in the economy,[21] Cuba enacted a new foreign investment legislation in June 2014 to replace Law 77. Passed in March 2014, Law 118 beefed up investment security, streamlined the approval process for new projects, and granted more attractive fiscal benefits to foreign investors.[22] In late 2013, Cuba also opened its first special development zone (*Zona Especial de Desarrollo Mariel*, ZEDM) around a container terminal at the port of Mariel, located approximately 30 miles west of Havana. Decree Law 313, which regulates operations in the ZEDM, offered overseas investors even better tax breaks and other incentives than those under Law 118.[23] Tourism was one of the sectors prioritized by the Cuban government in its search for new foreign investment projects (Salinas, Mundet, and Salinas 2018, 225).

Flows of International Visitors

The Cuban tourism industry, which was relatively small prior to 1990, has grown dramatically (as measured by the annual number of international visitors to Cuba) since the legalization of the dollar sector of the economy in 1993, with a small decline in the immediate aftermath of the September 11, 2001, terrorist attacks on the United States and another slight drop in 2006–7. According to official data displayed in Figure 2.1, international tourist arrivals in Cuba rose from 340,000 in 1990 to over 1 million in 1996 and broke the 2 million mark in 2004 and the 3 million mark in 2014. Although the Cuban government had even more ambitious expansion plans for the island's tourism sector,[24] tourist arrivals to Cuba increased on average at an impressive 9.9 percent annually in 1990–2014. They also increased in the aftermath of the 2007–8 global financial and economic crisis that affected many countries simultaneously and slammed the tourism industry worldwide. Most notably, Cuba recorded the highest rate of growth in tourist arrivals of all Caribbean countries in the decade or so following the collapse of the former Soviet Union. By the mid-2000s, Cuba had become the third most popular tourism destination in the Caribbean region behind Puerto Rico and the Dominican Republic and the second destination in the region for European travelers (Cerviño and Cubillo 2005, 223).

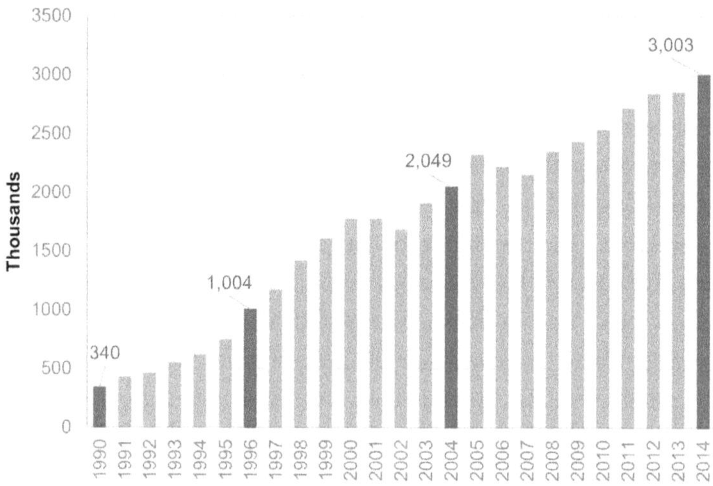

Figure 2.1 International visitors to Cuba, 1990–2014. Generated by the authors, based on ONEI 2015a, 2009a, 2004, 1998.

In reality, as reported in Table 2.1, during the 1990s Canada consolidated its role as the top source of tourists to Cuba (except in 1996–7 when it was surpassed by Italy). In 1996, more than fifty planes leaving from various parts of Canada every week transported around 160,000 Canadian tourists (the majority from Quebec) to Cuba.[25] By the end of the decade, Canada contributed nearly 300,000 annual visitors to the island or approximately 17 percent of total international arrivals. Nevertheless, arrivals from several European countries also increased markedly. Germany, Italy, Spain, and France each contributed well over 100,000 annual tourists to Cuba by 1998–9, and the UK sent a sizable number of visitors as well. As the 1990s came to a close, European travelers accounted for more than half of all foreign visitors to Cuba.[26] Other major source markets were Mexico, the United States, and, to a lower extent, Argentina, Venezuela, and Russia. Furthermore, Cuban overseas migrants, especially those living in South Florida, were making a growing contribution to total inbound arrivals.

In 1998 and 1999, fueled primarily by rising numbers of Cuban American trips to the island, Cuban residents abroad ranked fifth among Cuba's largest source markets for overseas visitors (ONEI 2000). And if we add US citizens of non-Cuban origin (62,345) to the number of Cuban Americans (an estimated 107,000) who traveled to Cuba in 1999 (a total of about 169,000 people), by the end of the 1990s, the United States was already the third largest source of international visitors to Cuba after Canada and Germany, despite a continuous ban on US-based travel to the island for tourism purposes, which remains in place today, and additional travel restrictions imposed by the US government (Spadoni 2010, 136). As a matter of fact, many individuals from the United States had been visiting Cuba through third countries without US travel permits for quite some time, technically violating US economic sanctions prohibiting the spending of money for unlicensed purposes. Between 1994 and 2002, the number of individuals subject to US law traveling to Cuba, but not authorized to do so, increased on average 19 percent to 21 percent, while legal visits rose by just 9 percent to 11 percent (USCTEC 2003). Cuban authorities, eager to accept US visitors paying in hard currency, did not stamp the passports of Americans (this practice ended in 2014 though), leaving no official trace of their presence.[27]

Although they have changed many times since 1963, restrictions on travel from the United States to Cuba have been a key component of US policy toward the Cuban revolutionary government for most of the last sixty years. During the 1990s, President Bill Clinton made several changes to travel regulations in response to actions by the government of Fidel Castro. As a reaction to

Table 2.1 International Visitors by Origin, 1990–9

	1990	1991	1992	1993	1994	1995	1996	1997	1998	1999
Canada	74,435	81,073	94,068	114,849	109,731	143,541	162,766	169,686	215,644	276,346
Germany	64,165	63,920	58,131	63,243	60,598	57,487	80,185	86,509	148,987	182,159
Italy	16,924	24,916	39,750	49,180	66,367	114,767	192,297	200,238	186,688	160,843
Spain	33,946	40,685	49,837	54,133	62,179	89,501	117,957	116,606	140,435	146,978
Cuban residents abroad[a]	2,900	5,100	16,200	21,500	37,200	43,700	64,800	79,792	105,467	133,463
France	8,244	9,864	13,177	23,352	33,910	34,332	62,742	93,897	101,604	123,607
UK	–	–	–	6,225	25,425	19,614	28,077	46,215	64,272	85,829
Mexico	34,465	48,771	44,644	56,254	49,096	32,069	37,229	52,712	61,589	70,983
United States	7,375	11,233	10,050	14,715	17,937	20,672	27,113	34,956	46,778	62,345
Argentina	5,450	16,236	19,478	29,254	36,414	32,583	31,331	41,511	47,579	42,612
Venezuela	17,306	10,280	11,369	14,444	9,026	8,621	8,270	8,558	9,986	12,723
Russia	–	–	–	9,350	9,038	8,263	9,116	11,406	11,711	11,200
Others	75,119	111,963	103,906	89,524	102,297	140,345	182,453	227,997	275,092	293,693
Total	340,329	424,041	460,610	546,023	619,218	745,495	1,004,336	1,170,083	1,415,832	1,602,781

Sources: ONEI 2000, 1998; UNWTO 2001; Alvarez and Amat 1996.
Note:
[a] Mostly Cuban Americans.

the *balsero* crisis of 1994,[28] Clinton banned family visits to Cuba by Cuban Americans except in cases of "extreme humanitarian need." In 1996, after the shooting down of two US planes flown by members of the Miami-based Cuban exile group, Brothers to the Rescue, he suspended direct flights between the two countries (Robyn, Reitzes, and Church 2002, 2). Yet, in 1999, as part of a policy aimed to promote people-to-people contacts, he streamlined travel procedures for students, athletes, artists, and other groups and individuals to visit Cuba. Clinton's policy, inaugurated in the wake of Pope John Paul II's trip to Cuba, also allowed resumption of charter flights from Miami to Havana as well as new direct flights from New York and Los Angeles. These changes were mainly intended to facilitate family reunions between Cuban Americans and their relatives on the island (Eckstein and Barberia 2001, 12).

Until recently, Cuba's official statistics published by the Office of National Statistics and Information (*Oficina Nacional de Estadística e Información*, ONEI) did not include data on arrivals to the island of Cuban residents abroad, let alone on arrivals of Cuban Americans. Stretching back to the year 2010, the annual numbers of visits to Cuba by Cuban residents abroad were first reported by ONEI in its *2015 Statistical Yearbook* (ONEI 2016a). Nevertheless, pre-2010 numbers can be compiled from the "other Caribbean arrivals" category in the country data provided by the Cuban government to the United Nations World Tourism Organization (UNWTO). This special category refers only to Cuban visitors because arrivals from each Caribbean country are reported separately in the UNWTO's tourism statistics. Data on Cuban American trips to Cuba (annual numbers for 2017 and 2018) instead appeared for the first time in early 2019 (ONEI 2019a). Previously ONEI simply published data on US citizens of non-Cuban origin traveling to Cuba with a US passport (reported as international visitors from the United States) but not on Cuban Americans entering the island with a regular Cuban passport.[29] Anyone who left Cuba after December 31, 1970, is considered a Cuban citizen by Cuban authorities and required to travel to his or her country of origin with a Cuban passport. At any rate, Paolo Spadoni (2010, 133–136) estimated that Cuban Americans accounted for around 90 percent of all Cuban residents abroad who visited Cuba in 1990–8 and for approximately 80 percent of those who visited in 1999 and in the 2000s. García Jiménez et al. (2006, 14) and the United States International Trade Commission (USITC 2007, 3–14) had similar estimates for the 2000s.[30] As for the annual flows of Cuban American travelers to Cuba in the 2010s, the authors of this book used official data from the Cuban Ministry of Tourism.

During the 2000s, Canada sustained the Cuban tourism industry by sending more tourists to Cuba than any other country. Canada had the highest degree of seasonality among Cuba's largest tourism source markets, with a strong concentration of tourist flows between December and March. It also had the heaviest focus on sun-and-beach tourism since more than 80 percent of its travelers to the island stayed at all-inclusive beach resorts, primarily in Varadero, Jardines del Rey, and Northern Holguín. Yet the average length of stay of Canadians in Cuba generally was shorter than that of Europeans (Quintana et al. 2004, 129–38).[31] As exhibited in Table 2.2, annual Canadian visitors to Cuba skyrocketed from 307,725 in 2000 to 818,246 in 2008. During that period, tourists from most European countries, especially Germany and Italy, declined substantially. A notable exception was the UK whose annual travelers to the island more than doubled to a peak of 211,075 in 2006 and to some 194,000 two years later, making British tourists the largest group among Europeans. Leaving aside arrivals of Cuban residents abroad that increased markedly from 137,158 in 2000 to 226,857 in 2008, tourists from the UK were the only major group of international visitors to Cuba that did not suffer a drop in the wake of the 9/11 terrorist attacks on the United States (ONEI 2003). In 2008, Canadian citizens accounted for about 35 percent of total international arrivals to Cuba, a bigger share than all European tourists combined (only 32 percent). That year, in terms of visits, Cuba was the fourth most important international destination for Canadians after the United States, Mexico, and the UK, and the most important destination in the insular Caribbean followed by the Dominican Republic. In 2008, according to Canada's official sources, Canadians spent an estimated 791 million Canadian dollars in Cuba (Statistics Canada 2010, 411).

Between 2000 and 2003, arrivals to Cuba from the United States also grew significantly, making US-based visitors the second-largest group among international travelers to the island. An estimated 135,000 Cuban Americans and 84,529 US citizens of non-Cuban descent (a total of about 220,000 people) visited Cuba in 2003, representing nearly 12 percent of total arrivals (ONEI 2004). Many of these visitors traveled to Cuba via third countries with or without their government's authorization.[32] In order to travel to Cuba, individuals subject to US law must be authorized by a general license (which requires no written authorization) or a specific license (which requires approval) from the Office of Foreign Assets Control (OFAC) of the Department of the Treasury. In 2003, the list of individuals who could travel without specific documentation from the OFAC included Cuban Americans on family trips (only once a year),

Table 2.2 International Visitors by Origin, 2000–2008

	2000	2001	2002	2003	2004	2005	2006	2007	2008
Canada	307,725	350,426	348,463	452,438	563,371	602,377	604,263	660,384	818,246
Cuban residents abroad[a]	137,158	139,099	157,501	168,251	125,442	148,891	158,463	192,847	226,857
UK	90,972	94,794	103,741	120,866	161,189	199,399	211,075	208,122	193,932
Italy	175,667	159,423	147,750	177,627	178,570	169,317	144,249	134,289	126,042
Spain	153,197	140,125	138,609	127,666	146,236	194,103	185,531	133,149	121,166
Germany	203,403	171,851	152,662	157,721	143,644	124,527	114,292	103,054	100,964
France	132,089	138,765	129,907	144,548	119,868	107,518	103,469	92,304	90,731
Mexico	86,540	98,495	87,589	88,787	79,752	89,154	97,984	92,120	84,052
Argentina	54,185	40,964	9,389	13,929	23,460	24,922	30,383	37,922	47,405
United States	76,898	78,789	77,646	84,529	49,856	37,233	36,808	40,521	41,904
Russia	14,429	14,391	10,653	12,610	17,457	20,711	27,861	29,077	40,621
Venezuela	12,481	16,375	10,977	15,228	86,258	185,157	83,832	33,593	31,931
Others	329,242	331,044	311,270	341,482	353,469	416,025	422,357	394,839	424,489
Total	1,773,986	1,774,541	1,686,162	1,905,682	2,048,572	2,319,334	2,220,567	2,152,221	2,348,340

Sources: ONEI 2009a, 2003; UNWTO 2010, 2009, 2004.

Note:

[a] Mostly Cuban Americans.

government officials, journalists regularly employed by news organizations, full-time academic researchers, and employees of US firms selling eligible agricultural and medical goods to Cuba. In contrast, a specific license was required for Cuban Americans visiting relatives in Cuba more than once a year, freelance journalists, members of religious and humanitarian organizations, university students, and athletes. Moreover, US rules of March 2003 forbade trips to Cuba that combined non-credit educational activities with people-to-people contacts. The latter had become a loophole for groups to travel to the island when the educational aspect was barely evident.[33] For the purpose of US policy toward Cuba, it is worth emphasizing that US-based visitors had become a valuable source of hard currency for the Cuban tourism industry. By 2001, even with sanctions in place, American travelers reportedly spent some $200 million a year in Cuba, thus contributing to more than 10 percent of the Castro government's revenues from tourism activities.[34]

On June 30, 2004, US president George W. Bush implemented tougher measures aimed at stemming the flow of hard currency reaching Cuba and hastening the end of the Castro government. Along with further restrictions on US-based educational travel to the island, these measures prohibited Cuban Americans from visiting relatives in Cuba more than once every three years (a specific license was required for every trip), limited their visits only to immediate family members (grandparents, grandchildren, parents, siblings, spouses, and children), limited their stay to fourteen days, and reduced the amount they could spend during their visits from $167 to $50 per day.[35] Trips to Cuba by US citizens of non-Cuban descent plunged to just 37,233 in 2005 (ONEI 2006) and remained virtually stagnant in 2006–8 at less than half their level of 2003. Cuban American visits instead suffered a less marked drop despite being the main targets of Bush's travel restrictions. They shrank by around 25 percent in 2004 but steadily recovered afterward to reach some 181,000 in 2008. Illegal trips sustained by family ties probably had a lot to do with the stark difference between Cuban American and other US citizens' travel activities with respect to Cuba in the period after 2004. Cuban American trips to Cuba in 2004–6 were more than three times the number of annual licenses issued by the OFAC for family visits (USITC 2007, 3–4).[36] After dropping to fifth place in 2005, the United States was again the second-largest source of foreign visitors to Cuba behind Canada in 2008.

Meanwhile, arrivals from Mexico remained sluggish, while visitors from nontraditional source markets such as Argentina and Russia experienced some notable growth. Around the mid-2000s, there was also a major increase in Venezuelans traveling to Cuba to undergo free eye surgeries under the Caracas-Havana joint health program Operation Miracle. The number of arrivals from Venezuela rose from 15,228 in 2003 to 185,157 in 2005. Later, beginning in 2007, this number fell drastically as Venezuelan and Cuban authorities chose to have Cuban physicians attend to the majority of Venezuelan patients in their own communities rather than flying them to the island to be treated. The move apparently was intended to ensure a more comfortable experience for Venezuelan patients as well as to reduce transportation, accommodation, and other costs assumed by the Cuban government (Kirk 2011, 369). Some Cuban economists maintained that health-related concerns were an additional factor. Many Venezuelans came from poor areas with a high prevalence of the dengue virus and inadequate sanitary controls, thus raising the possibility of infected travelers causing second-level outbreaks of the disease in Cuba at a time when Cuba was already battling an outbreak within its own territory.[37]

The global financial and economic crisis that began in 2007 and erupted with full force toward the end of 2008 dealt a serious blow to Cuba, which was already reeling from extensive damages caused by two powerful hurricanes, a weak macroeconomic situation, and the cumulative consequences of the long-standing US embargo (Vidal Alejandro 2010, 28).[38] The crisis drove prices of Cuba's food and oil imports upward, pushed prices of key export products like nickel downward, curbed demand for Cuban professional services abroad (the top generator of hard currency for the Cuban government), made credit tighter, and led to a drop in tourism revenues and possibly to a decline of family remittances (Mesa-Lago and Vidal Alejandro 2010, 695–8; CEPAL 2010; Pérez 2009). But total arrivals of international visitors to Cuba continued to increase.

It seems fair to argue that Canadian tourists enabled Cuba to weather the global economic crisis that had a strong negative impact on the tourism industry of other Caribbean countries (Maliska 2009). Among Cuba's top tourist supplier markets, Canada was the only country whose visitors to the island rose significantly in 2009 (almost 12 percent or around 100,000 more visitors than in 2008).[39] The annual number of Canadian tourists to Cuba increased an additional 28 percent in the next five years to reach roughly 1.2 million in 2014, when Canadians accounted for nearly 40 percent of all international arrivals on the island (Table 2.3). During the same period, total tourist arrivals from Europe fell slightly (with British and Spanish travelers suffering big drops) and in 2014 they represented just 23 percent of all arrivals from abroad. However,

Table 2.3 International Visitors by Origin, 2009–14

	2009	2010	2011	2012	2013	2014
Canada	914,884	945,248	1,002,318	1,071,696	1,105,729	1,175,077
Cuban residents abroad	296,181	375,431	397,873	384,188	373,427	361,228
Cuban Americans[a]	163,019	262,963	284,942	268,803	261,084	258,837
Germany	93,437	93,136	95,124	108,712	115,984	139,138
UK	172,318	174,343	175,822	153,737	149,515	123,910
Italy	118,347	112,298	110,432	103,290	95,542	112,076
France	83,478	80,470	94,370	101,522	96,640	103,475
United States	52,455	63,046	73,566	98,050	92,348	91,254
Mexico	61,487	66,650	76,326	78,289	84,704	82,820
Venezuela	28,657	30,965	34,096	36,373	45,943	78,839
Spain	129,224	104,948	101,631	81,354	73,056	77,099
Russia	37,391	56,245	78,472	86,944	70,401	69,237
Argentina	48,543	58,612	75,968	94,691	90,084	68,849
Others	393,407	370,353	400,319	439,761	459,199	519,743
Total	2,429,809	2,531,745	2,716,317	2,838,607	2,852,572	3,002,745

Sources: ONEI 2016a, 2011a; UNWTO 2011.
Note:
[a] MINTUR data obtained by the authors in February 2015.

then Cuban tourism minister Manuel Marrero stated in August 2009, "Tourism is growing, tourists keep coming, but they have changed the way they travel. . . . They are coming for less time, trying to come with a lot of discounts, and that has caused a decline in profits."[40]

In effect, Canada's official statistics reveal that Canadians spent less money in Cuba in 2014 (an estimated 780 million Canadian dollars) than in 2008, even if trips to the island had grown considerably.[41] Although they usually have tended to have shorter vacations and spend less money in Cuba than European travelers (Espino 2010, 368), Canadians were, until recently (until November 2022, when new rules allowed all foreign tourists to remain in Cuba for a total of 180 days)[42] the only group of tourists who could legally stay in Cuba for up to six consecutive months, as opposed to citizens of other countries whose maximum stay in a single trip could not exceed sixty days.[43] This was part of an effort by the Cuban government to attract Canadian "snowbirds" that spent a large portion of the winter in warmer places in the southern regions of the United States, Mexico, and the Caribbean.[44] Furthermore, it is not a secret that many Canadians, and Europeans as well, have always liked to go to Cuba for the simple reason that relatively few Americans go there.

Regarding visitors from the United States, the Cuban American segment experienced notable dynamism following President Barack Obama's decision in April 2009 to eliminate all restrictions on family visits (and family remittances) to Cuba.[45] Regulations enacting these changes were issued in September 2009.[46] The annual number of Cuban Americans traveling to Cuba rose from 163,019 in 2009 to 258,837 in 2014.[47] By 2010, Cuban Americans alone were the second-largest group of international visitors to Cuba (Perelló Cabrera 2011b, 27; ONEI 2011a). In early 2011, Obama also eased restrictions on travel to Cuba related to religious, educational, and other people-to-people exchanges and permitted all US international airports to apply for licenses to establish charter flights to Havana and other Cuban cities. These measures practically reinstated the policies of President Bill Clinton in the late 1990s (Sullivan 2012). Even so, in May 2012 the Obama administration issued stiffened rules on educational and cultural travel to Cuba, in part to prevent potential abuses of existing programs but also to placate Cuban American hardliners in the US Congress who had denounced people-to-people exchanges as just another form of tourism filling the coffers of the Raúl Castro regime.[48] After three years of virtual stagnation as a result of George W. Bush's restrictive policies, annual trips to Cuba by US citizens of non-Cuban descent resumed growth in 2009 and reached 98,050 in

2012 (a level well below their potential yet the highest since the 1950s) before dropping slightly to 91,254 in 2014. As the year 2014 neared its end with the momentous announcement of a thaw in relations between Washington and Havana, whose effects on travel and business activities will be analyzed in the next chapter, Cuba's largest visitor source markets were, in descending order, Canada, United States, Germany, UK, Italy, France, and Mexico.

Tourism Infrastructure and Foreign Participation

Since the beginning of the Special Period in the early 1990s, Cuba has worked actively to keep pace with tourist demand and accommodate a rapidly growing number of visitors from abroad. The Cuban government's efforts to cope with the tourist boom resulted in the widespread construction of new hotels (and repairs of some existing installations) with a great deal of new supplementary services and in substantial improvements to Cuba's infrastructure and support services such as airports, ports, roads, electricity, sewage, and telecommunications facilities that serve as the arteries of the tourism industry. Nevertheless, whereas the presence of foreign capital in key sectors of the Cuban economy closely linked to tourism has been considerable, the amount of foreign direct investment in Cuba's accommodation sector was modest as almost all hotel investments on the island were financed with domestic resources. Between 1990 and 2019, more than $5 billion was invested in hotel construction and renovation in Cuba. Foreign capital reportedly accounted for a meager 4 percent of that amount and for just 2 percent of all investment in Cuban tourism (Figueras Pérez 2020).

The Cuban tourism market is rather unique. Unlike leading tourism destinations in the Caribbean (Dominican Republic, Puerto Rico, Jamaica, and the Bahamas in particular) and elsewhere where most hotels are owned by international chains, Cuba is one of the world's few remaining countries where tourism establishments by and large are owned and controlled by the state. Furthermore, despite the introduction of piecemeal liberalizing measures to promote a transition to a mixed economy, Cuba is the only country in the world that boasts a significant international tourism sector within the framework of a centrally planned economy. Cuba is also unique in that it is a destination country subject to a comprehensive economic embargo imposed by the United States that prohibits US tourism on the island, severely restricts US-based travel to Cuba for other purposes, and makes most financial and commercial transactions with

Cuba illegal for US citizens and companies. This is all the more damaging given that the United States is the Caribbean region's principal tourism source market. And yet Cuba features a commercialization scheme based on a primary product like sun and beach marketed through the same distribution channels (travel agencies, tour operators, airlines, international hotel chains) that promote tourism in other Caribbean countries and help develop the region's tourism destination image (Perelló Cabrera 2019).

Since 1990, stimulated by the reopening of the island to international tourism, Cuba has witnessed significant hotel construction. As exhibited in Figure 2.2, Cuba's total hotel rooms doubled from 18,565 in 1990 to 37,178 in 2000 as several new hotels and resorts catering primarily to Canadian and European visitors were built in major tourism poles. In reality, if we exclude low-end establishments providing lodging for national tourists, Cuba's hotel capacity nearly tripled from 12,900 rooms in 1990 to about 36,000 in 2000 as close to 2,300 new hotel rooms were added each year during that period (Pérez Villanueva 2012, 221; Peters 2002).[49] The total number of hotel rooms has continued to increase at a rather steady pace since then, reaching 56,432 in 2010 and 62,090 in 2014. There is no question that over the past three decades the Cuban government has made tremendous financial efforts to expand the island's hotel capacity (González Núñez 2022, 124–5). Albeit at a level below the average annual growth rate in tourist arrivals (10.1 percent), Cuba's total hotel

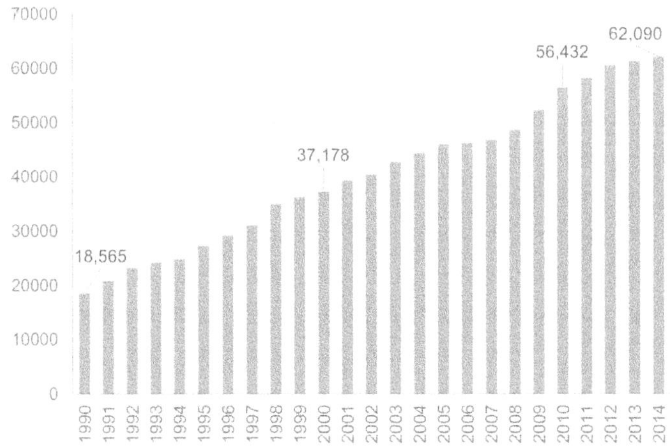

Figure 2.2 Cuba's total hotel rooms, 1990–2014. Note: Includes rooms for national and international tourism. MINTUR data obtained by the authors in May 2017 and May 2015.

rooms grew at an annual average of 5.1 percent during the 1990–2014 period, which translates into approximately 1,800 new rooms being added each year. Regarding the distribution among the various Cuban state entities of the 62,090 hotel rooms that existed in Cuba at the end of 2014, the military-run Gaviota group controlled 22,903 rooms (mainly in four- and five-star hotels) or about 37 percent of the total. As for the remaining rooms, 15,368 belonged to Cubanacán, 12,250 to Gran Caribe, 598 to Habaguanex, and 10,971 to other entities. In 2014, there were also 18,742 rooms in privately operated bed-and-breakfast lodging establishments (*casas particulares*), putting the total number of available hotel rooms in Cuba to 80,832 (Perelló Cabrera 2015a).

To highlight the transformation of Cuba's hotel room supply fueled by government efforts to develop upscale lodging accommodations for international tourists, Table 2.4 displays the annual percentage distribution by hotel rating of hotel rooms in Cuba during the 1990–2014 period. It must be underscored, though, that although Cuba has adopted a star rating system for its hotels that mirrors classification schemes used almost everywhere in the world, Cuban star ratings are widely considered to be overly generous, as many high-end hotels and resorts on the island remain somewhat below international standards (Feinberg and Newfarmer 2016).[50] In 1990, there were only three five-star and nineteen four-star hotels in Cuba with a combined 4,570 hotel rooms, representing 24.7 percent of the country's total hotel rooms. That year, nearly 40 percent of all rooms were in three-star hotels and over 20 percent in one-star hotels. A decade later, things had changed quite significantly. In 1999, more than half (55.6 percent) of Cuba's rooms were in seventeen five-star hotels and fifty-eight four-star hotels. Several of these properties were all-inclusive beach resorts catering to Western European and especially Canadian tourists. By 2013, forty five-star hotels and seventy-nine four-star hotels had a combined stock of nearly 40,000 rooms representing more than 60 percent of all hotel rooms available in Cuba. The bulk of Cuba's total hotel rooms (71.6 percent) in 2013 were in beach areas, 22.8 percent were in cities, and the remainder (5.6 percent) were in natural reserves and other areas.[51] At the end of 2014, the combined room stock of five- and four-star hotels accounted for over 70 percent of Cuba's total hotel rooms.

Table 2.5, which provides data on hotel rooms in Cuba's principal tourism poles between 1990 and 2014, shows how the Cuban government's policy for international tourism, despite a considerable growth of urban-based tourism and hotel accommodations in Havana in the post-1989 period (Colantonio and

Table 2.4 Hotel Rooms by Category, 1990–2014 (percentage distribution)

Category	1990	1993	1996	1999	2002	2005	2008	2011	2014
Five-star	5.6	13.4	13.1	16.6	18.5	18.4	22.4	30.2	36.9
Four-star	19.1	28.1	33.7	39.0	39.6	41.4	43.1	33.7	37.6
Three-star	38.2	33.3	30.1	19.7	18.2	18.9	22.4	22.8	16.7
Two-star	7.2	4.5	6.3	9.8	9.0	8.9	7.1	8.1	6.8
One-star	21.6	14.5	9.3	8.5	8.5	7.1	3.9	4.2	1.2
Unrated	8.3	6.2	7.5	6.4	6.2	5.3	1.1	1.0	0.8

Sources: ONEI 2015a, 2009a, 2003, 1998.

Table 2.5 Hotel Rooms in Principal Tourism Poles, 1990–2014 (percentage distribution)

Location	1990	1993	1996	1999	2002	2005	2008	2011	2014
Varadero	25.5	25.8	27.2	31.2	31.1	30.4	32.2	33.1	32.3
Havana	38.5	36.8	31.9	29.5	28.3	26.7	24.4	22.0	18.4
Holguín	5.6	5.5	6.5	7.3	8.6	9.5	9.6	8.3	9.5
Ciego de Ávila	–	–	4.2	6.7	8.3	8.7	8.2	8.4	7.0
Santiago de Cuba	5.5	7.3	7.0	5.2	4.9	4.8	3.9	3.2	3.7
Cayo Largo del Sur	1.3	1.4	1.8	1.8	2.1	1.8	2.5	2.2	2.2
Others	16.8	17.0	21.4	18.3	16.7	18.1	19.2	22.8	26.9

Sources: ONEI 2015a, 2009a, 2003, 1998.

Potter 2006b, 26–9), prioritized the development of coastal areas to create beach resorts. In 1990, according to official sources, 8,017 hotel rooms or 38.5 percent of Cuba's total rooms were located in Havana. With the construction of several new hotels and the refurbishment of some existing facilities, Havana's room stock for tourists rose notably in the next decade to reach 11,781 in 1999, yet its share of the island's total stock fell to 29.5 percent. The relative contribution of beach areas instead progressively grew in importance. Varadero, the most famous beach resort area in Cuba, had 5,299 rooms in thirty-three establishments in 1990, but apart from a handful of new four-star hotels,[52] these were old properties in poor condition that could not compete in the international tourism market (Woodard 2018, 197). Varadero expanded its hotels, including several five-star resorts, at a rapid pace during the 1990s (especially in the second half) and by the end of the decade had more than 12,000 rooms accounting for almost one-third of Cuba's total hotel rooms. Meanwhile, nontraditional beach areas also were developed as all-inclusive tourism destinations, in particular Guardalavaca on the northern coast of the Holguín province and the group of keys or *cayos* (Cayo Guillermo, Cayo Coco, Cayo Santa Maria, and others) collectively known as Jardines del Rey off the northern coast of the provinces of Ciego de Ávila and Camagüey. In 1999, the tourism poles of Holguín and Ciego de Ávila had a combined 5,600 rooms or about 14 percent of Cuba's total hotel rooms.

Besides major efforts to promote further growth of the aforementioned tourist places, Cuba developed additional beach areas (for some, development had already started in the 1990s) as all-inclusive resort destinations for international tourists during the 2000s. Among them were Guardalavaca's neighboring playas Esmeralda, Pesquero, and Turquesa, the small island of Cayo Largo del Sur off Cuba's south coast, Playa Santa Lucia on the northern coast of the Camagüey province, Cayo Ensenacho off the northern coast of the province of Villa Clara (near Cayo Santa Maria), and Playa Jibacoa on the north coast of the Mayabeque province. In 1998, there were thirty-three all-inclusive hotels in Cuba with 8,900 rooms, or approximately 30 percent of total rooms for international tourism. By early 2006, seventy-six resorts under such modality had a total of 26,000 rooms, representing close to 60 percent of the island's hotel room capacity,[53] and this trend has certainly intensified since then. At the end of 2014, hotel rooms in the urban tourism poles of Havana and Santiago de Cuba (the island's second-largest city) accounted for about 22 percent of Cuba's total rooms. Havana's share (18.4 percent), most notably, was less than half its level in 1990. On the other hand, the combined relative contribution of Varadero (32.3 percent), Holguín

(9.5 percent), Ciego de Ávila (7 percent), Cayo Largo del Sur (2.2 percent) and other poles (26.9 percent) to Cuba's total room stock was nearly 80 percent. The share of total rooms of each one of these poles was significantly higher than their level in 1990.

As noted before, foreign investment inflows have played a marginal role in financing hotel development for international tourism in Cuba. Joint ventures with overseas firms were the main form of foreign participation in the early 1990s when the Cuban economy was in free fall and domestic capital was in short supply. Typically, the foreign partner contributed capital for building the hotel and owned up to 50 percent interest in the joint venture, the Cuban partner provided the land and labor, and both parties shared management, marketing, and profits. The operational period of these arrangements was twenty-five years, renewable for another twenty-five years. While hotels under joint ventures were built in various tourist areas across the island, Varadero initially took center stage, with Cubanacán being the most active Cuban company pursuing these projects. The first joint venture with foreign capital in Cuba was between Cubanacán and Spain's Corporación Interinsular Hispana S.A. (CIHSA), in which the Spanish hotel chain Sol Meliá had a 25 percent stake. It completed three hotels in Varadero: the four-star Sol Palmeras in 1990, the five-star Meliá Varadero in 1992, and the five-star Meliá Las Américas in 1994. In 1990, a joint venture between Cubanacán and the Mexican group Desarrollo de Servicios Constructivos (DSC) completed the four-star Tuxpán hotel in Varadero that was managed by the German firm LTI International Hotels (Pérez-López 2004, 109–10; CEPAL 1995; Espino 1993, 57). Among the projects of Cubanacán in the early 1990s, the five-star Meliá Cohiba hotel in the Vedado district of Havana was supposed to be a joint venture with the Spanish consortium Ibercusa, but the partnership dissolved in late 1991, apparently due to the Spanish partner's inability to make payments for supplies. Cubanacán completed the project alone and Sol Meliá assumed the management of the hotel, which opened in 1994 (De las Cuevas Toraya 2001, 351).

Albeit no longer the predominant type of business partnership in hotel projects on the island as Cuba increasingly turned to foreign companies only for management activities, hotel joint ventures with overseas partners continued to be promoted in the second half of the 1990s and until around the mid-2000s, when a process of economic recentralization led to plummeting foreign investment inflows in nearly every sector of the Cuban economy (Spadoni 2014, 84). The number of Cuba's hotel rooms under joint ventures with foreign firms jumped from just 607 in 1990 to around 1,800 in 1995 and reached 5,207 in thirteen hotels in 2005 before experiencing a decade of almost no growth (MINTUR 2015a; CEPAL 2006, Pérez Villanueva 2004, 179). As for the ownership structure of these establishments, some international hotel chains, especially from Spain, acquired equity interests (never more than 50 percent) in certain hotels they managed in Cuba, while others invested in hotel joint ventures and later sold their shares to other foreign investors. Most hotel joint ventures were with Cubanacán, a few with Gran Caribe, and none with Gaviota.

At the end of 2014, foreign chains had equity interests in the following hotels in Cuba (still open today): Spain's Sol Meliá, which changed its name to Meliá Hotels International in 2011, had stakes in the hotels Sol Palmeras, Meliá Varadero, Meliá Las Américas, Meliá Habana, and Meliá Cayo Coco; Spain's Iberostar in the hotels Iberostar Cayo Coco-Mojito, Iberostar Trinidad, Iberostar Varadero, and Iberostar Parque Central (Havana); Spain's Blau in the hotel Blau Varadero; Spain's Roc Hotels in the hotel Arenas Doradas (Varadero); and Canada's Blue Diamond in the hotels Royalton Royal Hicacos (Varadero) and Memories Jibacoa, previously operated, respectively, by Jamaica's Sandals Resorts and Jamaica's SuperClubs.[54] Canada's Sherritt International, one of the largest foreign investors in Cuba with a diversified portfolio in mining, oil exploration, and power generation, held 25 percent of the hotel Meliá Las Americas and 12.5 percent of the hotel Meliá Habana until late 2003 when it sold its stakes in those hotels.[55] Another Canadian company, Leisure Canada, sought for several years to develop a luxury hotel in the Monte Barreto neighborhood of Havana, and a resort at Jibacoa, east of Havana. Leisure Canada created a hotel brand (Mirus Resorts and Hotels) to use on properties within Cuba and in mid-2003 supposedly was given the right to manage the Monte Barreto hotel under its own new brand (Spadoni 2010, 202), and yet it never broke ground on its projects.[56] Since 2011, Cuba also has received a number of proposals from foreign investors to develop real estate with modern golf courses on the island and some joint ventures reportedly were signed, but none of these projects have been completed so far.[57]

In general, negotiating joint venture agreements to build hotels in Cuba has always been difficult given the Cuban government's emphasis on state ownership of tourism facilities (Feinberg 2016, 111). Since the mid-1990s, as more domestic capital became available, Cuba's partnerships with international hotel chains in the tourism sector have prioritized management contracts rather than mixed enterprises. The duration of these contracts is usually five to seven years and can be prolonged if the partners wish to do so (Cárdenas O'Farrill 2021, 244). International hotel chains improved the quality of Cuba's tourism services, attracted large tour operators and major airlines interested in the island as a new tourist destination in the Caribbean, and played a significant tutorial role by transferring not only management but also marketing know-how to Cuban corporations. They also created substantial employment. As foreign partners do not receive a share of ownership like in joint ventures, management contracts ensure that more profits remain in Cuba. One obvious disadvantage for Cuba, though, is that it has to make the entire initial capital expenditure for the construction of the hotel. At the same time, Cuba has been a profitable market for foreign hoteliers in terms of management fees. The first contracts in the early 1990s established that the foreign partner's compensation consisted of a basic fee based on a share (3 percent to 5 percent) of the hotel's gross revenues and an incentive fee based on hotel occupancy. Later, the incentive fee was tied to overnight stays, and finally to profits. In their current form, management contracts no longer include basic fees and establish that payments to foreign companies consist only of incentive fees linked to levels of gross operating profits to encourage operators to perform well (D'Meza Pérez, Zaldívar Puig, and Martín Fernández 2017, 144; Cerviño and José María Cubillo 2005, 239).

As shown in Table 2.6, the number of hotel rooms under management contracts with overseas firms rose from only 1,330 in three hotels in 1990 to 14,221 in forty-six hotels in 1999, when they accounted for around 40 percent of Cuba's total rooms. By the end of the 1990s, there were already nearly five times as many rooms under management contracts as rooms under joint ventures. And while the latter barely increased after 2005, the former grew steadily to 22,705

Table 2.6 Hotel Management Contracts with Foreign Companies, 1990–2014

Year	Cuba's Total Hotel Rooms	Under Management Contract		
		Contracts	Rooms	% of Total Rooms
1990	18,565	3	1,330	7.2
1995	27,256	29	8,424	30.9
1997	31,043	33	10,037	32.3
1999	36,252	46	14,221	39.2
2004	44,268	54	19,124	42.3
2007	46,862	57	22,705	48.4
2013	61,250	63	31,573	51.5
2014	62,090	68	32,415	52.2

Sources: MINTUR data obtained by the authors in May 2017 and May 2015; Pérez Villanueva 2012.

rooms in fifty-seven hotels in 2007 and to 32,415 rooms (representing more than half of Cuba's total rooms) in sixty-eight hotels in 2014. Virtually all of these facilities were four- and five-star hotels (D'Meza Pérez, Zaldívar Puig, and Martín Fernández 2017, 143). Since 1990, many foreign hotel companies have entered and left the Cuban market. At the end of 2014, the list of international hotel operators with a Cuban portfolio included France's Accor and Warwick Hotels, Canada's Blue Diamond Resorts, Portugal's Pestana, Jamaica's SuperClubs, and the Spanish chains Meliá Hotels International, Iberostar, Barceló, BlueBay, Blau, Be Live, H10, Roc Hotels, NH, and Valentin.[58] Meliá was by far the leading operator in Cuba as it managed 12,310 rooms in twenty-nine hotels (with equity interests in five of them), representing one-fifth of all hotel rooms on the island.[59] Remarkably, between 1990 and 2010, Meliá's hotels generated close to $3.7 billion in revenues for Cuba, or an average of approximately $180 million per year.[60]

During the 1990–2014 period, Cuba built or refurbished about eighty hotels of category four and five stars with some 36,000 rooms combined.[61] Until the mid-2000s, the capital for such a massive construction effort came primarily from a tourism development fund administered by MINTUR, with a portion of the funds secured through international credits backed by the foreign chains that operate in Cuba. This financing scheme is still in place today, and new lines of credit have recently been extended for the renovation of some hotels of Cubanacán, Gran Caribe, and Islazul. Yet since around 2004, the Cuban military's tourism business arm, Grupo de Turismo Gaviota S. A., has been the most active player in pursuing tourism development projects in Cuba, using its own resources to finance hotel construction in Havana and in other choice locations (Figueras Pérez 2020). In addition to a sizable number of hotels and resorts (mainly five-star establishments), Gaviota's portfolio includes marinas, a tour company, a transportation service company, and other businesses.

It must be acknowledged that the extensive construction of new hotels and resorts in Cuba in the post-1990 period was carried out amid serious difficulties. It soon became evident that the inefficiencies and limitations of Cuba's construction firms constituted a major obstacle to the improvement of tourism services on the island, as the financial resources for the expansion of Cuba's tourism infrastructure clearly exceeded the construction capacity of the country. The Cuban construction industry (like other Cuban industries) suffered, among other things, from excessive bureaucracy and inadequate management, shortages of tools, equipment, materials, and fuel, transportation problems, reiterated delays in completing projects, poor workmanship and low quality materials (Miller and Henthorne 1997, 130; CEPAL 1995). By the end of the 1990s, in response to the growing demand from the tourism industry, Cuba had initiated an integrated partnership with the French company Bouygues Construction for the development of luxury complexes on the island. A global leader in construction with operations in over sixty countries, Bouygues has taken numerous orders for building complexes in Cuba over the past two decades, completing hotels in Havana, Varadero, the northern keys, and other popular tourist areas. In early 2003, most notably, Fidel Castro presided over the opening of the Gaviota-owned hotel Playa Pesquero in the eastern province of Holguín, a massive $100 million resort that was completed in less than two years using Cuban capital and labor with technical assistance by Bouygues.[62] As evidence of its substantial tourism presence in Cuba, Bouygues revealed in October 2017 that since it signed the first contract for a Cuban hotel in 1999, the company had delivered more than 30,000 rooms, or 50 percent of total hotel rooms available in Cuba and 100 percent of those in five-star establishments.[63]

In contrast to their modest contribution to the expansion of hotel capacities and supplementary hotel services, foreign investors have played a critical role in developing Cuba's infrastructure to support tourism growth and competitiveness. Designed by Canada's Intelcan Technosystems and built with a loan from the Canadian Export Development Corporation, the new terminal three at Havana's José Martí international airport was inaugurated in April 1998. That same year, the Cuban Ministry of Transportation and the Italian port operator Silares Terminales del Caribe NV formed the joint venture Cubanco to manage the cruise terminals in the ports of Havana, Santiago de Cuba, and Punta Francés on the Isle of Youth (Wright 2009, 208; Babun, Jr. 2007, 54). Canada's Sherritt International entered the power business in the late 1990s by signing a joint venture (Energas S. A.) with the Cuban government to build gas-fired electricity generation facilities at Varadero and Boca de Jaruco. In 1999, a joint venture between the Cuban state and the Spanish company Aguas de Barcelona created the firm Aguas del Oeste to manage the water service for the city of Havana. Furthermore, in the mid-1990s, Italy's Telecom Italia entered into a joint venture (ETECSA) with the Cuban state-run telephone firm Empresa Telefónica de Cuba (EMTEL) for the modernization and expansion of Cuba's telephone system. In early 1998, Sherritt also bought a 37.5 percent interest in the Cuban cellular carrier Teléfonos Celulares de Cuba (Cubacel), which had exclusive rights to frequencies in Cuba's dollar-priced cellular phone market available only to foreign visitors. Although both Sherritt and Telecom Italia later exited the telephone business on the island,[64] their presence and operations contributed significantly to the improvement of Cuba's telecommunications services (Spadoni 2014, 140–3; Pérez Villanueva 1999).

Finally, the enactment by the United States of the extraterritorial Helms-Burton law in March 1996 deserves a special mention since it complicated the business operations of foreign investors in Cuba and produced some negative effects on the Cuban economy. The final text of this law is composed of thirty-three sections grouped into four titles. Title I codifies all US embargo restrictions against Cuba in effect as of March 1, 1996. Title II lays down a series of conditions that would have to be met before Washington's reengagement with any future Cuban government. Titles III and IV are the aspects of the law aimed at Cuba's commercial partners. Title III of the legislation enables Americans, and Cubans who later became US citizens, to bring lawsuits in US courts against foreign firms and individuals who "traffic" in properties confiscated from them by the Castro government in the early 1960s. Title IV empowers US authorities to bar entry into the United States to senior officials of those firms and their immediate family members, such as spouses and any dependent children. To make it more difficult for foreign companies to evade Helms-Burton's reach, the authors of the legislation consciously left a margin of uncertainty in the interpretation of trafficking. This is broadly defined and includes selling, leasing, managing, and purchasing expropriated properties. It also includes the use of trademarks or licenses claimed by US firms.[65] Due to come into effect on August 1, 1996, Title III was waived every six months and thus never implemented until US president Donald Trump lifted its suspension in May 2019. An examination of the effects of the activation of Title III will be provided in Chapter 5.

While a thorough analysis of the impact of the Helms-Burton law on foreign investment in Cuba is beyond the scope of this study, business activities in the tourism sector demonstrate how the broad definition of trafficking added pressure on foreign firms to end their operations in Cuba. Several European and Canadian firms operating in the Cuban tourism sector have been targeted by Helms-Burton not only for equity positions in confiscated properties but also for management activities. Between 1996 and 1999, the US State Department sent "warning letters" to Leisure Canada and Air Transat (Canada), Club Med (France), LTI (Germany), and Sol Meliá (Spain) advising them that their operations in Cuba might constitute "dealing in expropriated goods" and they could face penalties under Title IV of Helms-Burton. Sol Meliá, in particular, was repeatedly threatened. In the mid-2000s, some Spanish hotel chains based in Majorca received similar warnings because of their management and investment activities in Cuban hotels allegedly built on expropriated properties (United Nations 2006, 23; Spadoni 2001, 29). And yet there is no evidence that Helms-Burton forced any of these firms to abandon Cuba or to refuse the renewal of a contract. In June 2004, Jamaica's SuperClubs pulled out of two hotel contracts in Cuba after receiving a US notice that it had a forty-five-day grace period before the imposition of visa restrictions on its top executives for travel to the United States, but the company continued to manage two other hotels on the island.[66] The US action against SuperClubs was the last application of Title IV to date. Between 1996 and 1997, Washington had already sanctioned three foreign investors in Cuba under this provision, Canada's Sherritt International for its nickel activities, Mexico's Grupo Domos for a telecommunications joint venture, and Israel's BM Group for its citrus operations (Pérez-Stable 2011, 69).

Overall, the Helms-Burton law caused losses for Cuba by hindering certain ongoing foreign investment projects, raising the cost of external financing due to a higher risk of doing business with the Cuban government, and producing a chilling effect on prospective investors. In the tourism sector, it is conceivable that some foreign firms with operations in the United States stayed out of Cuba because of the US policy toward the island that was reinforced by Helms-Burton (Spadoni 2010, 126–7). Oddly enough, even if Canada has been almost uninterruptedly the largest source market for international tourism in Cuba since 1985, the presence of Canadian firms (equity investors and management contractors) in Cuba's hotel business remained quite limited for many years, most likely because of their exposure to the US market and potential problems with Helms-Burton (Spadoni and Sagebien 2013, 86). This situation changed in the early 2010s when the Canadian chain Blue Diamond Resorts entered the Cuban tourism market and quickly became a major player. Blue Diamond, whose activities will be analyzed in the next chapter, manages several hotels (with equity positions in a few of them) in Cuba and has properties in the Dominican Republic, Jamaica, Mexico, and other Caribbean countries, but not in the United States.[67] However, Blue Diamond is the hotel division of Canada's Sunwing Travel Group, which does business in the United States, and has a partnership agreement with the US hotel giant Marriott International. In short, the Helms-Burton law might have discouraged some new foreign investment in the Cuban tourism market but was unable to stifle the growing presence of international hotel chains in Cuba and their contribution to the development of the island's tourism sector.

Economic Impact of Tourism

The main goals of any country that seeks to develop its tourism sector are fundamentally economic. Above all, tourism is promoted because of its strong potential to fuel economic growth, drive exports, generate hard currency earnings, stimulate employment, and open up investment and business opportunities that contribute to national development and prosperity. Along with these direct effects, tourism makes a wider contribution by establishing linkages with other areas of the economy and creating spillover benefits (indirect effects) for a number of sectors. To give a sense of the importance of tourism activities in today's world, in 2019 (prior to the Covid-19 pandemic) the travel and tourism sector accounted for 10.4 percent of the global gross domestic product (GDP), 6.8 percent of total global exports, and 27.4 percent of total global service exports. It also accounted for 334 million jobs, or 10.6 percent of total global employment in 2019, and was responsible for creating one in four of all new jobs in the world that year (WTTC 2021).

Cuba's efforts to turn international tourism into a key driver for the country's economic progress began in the 1980s, yet gathered tremendous pace in the 1990s after the demise of the Soviet Union wreaked havoc on the Cuban economy. Previously, during the 1950s, Cuba had worked actively to attract foreign visitors (mainly from the United States) and boost hard currency revenues from tourism activities, but tourism did little to foster economic development since the island essentially became a haven for prostitution, brothels, and gambling activities that benefited above all the American mafia, then-president Batista, and his cronies. Although tourism earnings and their contribution to Cuba's GDP increased significantly toward the end of the 1950s, the financial balance of the tourism sector was negative for most of the decade as Cuban nationals spent more money abroad than foreign tourists and excursionists spent in Cuba, and only growing revenues from gambling allowed the balance to record surpluses in 1956–8 (Durán Cárdenas 2000, 32–3). Furthermore, revenues from tourism activities paled in comparison with those from sugar exports. During the 1990s, on the other hand, tourism assumed a much more vital role in the Cuban economy. As Fidel Castro stated in a speech in July 1993, "Tourism is an extremely important source of convertible currency, which reaches the people in many ways."[68] Later that year, reflecting on the "exceptional circumstances" ushered in by the Soviet collapse that forced Cuba to vigorously promote international tourism and place it among the prioritized work plans, Fidel added,

"It was not the only program, nor is it the only program, but it is definitely the one with the best prospects."[69]

As a matter of fact, the Cuban economy experienced a crucial transformation during the 1990s from an economy centered on agriculture, and especially sugar production, to one based on services such as international tourism. It witnessed yet another new dramatic change in the post-2004 period, principally as a result of Venezuela's financial largesse and its booming ties with Cuba (Sánchez Egozcue and Triana Cordoví 2008). In late 2000, the governments of Venezuela and Cuba signed the Integral Cooperation Agreement (Convenio Integral de Cooperación, CIC) under which Venezuela agreed to sell 53,000 barrels per day of crude oil and derivatives to Cuba at preferential conditions. In exchange, Cuba agreed to send 13,000 workers to Venezuela, mostly physicians, nurses, and paramedics along with a smaller number of sports trainers, teachers, and advisers. Essentially a barter deal at first, the CIC was amended in 2003 to include hard currency payments for professional services provided by Cuban personnel. This led to growing cooperation in the area of energy and boosted Cuban involvement in various social initiatives in Venezuela. In April 2005, the CIC was renewed and expanded. Venezuelan oil shipments to Cuba increased to over 90,000 bpd of crude oil and derivatives while Cuba agreed to send 30,000 specialists to staff medical facilities and rehabilitation centers in Venezuela (Romero 2010, 128; Feinsilver 2008, 278–9).

Exports of professional services mainly offered by Cuban physicians and nurses in Venezuela quickly became (and still are) the top generator of hard currency for Cuba and reinforced the process of "tertiarization" of an economy whose export activities were once dominated by sales of primary goods. In recent years, Cuba also has provided for-profit medical services to other countries, among them Brazil and various Latin American nations, Portugal, oil-rich nations in Africa and the Middle East, and several additional countries during the Covid-19 pandemic (Luis Luis 2020; Torres Pérez 2015, 168).[70] Whereas in 1990 over 90 percent of Cuba's total exports were accounted for by goods, by 2014 Cuba's exports of services generated over twice as much revenues as its exports of goods (ONEI 1998, 2015a). Today that ratio is even more skewed in favor of services even if both tourism revenues and exports of medical services have declined, the latter mainly because of the Venezuelan economic crisis, and the former because of Trump's tougher economic sanctions against Cuba and the Covid-19 outbreak that negatively impacted international travel (Spadoni 2021, 247–8).

Table 2.7 highlights the aforementioned changes by presenting annual data on the sectoral composition of Cuban exports (percentage distribution in terms of earnings) between 1990 and 2014. In 1990, when the Soviet Union was about to disintegrate, Cuba's sugar industry brought in over $4 billion worth of exports, representing 72.6 percent of the island's total hard currency revenues. Nickel exports and an incipient tourism industry also made a modest contribution. While sugar fell on hard times during the 1990s, gross revenues from international tourism grew noticeably and by 1996 had surpassed those from sugar exports, giving tourism the top spot among Cuba's hard currency sources. Overseas sales of nickel also expanded during this period. By the early 2000s, nickel had become the leading foreign exchange earner among Cuban goods (Pérez-López 2011a, 439). The share of tourism revenues in Cuba's total exports of goods and services reached a peak of 45.7 percent in 2002,[71] and it was still 37.5 percent in 2004, right before Havana's government stepped up the deployment of medical personnel and other professionals abroad. In 2004, nickel exports accounted for 19 percent of total earnings, while the share of sugar, the mainstay of the Cuban economy for most of its history, was less than 5 percent. The share of tobacco, another traditional export sector, did not reach 4 percent (ONEI 2005).

But since then, almost everything has changed due to the Venezuelan factor. Mostly driven by Cuba's special deals with Venezuela in the area of medical assistance, the percentage of Cuba's total exports of goods and services accounted for by services other than tourism almost tripled between 2004 and 2014, from 21.0 percent to 58.4 percent. Meanwhile, revenues from international tourism dropped to 14.3 percent of the total. The relative contribution of nearly every other sector of the Cuban economy underwent a similar downward trend.[72] In 2014, nickel exports generated 4.2 percent of Cuba's foreign exchange income from all trade activities, and sugar exports accounted for just 2.3 percent. International sales of tobacco products (mainly cigars) provided only 1.3 percent of all hard currency earnings (ONEI 2015a). At any rate, it must be stressed that a significant portion of Cuba's revenues from the export of professional services to Venezuela and other nations, and even the government-backed credits the island receives from China and other strategic partners, are funds committed to the purchase of products from these countries. Thus, both before and after healthcare professionals became Cuba's biggest export, the Cuban government has had to rely primarily on international tourism, along with overseas remittances (captured in state-owned hard currency stores), and exports of

Table 2.7 Exports of Goods and Services by Sector, 1990–2014 (percentage distribution)

Year	Sugar	Nickel	Tobacco	Pharmaceuticals	Other Goods	Tourism	Other Services[a]
1990	72.6	6.7	1.9	1.4	8.6	4.1	4.7
1993	38.2	8.1	3.6	0.3	8.6	36.6	4.6
1996	25.8	11.4	2.9	1.4	8.8	36.0	13.7
1999	10.6	9.5	4.8	0.7	9.1	44.1	21.2
2002	11.4	11.2	3.7	1.3	9.1	45.7	17.6
2004	4.8	19.0	3.8	2.9	11.0	37.5	21.0
2008	1.8	11.5	1.9	2.4	11.7	18.8	51.9
2011	2.1	8.3	1.3	3.0	19.5	14.6	51.2
2014	2.3	4.2	1.3	3.1	16.4	14.3	58.4

Sources: ONEI 2015a, 2009a, 2003, 1998; CEPAL 2009, 2006, 2000.
Note:
[a] Mainly medical and other professional services after 2004.

nickel and a few other products, to build foreign exchange liquidity and expand international reserves. Liquidity is essential to pay unavoidable short-term debts with high interest rates, import foodstuffs and other goods, finance domestic business activities, and stabilize price distortions that put pressure on salaries (Spadoni 2017a, 107).

As presented in Table 2.8, gross revenues from international tourism rose steadily from $243 million in 1990 to about $2.1 billion in 2004 (with a small decrease of 5.5 percent in 2001 and another small drop of 3.9 percent in 2002), making the tourism industry, as Cuban officials often described it, the "locomotive" of the island's economy (Gutiérrez-Castillo and Gancedo Gaspar 2002). Approximately 90 percent of tourism earnings reportedly came from foreign visitors' expenses in Cuba, including in private establishments, and from

Table 2.8 Gross Tourism Revenues, 1990–2014 ($US millions)

Year	International Tourism	International Transportation Services	Total Revenues
1990	–	–	243.0
1991	–	–	402.1
1992	–	–	549.5
1993	635.5	84.5	720.0
1994	763.2	86.8	850.0
1995	962.6	137.4	1,100.0
1996	1,170.6	162.5	1,333.1
1997	1,310.7	204.3	1,515.0
1998	1,556.3	203.0	1,759.3
1999	1,695.0	206.0	1,901.0
2000	1,737.4	210.8	1,948.2
2001	1,692.2	148.2	1,840.4
2002	1,633.0	136.0	1,769.0
2003	1,846.3	152.9	1,999.2
2004	1,914.7	198.9	2,113.6
2005	2,149.9	249.0	2,398.9
2006	1,969.0	265.9	2,234.9
2007	1,982.2	254.2	2,236.4
2008	2,090.4	256.5	2,346.9
2009	1,899.2	183.2	2,082.4
2010	2,025.1	193.3	2,218.4
2011	2,282.7	220.4	2,503.1
2012	2,325.6	287.7	2,613.3
2013	2,325.1	282.7	2,607.8
2014	2,367.3	178.8	2,546.1

Sources: ONEI 2015a, 2009a, 2004, 1998.

sales of raw materials and recovered materials as well as earnings from wholesale operations, among other concepts (ONEI 2009a). The rest were payments to national carriers for international transportation. After a difficult period in 2008–9 during which Havana's authorities delayed payments to foreign creditors and froze hundreds of millions of dollars in the Cuban bank accounts of many foreign suppliers and partners in joint ventures, Cuba's liquidity problems began to improve thanks above all to record numbers of international arrivals and associated revenues and to a spike in remittances from abroad.[73] While fluctuating widely, annual gross revenues from tourism activities reached $2.2 billion in 2010 and some $2.5 billion in 2014. Food and accommodation services accounted for the majority of those earnings, followed by transportation services, retail trade, and recreation activities (ONEI 2015b). Remittances, according to unofficial calculations, more than doubled from $1.6 billion in 2009 to over $3 billion in 2014 and possibly emerged as Cuba's second-largest source of hard currency revenues after exports of professional services (Morales 2016).

Albeit far less important as a hard currency generator than international tourism, national tourism in Cuban convertible pesos or CUCs (the CUC was pegged 1:1 to the US dollar until it was removed from circulation at the beginning of 2021) should be discussed. This kind of tourism expanded considerably after Raúl Castro's decision in late March 2008 to allow Cubans to stay at hotels until then designated only for foreigners, rent cars, and use facilities that were formerly off-limits to them, including the best beaches frequented by overseas tourists in Varadero (Salinas, Mundet, and Salinas 2018, 224; Ayala Castro 2014).[74] Data from the Cuban Ministry of Tourism reveal that the annual number of national tourists in Cuban hotels shot up from 61,000 in 2008 to over 800,000 in 2014. During this period, annual gross revenues generated by national tourists increased steadily from just 5 million CUCs to 84.2 million CUCs.[75] Cuba's National Statistics Office (*Oficina Nacional de Estadística e Información* or ONEI) put the same indicator at 147.3 million CUCs in 2014 (representing nearly 6 percent of income from international tourism), of which 58.1 million CUCs was for lodging, 48.3 million CUCs was for gastronomy, and 38.3 million CUCs was for transportation. While national tourism remains concentrated at tired lower-end state-run establishments and at government camping facilities, around 20 percent of Cuban tourists stayed at four- and five-star hotels in 2014 (ONEI 2015a, 2015c). In short, whatever the exact amount of hard currency income from national tourism, there is no doubt that the latter has become a significant segment of the Cuban tourism sector whose importance is magnified during the summer off-season when beach hotel operators look to Cuban guests to boost occupancy rates (Feinberg and Newfarmer 2016, 28).

Although it is sensitive to the economic conditions in its source markets, international tourism is critically important for the economy of many developing countries since it provides more stable hard currency earnings than primary products whose prices are often volatile. The international prices of Cuba's key export (and import) goods indeed tend to experience wide swings and, to further complicate things, major Cuban export sectors like nickel and sugar lack the ability to boost production to react to higher prices (Pérez-López 2017, 32; Sánchez Egozcue 2011). While there are today more comprehensive, and arguably more accurate, ways to assess the extent to which a country is dependent on tourism,[76] the relation between tourism revenues and exports of goods has traditionally been used to evaluate the economic dependency of small countries on tourism activities (Cleverdon 1979, 56). With a feeble production base and serious systemic flaws in its economy, Cuba became considerably dependent on tourism during the 1990s, and by 1998 the tourism sector's revenues were higher than exports of all goods combined. Despite some significant variations, this trend continued in the first half of the 2000s but experienced a reversal in the second half of the decade, and especially in the early 2010s, as exports of goods grew markedly. By 2014, Cuba's tourism revenues were about half the total value of the country's merchandise exports, and yet the next five years saw a trend similar to that of the 1990s (ONEI 2020a, 2015a). Put simply, Cuba's crucial trade vulnerabilities, such as the failure to diversify export commodities and an acute sensitivity to international price fluctuations, magnify the country's dependency on tourism as a source of foreign exchange.

Moreover, given that income from tourism helps Cuba satisfying part of its import needs, Table 2.9 examines the import and trade deficit coverage provided by tourism revenues between 1990 and 2014. Besides oil products that at least since the mid-2000s have been supplied largely by Venezuela and financed with the services of Cuban professional workers stationed in that country, Cuba generally has relied heavily on imported foodstuffs, machinery and equipment, and manufactured goods to supplement insufficient domestic production and alleviate the needs of its society (Spadoni 2012, 174). It will likely continue to do so in the future because the stimulation of productive forces and the recapitalization of key national industries require considerable purchases from abroad. Despite significant economic reforms aimed at bolstering agricultural output and substituting expensive food imports, Cuba still imports more than

Table 2.9 Import and Trade Deficit Coverage by Tourism Revenues, 1990–2014 (percentage)

	1990	1993	1996	1999	2002	2005	2008	2011	2014
Import coverage[a]	3.3	34.1	36.4	43.3	42.2	31.5	16.5	17.9	19.5
Trade deficit coverage[b]	12.1	75.5	74.4	65.7	63.9	44.1	22.2	31.0	31.1

Sources: Calculations of the authors from ONEI 2015a, 2009a, 2003, 1998.
Notes:
[a] Gross tourism revenues/imports of goods
[b] Gross tourism revenues/trade balance for goods

60 percent of the food it consumes at a cost of roughly $2 billion each year.[77] The country's industrial sector has severe problems as well and has yet to recover from its collapse in the early 1990s. Albeit with significantly lower import levels than in the late 1980s when Cuba bought annually $7–8 billion worth of goods worldwide, the ratio of Cuba's tourism revenues to the total value of merchandise imports increased from just 3.3 percent in 1990 to a peak of 43.3 percent in 1999 and to 42.2 percent in 2002, when the island imported approximately $4.2 billion worth of goods. Since then, due to a rapid expansion of imported goods that reached $14.2 billion in 2008 and remained above $13 billion a year in 2011–14, this ratio has plunged considerably. It was just 16.5 percent in 2008 and 19.5 percent in 2014 (ONEI 2015a, 2009a).

With respect to the external trade in goods, Cuba has experienced a chronic deficit since the victory of the revolution led by Fidel Castro in 1959.[78] Yet the country's annual trade deficit, which was less than $1 billion in 1992–4, has soared dramatically since the mid-1990s. It was roughly $2–3 billion a year in the second half of the 1990s and in the first half of the 2000s, a level similar to that of the late 1980s. It increased even more after 2005 as booming revenues from exports of professional services and new credit lines from Venezuela and China triggered unprecedented levels of merchandise imports. Owing primarily to rapidly worsening terms of trade in connection with the global economic crisis, Cuba's balance of goods posted a deficit of more than $10 billion in 2008, the largest ever. A combination of significant import cuts and growing export receipts drove the trade deficit down to about $6 billion in 2010 before it climbed back to over $8 billion in 2014 (Pérez-López 2017, 33). Meanwhile, the ratio of Cuba's tourism earnings to the country's merchandise trade balance rose from 12.1 percent in 1990 to nearly 75 percent in 1996 and around 64 percent in 2002. This ratio fell noticeably after 2005 but was still over 30 percent in 2014. In sum, between 1990 and 2014, cumulative revenues from exports of goods were less than half (only 39 percent) of what Cuba had to pay to satisfy its import needs.[79] With such a weak export performance, Cuba has struggled to finance its trade deficit and frequently fell behind on its payments to foreign business partners. Hard currency income from tourism, especially until the early 2000s but afterward as well, provided enough liquidity to finance a sizable portion of Cuban imports of goods and to cover an even greater share of the country's trade deficit.[80]

Tourism can also become a key pillar of a country's economy through its direct and indirect contributions to economic growth and employment generation. According to the standard statistical framework for the economic measurement of tourism developed by the United Nations Statistics Divisions (Tourism Satellite Account or TSA),[81] the direct contribution of tourism to a country's GDP takes into account the gross value added (the value of output less the value of intermediate consumption) generated by industries that directly serve tourists such as lodging services (hotels, resorts, bed and breakfasts), food and beverage services (bars and restaurants), transportation services (airlines, cruises, car rentals, taxis), reservation services (travel agencies and tour operators), retail trade (tourist shops), and recreational services, among others. It is equivalent to total visitor spending less the purchases made by tourism industries, including imports.[82] As for tourism's contribution to employment, the TSA recommends only the estimation of jobs in the tourism industries because "employment strictly related to the goods and services acquired by visitors and produced either by tourism industries or other industries cannot be directly observed" (United Nations 2010a, 48). Whereas the TSA recognizes only the direct contributions of tourism, the World Travel & Tourism Council, or WTTC, quantifies also the indirect and induced impacts. The indirect impacts include private and government investment in the travel and tourism sector and the effects of supporting businesses in the supply chain. The induced impacts refer to direct and indirect employee spending (WTTC 2015). In brief, tourism activities have significant multiplier effects as the flow of money generated by visitor spending passes through and stimulates many sectors of the national economy.

In the case of Cuba, prior to the monetary unification that started in January 2021, an accurate calculation of the contribution of tourism to the country's GDP was complicated by the use of the official exchange rate of one Cuban peso for one US dollar that undervalued the impact of tourism,[83] and by the existence of a dual currency system with multiple exchange rates that produced distortions in almost all economic measurements, including the value of purchases made by tourism industries (Vidal Alejandro and Pérez Villanueva 2013, 11; Vidal Alejandro 2012, 107; Quintana et al. 2004, 177). In 2005, Cuba also adopted a new GDP formula and revised previously reported figures back to 1996. The new formula inflated the size of the Cuban economy by recognizing the value added of subsidized social services (mostly in the fields of health care, education, and sports) provided by the Cuban state to its population and to citizens from other countries (Spadoni 2014, 7–9). This altered the relative weight of tourism-related services (commerce, restaurants, and hotels) in Cuba's GDP. Furthermore, according to the TSA methodology, to appropriately measure the direct economic contribution of tourism to GDP, the gross value added generated by tourism industries (and by non-tourism industries as long as it is linked to visitor use) needs to be refined, through special calculations, to include only the part of value added that responds to visitor consumption, and leave aside the indirect and induced effects created by this consumption. For instance, with respect to restaurants, this measurement would include the portion of value added generated by meals for tourists and exclude the portion corresponding to meals for locals and to any other output not delivered to visitors (United Nations 2010a, 46–47). It goes without saying that carrying out these special calculations to build aggregate indicators to gauge the true size of tourism in the Cuban economy is a daunting task.

Nevertheless, Table 2.10 exhibits statistics of the World Travel and Tourism Council on the direct and total contributions of travel and tourism to Cuba's GDP and to employment generation between 1995 and 2014. Calculated in a way reportedly consistent with the TSA methodology, the annual share of GDP generated by industries that deal directly with tourists increased from 2.7 percent in 1995 to 3.7 percent in 2005, and then steadily declined to 2.3 percent in 2014. During the same period, and using the same years as a reference, the total contribution of travel and tourism to GDP, which includes wider effects from investment, the supply chain, and induced income impacts, rose from 9.2 percent to 13.4 percent, and then plunged to 8.8 percent. It comes as no surprise that both indicators started to plummet in the mid-2000s because that was the time when Venezuela and Cuba deepened their special economic and political relationship centered on Caracas providing critical oil supplies on subsidized terms to Havana in return for the services of thousands of Cuban doctors and other professionals. Tourism, though, has continued to play a crucial role in the Cuban economy. With respect to job creation, as tourism expanded, the number of Cubans who worked directly in the sector grew from 95,800 in 1995 to a peak

Table 2.10 Contribution of Travel and Tourism to GDP and Employment, 1995–2014

	GDP (%)		Employment (thousands)	
Year	**Direct**	**Total**	**Direct**	**Total**
1995	2.7	9.2	95.8	330.3
1996	2.9	10.2	104.4	373.9
1997	3.2	11.3	116.0	420.3
1998	3.6	12.8	134.3	481.7
1999	3.6	12.9	135.7	489.1
2000	3.3	12.0	126.8	464.0
2001	3.3	12.1	128.5	480.1
2002	3.2	12.0	130.0	487.2
2003	3.3	12.3	135.0	508.9
2004	3.3	12.3	138.6	520.6
2005	3.7	13.4	158.6	574.1
2006	3.2	11.4	139.9	496.3
2007	2.9	10.1	128.1	451.6
2008	2.8	10.8	130.3	495.1
2009	2.6	10.1	120.4	458.8
2010	2.6	9.5	116.4	432.4
2011	2.6	9.6	123.6	451.5
2012	2.4	9.6	119.5	460.7
2013	2.4	9.3	116.9	446.9
2014	2.3	8.8	109.9	413.8

Source: WTTC (World Travel & Tourism Council) Data Visualization, https://wttc.org/research/economic-impact/data-visualisation.

of 158,600 in 2005. That year, if we add the indirect and induced effects, travel and tourism supported a record 574,100 jobs, representing over 12 percent of total employment in Cuba. By 2014, direct and total jobs linked to tourism had dropped to 109,900 and 413,800, respectively. At any rate, the tourism sector requires a considerable workforce, has significant employment multiplier effects on other sectors (Figueras Pérez and Pérez Cárdenas 2015, 184),[84] and, to meet quality standards and tourist demand, needs a certain amount of skilled labor along with less-skilled workers. The seasonal nature of work in tourism, however, means that employment does not continue year-round and that many positions are available only during certain times of the year.

A few additional positive effects of tourism should be acknowledged. Since they are classified as exports of services, tourism revenues may improve a country's balance of payments (BOP). In the second half of the 1990s and early 2000s, the deficit in the current account of Cuba's BOP was partially mitigated by tourism earnings along with remittance inflows. Since then (at least until

2015 and except in 2006 and 2008), the current account, which essentially determines the international exposure of an economy,[85] has recorded surpluses or remained balanced mainly because of thriving exports of professional services (ONEI 2018a). Income from tourism may also be used for the expansion and modernization of basic infrastructures (airports, hotels, roads, water supply, electricity generation, telecommunications services, and other public utilities) that are undertaken to attract international visitors, as well as of infrastructures for all kinds of economic activities, which in turn contribute to overall economic development (Crespo and Negrón Díaz 2007, 157). In other words, justified as part of efforts to foster tourism, the benefits from infrastructure investments may be shared by other sectors of the economy and by industries that do not serve the needs of tourism. And they may be shared by the resident population as well. Still presenting serious infrastructure deficiencies typical of developing countries, Cuba has nonetheless witnessed substantial infrastructure improvements over the past three decades as a result of the active promotion of international tourism.

Critical Challenges

Despite its remarkable achievements, the Cuban tourism industry has faced serious challenges since reopening to international tourism in the early 1990s. Among these challenges, it is worth emphasizing the re-emergence of prostitution that supposedly had been eradicated after the 1959 revolution, the displacement of many qualified workers from other sectors to non-professional jobs in tourism, and, most notably, critical efficiency problems that conspire against the industry unleashing its full potential as a driver of economic growth and of improved living standards. And even if sex tourism in today's Cuba is arguably less popular than it was some twenty years ago, by and large, the aforementioned challenges have yet to be properly addressed.

Sex tourism in Cuba exploded around the mid-1990s as many Cuban women turned to prostitution for survival, to supplement their income (especially professional women), or simply to live a better life at a time when economic conditions on the island were extremely dire and tourist dollars were creating new unofficial and unregulated opportunities for Cubans to make money.[86] Speaking before the National Assembly in July 1992, Fidel Castro acknowledged the existence of prostitution in Cuba but specified, "There are no women forced to sell themselves to a man, to a foreigner, to a tourist. Those who do so do it on their own, voluntarily, and without any need for it. We can say that they are highly educated hookers and quite healthy, because we are the country with the lowest number of AIDS cases." Yet, making a case against the use of US dollars by Cubans, he added, "Every whore [jinetera] around would be looking for dollars. They could come to the hotels if they did a trick with a tourist and were paid in dollars . . . the effect would be disastrous because it would promote all sorts of immoralities and shady deals."[87] Indeed, it was precisely after the legalization of the circulation and possession of the US dollar in Cuba in August 1993 that prostitution (and the infamous business of pimping associated with it) came back on the island in dramatic fashion. By the end of that year, the real average monthly salary (adjusted for inflation) in Cuban pesos (CUPs) in Cuba had plunged to about 10 percent of its value in 1989 and, based on the unofficial CUP-dollar exchange rate, that salary was the equivalent of less than two dollars per month (Vidal Alejandro 2009; Cepal 2000).[88] A survey carried out in 1995 by the Italian magazine *Viaggiare* rated Cuba the number one country in the world for sex tourism, ahead of popular destinations like Thailand, Brazil, and the Philippines.[89]

Technically not a crime, prostitution is persecuted in Cuba under certain provisions of the country's penal code that cover dangerous antisocial behavior. The government of Fidel Castro started to crack down on prostitution in 1996, mainly in Varadero where many Cuban women from all parts of the island had come to work as prostitutes. Cuban authorities returned thousands of women to their places of origin, investigated several suspected house brothels, slapped stiff penalties on anyone who housed a prostitute, and ultimately banned *casas particulares* (private homes with rooms for rent) in Varadero (Trumbull 2001, 363–4). Nevertheless, prostitution continued to be rampant in the city of Havana and its eastern beaches and, even in Varadero, the provision of sex to guests remained an integral feature of all-inclusive beach resorts that catered to foreign tourists (Cabezas 2006, 2009). Another major offensive against prostitution took place in 1998–9, this time in the Havana area. Along with random inspections of *casas particulares*, government tactics included the closure of bars and nightclubs, police raids to round up prostitutes, routine identification checks of Cubans seen in contact with tourists to fine, incarcerate, or deport pimps and jineteras who did not possess "legal residency" in Havana,[90] the sending of sex workers to rehabilitation schools, and new penalties for sex-related crimes like child abuse and pimping.[91] Since then, tourism-oriented prostitution in Cuba has likely diminished, in part because of the effectiveness of the government's responses, its efforts to diversify Cuba's tourism product, and improved economic conditions on the island. But it is also true that, to some extent, the phenomenon has simply become more discreet and less visible as many jineteras developed subtler techniques to interact with tourists and circumvent police control (Simoni 2016, 45–6, Valle Ojeda 2006, 242–3).

The spectacular growth of international tourism in Cuba in the 1990s, amid a plummeting purchasing power of Cuban salaries in regular pesos, also led to an exodus of public service workers, among them many professionals, into low-skill service jobs in the tourism sector that provided much-needed hard currency (Uriarte 2008, 326; Crahan and Harmony 2007, 147). Besides qualified people who secured licenses for self-employment (*cuentapropistas*) to open small restaurants or rent spare rooms in their homes to foreign tourists, several Cuban lawyers, physicians, engineers, technicians, scientists, and teachers abandoned the profession for which they were trained to find work as waiters, bartenders, and taxi drivers. They could obtain tips in US dollars and earn more in one day than what they earned in a week or even a month at their former job. While a formal training and hiring process for all tourism-related positions was reportedly implemented by MINTUR in 1995 in an effort to stem the flow of Cuban professionals making their way into the tourism sector (Henthorne 2018, 69–70), the latter has continued to attract qualified workers from other sectors seeking access to hard currency. What is more, Raúl Castro's decision in 2010 to lay off a large number of state employees and expand the legal categories of self-employment in Cuba gave further impetus to the misallocation of human capital since most Cuban professionals were not allowed to be *cuentapropistas* in jobs directly related to their careers (Ritter and Henken 2015, 174; Peters 2012, 20).[92] As the labor market in the Cuban tourism industry exemplified, monetary duality was the leading cause of the mismatch and underutilization of workers' skills so prevalent in Cuba. It also fueled income inequalities, inefficiencies, and corruption. The Cuban workforce by and large is still highly educated, but with every passing day, it is becoming less qualified since many professionals have not worked in their own fields for a long time.

Additional economic challenges must be examined. In terms of tourist expenditures, as shown in Figure 2.3, gross revenues per tourist per year more than doubled from $715 in 1990 to $1,475 in 1995. Since then, annual gross revenues per tourist have decreased significantly, and by 2014 they were just $848, about 42 percent below the level of 1995. Gross revenues per room grew

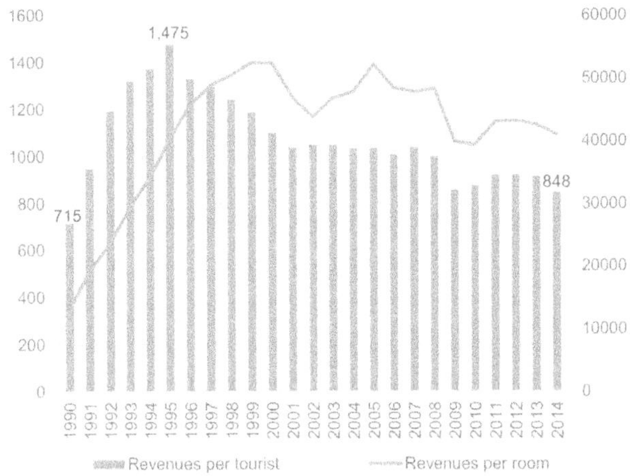

Figure 2.3 Gross revenues per tourist and per room, 1990–2014. Calculations of the authors from ONEI 2015a, 2009a, 2004, 1998, and from MINTUR data obtained in May 2017 and May 2015.

fourfold during the 1990s, yet they plummeted more than 20 percent during 2000–2014. Overall, these problems denoted an exhaustion of the model of tourism development based on massive construction of hotel rooms and suggested that the future contribution of tourism to economic growth could diminish greatly once the sector reached its maturity (Perelló Cabrera 2016, 230–1; Pérez Villanueva 2013, 26). Moreover, the beginning of a downward turn in gross income per tourist coincided with the decision by Fidel Castro's government in the mid-1990s to actively promote all-inclusive vacation packages, which tend to produce a smaller impact on a host country's economy than other accommodation subsectors. Overconcentration on all-inclusive sun-and-beach activities is still a key feature of the Cuban tourism sector, even though other forms of tourism like health, cultural, educational, and ecotourism are being promoted.

While there is no doubt that the Cuban tourism industry suffers from severe efficiency problems, performance indicators based on official tourism revenue data should nonetheless be taken with caution given the unreliability of Cuban statistics and the difficulties of measuring certain expenses of overseas visitors to Cuba. Above all, ONEI's figures on international tourism revenues reportedly include the earnings of the private sector, which are substantial, yet it remains unclear how those earnings are calculated. As Ricardo Torres Pérez, an economist from the island, wrote, "The statistical system is probably incapable, for various reasons, of detecting a part of the [tourist] expenditures made in the private and cooperative sectors."[93] Following recommendations by the World Tourism Organization (United Nations 2010b, 36–8), quite a few countries carry out surveys of inbound visitors (either at the border or at accommodation establishments and/or popular tourist sites) to estimate total tourism expenditure and they complement the results with other data sources such as credit card records.[94] Although to some extent it might be easier for Cuba than for other countries to measure international tourism receipts since the Cuban state controls most of the tourism sector, to the best of our knowledge Cuba does not conduct rigorous surveys of inbound visitor spending that would also be particularly useful to estimate the tourism-related revenues of *casas particulares* (bed-and-breakfasts), *paladares* (restaurants), taxi services, and other private businesses.

It is telling that ONEI reported for the first time the separate amount of income of the private sector (officially referred to as "non-state sector") within total tourism revenues only in its 2017 annual yearbook of statistics and added figures back to 2013 (ONEI 2018a), which coincidentally was the year when a new comprehensive tax code went into effect in Cuba.[95] The island's authorities most likely estimate (and probably underestimate) the tourism earnings of private entrepreneurs from their tax returns and from various assumptions linked to their volume of business. Along with the potential problems of those assumptions, a key issue is that Cuban *cuentapropistas* continued to engage in substantial tax evasion and income underreporting even after the new tax code introduced several additional tax deductions to stimulate their work (Sarduy González, Pons Pérez, and Traba Muñoz 2015, 96; Pons Pérez 2013, 49).[96] To further complicate measurements, both Cuban residents and visitors from abroad shop in Cuba's hard currency stores, and even *cuentapropistas* buy products there since there is no wholesale market for them. In Cuba, there were dollar stores from 1993 until November 2004, when the US dollar was withdrawn from circulation and the dollar-pegged Cuban convertible peso became the only accepted currency in these stores. Starting in late 2019, dollar outlets were revived with the opening of stores in freely convertible currency (*moneda libremente convertible*, MLC) where prices are set in US dollars and payments are only by card.[97] Moreover, it is complicated to track many transactions of foreign tourists in Cuba since credit card usage on the island is still very limited. In a nutshell, the degree of inaccuracy of Cuba's official tourism revenue figures (earnings of the private sector and total revenues) might be quite significant.

Another shortcoming of Cuba's leisure industry is exemplified by its excessive operational costs. In March 2001, then Cuban vice president Carlos Lage Dávila estimated the cost per dollar of gross income from tourism activities at $0.76.[98] This indicator is high and refers only to the direct cost in dollars, not the indirect cost incurred by the state in the tourism sector. Also consider that domestically produced goods for tourism have an imported (indirect) component in dollars, which implies that the cost per dollar of gross income would be even higher. Direct and indirect costs per dollar were estimated in the early 2000s at more than $0.80, which would mean for the country a net result of only $0.20 for every dollar of gross income from tourism.[99] This was confirmed by Marta Maíz, then Cuban vice minister of tourism, in May 2003. In an interview for the Cuban magazine *Bohemia*, Maíz said that in 2002 "income from tourism was $52.2 million less than in 2001, with a cost of USD 80 cents for every dollar captured by the country."[100] More recent official data for the City of Havana show that operational costs remained quite high until 2008, with some notable improvement afterward. The cost per dollar of gross income from tourism activities in Havana, which generate approximately 20 percent of Cuba's total tourism revenues, grew from $0.75 in 2003 to $0.81 in 2008. This indicator dropped to $0.70 in 2010 and $0.65 in 2013, but annual gross revenues (in particular from extra-hotel services) plunged 25 percent during this period, thus limiting net profits (ONEI 2015d, 2009b).[101]

At any rate, on the subject of efficiency, it is worth emphasizing that hotels managed by renowned international chains have exhibited much better operating results than those administered by Cuban state corporations. Citing official government statistics, Victor Alejandro Rodríguez García (2009) compared some key performance indicators of Cuban hotels by type of management in 2008. That year, with practically the same number of rooms as hotels under national administration, hotels under foreign management accounted for 77 percent of total gross revenues and a staggering 89 percent of profits after payment of fees. They also accommodated more than two-thirds of all hotel guests in Cuba and occupied nearly twice as many rooms as state-run hotels. As the portion of hotel rooms under foreign administration within the total stock of hotel rooms existing in Cuba increased considerably over the next decade to more than 60 percent, it is safe to assume that the relative shares of total revenues and profits generated by lodging facilities directly managed by international hotel groups on the island became even more substantial.

When it comes to improving the performance of the Cuban tourism industry and spreading its benefits to the overall economy through production linkages with other sectors (in particular agriculture and manufacturing), the industry's vast amount of importing remains a major shortcoming despite some positive results in the post-1990 period, stimulated above all by a growing influx of foreign investment. In effect, the share of domestically produced goods provided to the Cuban tourism industry reportedly increased from just 12 percent in 1990 to nearly 70 percent around the mid-2000s (García Jiménez 2010).[102] In the early 1990s, virtually all products for hotels and restaurants in Cuba had to be imported. The development of mixed enterprises with foreign firms in tourism stimulated the formation of new joint ventures in various industries, agriculture, and services to supply tourist establishments at low cost. Regarding domestic sales to the tourism sector, foreign capital and know-how assumed an important role in the markets of beers, soft drinks, mineral waters, processed meats, personal hygiene products, construction materials, tourist buses, air conditioning systems, wireline and wireless telephony, electronic equipment, and software solutions, among others (Figueras Pérez 2002, 111–13).

Nevertheless, since the mid-2000s, in the wake of a process of recentralization of the Cuban economy that led to a considerable reduction of foreign direct investment flows to Cuba, the share of total supplies to the island's tourism sector consisting of locally produced goods almost certainly has witnessed a sizable decline. Between 2003 and 2005, a series of administrative decisions by the government of Fidel Castro stripped certain Cuban corporations in tourism and other areas of the economy of the relative autonomy they had enjoyed until then, centralized the use of hard currency and its availability in the state's coffers, established that import plans had to be approved by officials at the Cuban Central Bank with little practical knowledge of the functioning of the economy, and reevaluated the dollar-pegged CUC by 8 percent against all international currencies (Spadoni 2014, 95–6).[103] Besides rendering Cuba less attractive to foreign investors, these measures weakened the integration of tourism with agriculture and various industries, made it more difficult and longer for local producers to obtain government approval for importing inputs, and made domestic production with substantial imported components more expensive.

The inevitable result was a reduction of domestic supplies to the Cuban tourism industry and a spike in imported goods that continued after Raúl Castro took over the country's leadership from his older brother in late July 2006. Hotel shops and hard currency stores began to import beers, soft drinks, various types of dairy products, chocolate, cookies, and clothing, among other items. The construction of new hotels led to an increase in imported furniture, decorations, and accessories. Moreover, Cuba gained access to new foreign markets where it could buy products at good prices. Imported foodstuffs from the United States substituted the local production of chicken for the island's tourism sector and part of the pork meat production.[104] After 2005, as generous soft credits from China became available, the Cuban tourism sector started to purchase from that country all uniforms and shoes for its workers and most of its hotel and restaurant linens.[105] Therefore, as it became increasingly difficult to cover the demand of the tourism industry with domestic production and to avoid sacrificing the quality of goods and services for international tourists as well as to prevent supply delays and defaults, the Cuban government practically decided to free tourism from the responsibility to stimulate the growth of other economic sectors through backward linkages (Figueras Pérez 2019).

It should also be emphasized that hotel occupancy rates in Cuba have traditionally been low, especially if compared to other Caribbean countries (Thomas, Kitterlin-Lynch, and Lorenzo Del Valle 2015, 9; Espino 2000, 367). Official statistics in Figure 2.4, which refer to total existing rooms for both national and international tourism, reveal that the average year-round hotel occupancy rate in Cuba rose from little more than 50 percent in 1994, during the worst days of the post-Soviet economic crisis, to nearly 80 percent toward

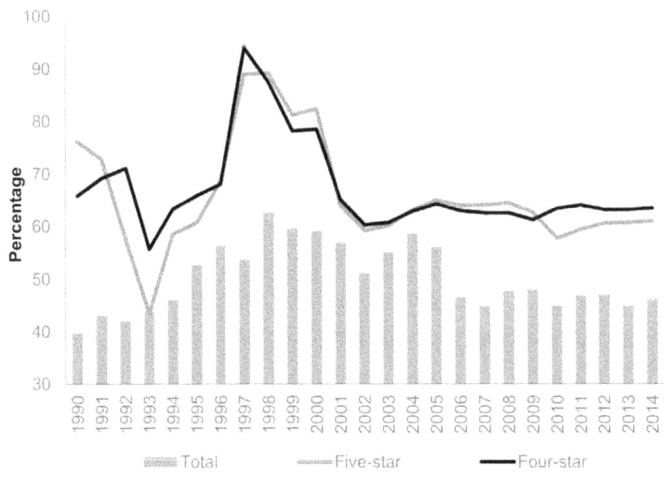

Figure 2.4 Hotel occupancy rates, 1990–2014. Note: Rates refer to existing rooms for national and international tourism. Generated by the authors, based on ONEI 2015a, 2009a, 2004, 1998.

the end of that decade. Since then, though, rates have dropped considerably to less than 60 percent in the first half of the 2010s and to even lower levels today. Accessible only to foreign tourists until Raúl Castro's government opened their doors to Cuban residents in March 2008,[106] four- and five-star hotels saw their annual occupancy rates grow markedly after 1994 to a peak of approximately 90 percent in 1997–8. But at that time the total number of rooms in these categories was just about one-third its current level (ONEI 2019b, 2000). Occupancy levels in Cuba's upper-tier lodging facilities witnessed a major drop in the years ahead (the average year-round rate was 61 percent in five-star hotels and 63.5 percent in four-star hotels in 2014) as many new luxury hotels and resorts were built across the island. Albeit incomplete and inconsistent, historical data on occupancy levels in Cuba's international tourism sector alone (excluding national tourism) indicate that annual rates hovered below 60 percent in the 1990s and in the early 2000s (except in 1998 when occupancy averaged 62.5 percent) and below 50 percent in 2006–14 (CEPAL 2004, 2000; ONEI 2015b, 2013, 2011b, 2009c, 2008).

To some extent, relatively low year-round occupancy rates in Cuban hotels are due to the strong seasonality of the tourism market, which is particularly challenging for a destination like Cuba, that mainly depends on the one-dimensional sun-and-sea product and has had limited success so far in diversifying its product to attract low-season tourists. International tourist flows (overnight visitors) tend to concentrate in the months from December to March, thus placing heavy pressure on Cuba's physical and social resources that are required to meet the demand for certain goods and services during the high tourism season. Conversely, the underuse of infrastructure and the reduced demand for resources during the low-season months of May-June and September-October negatively affect profitability and employment levels. However, the Cuban tourism sector's own problems greatly contribute to depressed occupancy levels. Cuba has a large number of rooms either out of service or in very poor condition and has devoted far more efforts on building new hotels than repairing existing facilities. Moreover, the island's tourism sector suffers from poor quality standards of accommodation, contributing to a low rate of repeat tourism, inadequate supporting infrastructure, and insufficient diversification of complementary services (Rutty and Richardson 2019, 2; Feinberg and Newfarmer 2016, 8; Sharpley and Knight 2009, 252; Elliott and Neirotti 2008, 389–90; Espino 2008, 136). Improvements in these areas and the overall development of the Cuban tourism industry are also limited by key structural problems of the Cuban economy that include a weak capacity to generate domestic savings to support investment, low efficiency levels of most state-run firms, relative price distortions stemming from government controls that hamper market mechanisms, heavy red tape and overcentralization, and insufficient levels of foreign direct investment.[107]

Conclusion

Along with the pursuit of foreign investment in the search for the markets, technology, and financing that disappeared with the demise of the socialist bloc in Eastern Europe, the main novelty of Cuba's response to the plunge in its economy brought about by the Soviet collapse in the early 1990s was the active promotion of international tourism after a long period of neglect. As a matter of fact, at that time, Cuba was the only Caribbean country without a sizable tourism sector and a strong tourism promotion program. The resurgence of tourism development in Cuba was a success story as the island rapidly became one of the Caribbean's most popular holiday destinations. Hotels and supporting tourism infrastructure, arrivals of international visitors, and tourism-related revenues all grew considerably during the 1990s. By 1996, Cuba already received more than one million foreign tourists per year and several new hotels with almost 10,000 rooms combined had been built, especially in Varadero. Meanwhile, tourism had replaced sugar as Cuba's top foreign exchange earner, generating more than $1 billion in gross revenues per year. An almost unrelenting expansion of the Cuban tourism industry continued over the next two decades or so. In 2014, Cuba attracted over 3 million visitors from abroad who generated $2.5 billion in gross hard currency earnings, and boasted about 62,000 hotel rooms (with an additional 19,000 private rooms), mainly in all-inclusive four- and five-star resorts in various beach areas across the island. Approximately 25 percent of available rooms were instead in urban areas, primarily in Havana. Renowned international hotel chains managed more than half of Cuba's rooms, with some establishments being run as joint ventures with Cuban corporations. The tourism sector contributed (direct and indirect effects) to 11 percent of Cuba's GDP in 2014, and featured around 520,000 direct and indirect workers.

Cuba's reopening to international tourism in the post-Soviet era, though, was not without serious challenges. Critical problems, particularly during the 1990s, included the proliferation of sex tourism and prostitution that gradually became a source of embarrassment for the Cuban government and led to repeated police crackdowns, and a massive exodus of medical, educational, and other professional workers from state employment into tourism and the informal sectors to access dollars. Additional problems of the Cuban tourism industry emerged over time, among them a declining trend in the average spending per tourist and per room, low levels of profitability and high operational costs, the industry's overreliance on imported products and thus its limited backward linkages with other economic sectors, and low hotel occupancy rates. Furthermore, by the mid-2000s, mainly boosted by Cuba's special agreements with Venezuela, exports of medical and other professional services had become Cuba's largest source of hard currency, bringing in twice as much revenues as tourism. Cuban exports of professional services grew steadily until the death of Venezuelan president Hugo Chávez in early 2013. Since then, while continuing to be Cuba's leading foreign exchange generator, they have declined notably as the beginning of a severe economic crisis in Venezuela, aggravated by a fall in the international price of oil, significantly weakened the economic ties between Caracas and Havana. As 2014 came to a close, Cuba's gross revenues from tourism activities remained substantial, but they had been practically stagnant since 2011. However, as we will see in the next chapter, a major change in the relationship between Cuba and the United States was about to open up new opportunities for the Cuban tourism industry and boost considerably both the industry's revenues and the arrivals of international visitors to the island.

Notes

1 Founded on August 15, 1511, by Diego Velázquez de Cuéllar, Baracoa is the oldest Spanish settlement in Cuba and the country's first capital.

2 In addition to "coordinated supply plans" and exports, Ernesto Hernández-Catá (2001, 4) estimates that Soviet subsidies and aid to Cuba averaged $4.3 billion a year between 1986 and 1990.

3 Fidel Castro's inaugural address to the fourth congress of the Cuban Communist Party, Santiago de Cuba, October 10, 1991, http://www.cuba.cu/gobierno/discursos/1991/esp/f101091e.html.

4 The resolution on the economy (*Resolución Sobre el Desarrollo Económico del País*) and other official documents of the 1991 PCC congress are available at http://congresopcc.cip.cu/congresos/iv-congreso-pcc.

5 "Constitución de la República de Cuba," *Gaceta Oficial*, August 1, 1992, art. 23.

6 The Cuban government also improved Cuba's overall business environment by exempting foreign investors from taxes on gross income, personal income, and the transfer of real estate and businesses, by lifting restrictions on hiring foreign executives and technical personnel, by eliminating custom duties for essential goods and equipment, and by allowing the unrestricted repatriation in hard currency of dividends and profits (Travieso-Diaz 1997, 106).

7 Between 1992 and 1998, FINATUR granted $437.2 million worth of credits to over 200 companies of different sizes that supplied goods to the Cuban tourism sector (Colantonio and Potter 2006a, 44).

8 The creation of MINTUR was part of a bureaucratic reorganization pursuant to Decree Law 147 of April 1994 that eliminated a number of state committees and national institutes belonging to Cuba's Council of Ministers and assigned their operations to existing and new ministries.

9 Kirby Jones (2008, 4) described this way the business activities of Cuba's tourist corporations: "These entities operate as free from day-to-day government control and oversight as any private sector firm in any country in the world. Some borrow on the international financial markets for their own account, many are audited by leading Western accounting firms, and they are flexible to undertake any business deal they want. But they only have one shareholder: the government of Cuba. If they make their profit targets and operate with acceptable margins, they keep going. If they do not, people are fired, new management is inserted, and changes are made. What could be more capitalistic than this?"

10 "Ley de la Inversión Extranjera en Cuba (Ley 77)," *Gaceta Oficial*, September 6, 1995.

11 Mauricio Vicent, "Cuba Investiga un Escándalo de Corrupción en el Sector Turístico," *El País*, December 6, 2003; Nancy San Martin, "Cuba's Tourism Chief Replaced by Army Colonel," *Miami Herald*, February 12, 2004.

12 Marc Frank, "Castro Reins in Cuban Tourism," Reuters, July 13, 2004; Marc Frank, "Cuba Tightens Up on Tourism," Reuters, October 16, 2004.

13 "Cuba Travel Agency Rumbos Cuba and Viajes Gran Caribe To Be Taken Over by Cubatur and Viajes Cubanacán," Havana Journal, November 2, 2004.

14 "Cuba Devalues Convertible Peso, at Dollar Parity," Agence France-Presse, March 14, 2011; Kaelyn Forde, "What Will Cuba's new Single Currency Mean for the Island?" Aljazeera, January 1, 2021.

15 Calculations of the authors from Perelló Cabrera 2015a and MINTUR data obtained in February 2017.

16 For a comprehensive review and assessment of Raúl Castro's economic reforms, see Mesa-Lago and Pérez-López 2013. For a more concise review, see Spadoni 2014, 165–197.

17 "Resolución No. 32/2010," *Gaceta Oficial*, October 8, 2010.

18 "Resolución No. 145/13," *Gaceta Oficial,* October 8, 2013; Marc Frank, "Cuba Allows Tourism Industry to Hire Private Contractors," Reuters, October 9, 2013; "Cuba Promotes Private Offers in Tourism," Xinhua, September 8, 2013.

19 "Cuba Allows Foreign Investment in Tourism," BBC News, May 5, 2010.

20 "Decreto-Ley No. 273/10," *Gaceta Oficial,* August 13, 2010.

21 "Exploran Nuevas Oportunidades," *Opciones*, January 17, 2014.

22 "Ley No. 118/2014," *Gaceta Oficial*, April 16, 2014.

23 "Decreto-Ley No. 313," *Gaceta Oficial*, September 23, 2013.

24 The Cuban Ministry of Tourism estimated in 1997 that the annual number of international visitors to Cuba would reach 2 million by the year 2000. Based on this official estimate and assuming little or no change to Cuba's political and economic systems, Crespo and Negrón Díaz (1997, 159) estimated that Cuba would double its annual number of tourist arrivals to 4 million by the year 2007.

25 Nomi Morris, "Can Cuba Change?" *Maclean's*, January 15, 1996.

26 In 1999, besides arrivals from Germany, Italy, Spain, France, and the UK, international visitors to Cuba from European countries came primarily from Switzerland (26,111), Portugal (25,474), Holland (22,365), Belgium (20,092), and Austria (16,119). For additional information, see ONEI 2000.

27 Kelly Sullivan, "Americans Defy Cuba Embargo," *Washington Post*, October 31, 2001.

28 In August 1994, after a series of violent boat hijackings and unprecedented street riots, Fidel Castro declared that anyone who wanted to leave Cuba on his own could do so. In the following weeks, some thirty-five thousand Cuban rafters left the island for the United States in what became known as the *balsero* crisis.

29 The category of international visitors to Cuba from the United States in Table 2.1 (also in Table 2.2 and Table 2.3) includes some arrivals of Cuban Americans who left Cuba before December 30, 1970, and entered the island with a US passport.

30 Such estimates seem reasonable if we consider that more than 1.4 million people of Cuban descent (nearly 85 percent of all Cuban nationals overseas) lived in the United States in 2007. That year, there were also 70,000 Cubans living in Spain, 66,000 in Venezuela, and 20,000 in Mexico (Martín Fernández et al. 2007; Gutiérrez Guerra and Gutiérrez Castillo 2011).

31 In 2003, the average length of stay of Canadian tourists in Cuba was 8.5 days, significantly shorter than that of German (13.6 days), British (11.6 days), Italian (11.5 days), French (10 days), and Spanish (9.5 days) tourists (Quintana et al. 2004, 130–7). A similar situation existed in 1996–8 (Durán Cárdenas 2000, 47).

32 Approximately 50,000 individuals subject to US law reportedly visited Cuba via third countries (legally or illegally) in 2002. "Embargo or Not, US Is Now Fidel

Castro's Economic Partner," *The Economist*, January 4, 2003. As recalled by a Canadian official some years ago, US tourists "are quite visible among visitors arriving in Cuba on flights from Montreal, Toronto, Kingston, Nassau, and Mexico City" (USITC 2001, 3–25).

33 "Cuban Assets Control Regulation," Federal Register 68(56): 14141–14148, March 24, 2003.

34 Andres Oppenheimer, "U.S. May Tighten Cuba Travel," *Miami Herald*, December 15, 2002.

35 "Cuban Assets Control Regulation," Federal Register 69(115): 33768–33774, June 16, 2004.

36 In March 2005, the US Treasury Department launched an investigation of virtually all religious Cuba-travel licensees in the United States after some Florida-based religious groups were accused of abusing their licenses to promote family visits to Cuba. Oscar Corral, "Groups Warned to Obey Travel Limits," *Miami Herald*, April 8, 2005. As a result, the OFAC implemented more stringent rules for these organizations by restricting the number of people they could send to Cuba and the frequency of their trips (Sullivan 2009, 6).

37 Cuban economists based in Havana, interviewed by the authors, Havana, December 6, 2006.

38 Hurricanes Gustav and Ike hit Cuba in August and September 2008, respectively. Along with hurricane Paloma, which hammered Cuba's southern coast later in November, they caused nearly $10 billion worth of material losses. About 530,000 homes across the island were either damaged or destroyed (MEP 2008).

39 Visitors to Cuba from Spain, the United States, and Argentina also increased slightly in 2009.

40 Will Weissert, "Castro Says Cuba to Cut Spending, Communism Secure," Associated Press, August 3, 2009.

41 Statistics Canada, "Travel by Canadians to Foreign Countries, Top 15 Countries Visited," https://www150.statcan.gc.ca/t1/tbl1/en/tv.action?pid=2410003701.

42 Roe Monique Varela Enrique, "Los Viajeros Podrán Permanecer Más Tiempo en Cuba Gracias a Este Cambio en la Visa," *El Nuevo Herald*, November 2, 2022.

43 Canadian tourists in Cuba could apply for a three-month extension of their stay prior to the ninetieth day in the country. All other foreign tourists could apply for a one-month extension prior to the thirtieth day in the country.

44 "Canadian Snowbirds Boost Cuban Tourism Recovery," Reuters, April 9, 2008.

45 President Obama's measures of 2009 also broadened the range of relatives that Cuban Americans could visit in Cuba and removed limitations on the duration of their trips and related expenditure amounts. Jeff Franks and Pascal Fletcher, "Cubans Say Obama Move Will Help Families, Budgets," Reuters, April 13, 2009.

46 "Cuban Assets Control Regulation," Federal Register 74(172): 46000–46007, September 8, 2009.

47 Data from the Cuban Ministry of Tourism in Table 2.3 indicate that Cuban Americans accounted for only 55 percent of all Cuban residents abroad who visited Cuba in 2009 and for about 70 percent of those who visited in 2010–14. This might be due to the fact that the MINTUR data reportedly refer only to direct arrivals from the United States and thus exclude arrivals of Cuban Americans via third countries that became less common after Obama allowed US citizens of Cuban descent to engage in unlimited family travel to Cuba.

48 Marc Frank, "U.S.-Cuba Travel Snarled by Regulations, Politics," Reuters, September 13, 2012.

49 During the 1990s (and most of the 2000s), since higher-end hotels and resorts in Cuba operating in hard currency were reserved exclusively for foreigners, Cuba's official data on its accommodation capacity included a separate series on hotel rooms available for international tourism (CEPAL 2000).

50 "Five-star Prices, Without the Service: Cuba's Hotel Problem," Agence France-Presse, February 15, 2017.

51 MINTUR data obtained by the authors in November 2015.

52 The small number of new four-star hotels that existed in Varadero in 1990 included the hotels Paradiso and Puntarena and the first joint venture hotels Sol Palmeras and Tuxpan (Woodard 2018, 199; De las Cuevas Toraya 2001, 350–1). Evaristo Villalba Garrido (1993, 201) wrote that the following hotels began operations in Varadero in 1990–1: Las Palmeras, Paradiso, Tuxpan, Puntarena, Conraforte, Cuatro Palmas, Club Varadero, Gaviota II y III, Caracol, Meliá Varadero, and Meliá Las Américas.

53 Miguel Alejandro Figueras Pérez, presentation at the Conference "Doing Business in Cuba," Orlando, FL, April 13, 2006. In 2006, Figueras Pérez was senior adviser to the Cuban minister of tourism.

54 MINTUR data obtained by the authors in June 2018. Cubanacán also had a joint venture (Shanghai Suncuba) in China with the Chinese company Suntime International. The joint venture completed a deluxe five-star hotel in Pudong, Shanghai in February 2010 that was managed by Sol Meliá as Gran Hotel Meliá Shanghai for about three years. In May 2013, the Swiss chain Kempinski Hotels took over the management of the hotel that reopened as Grand Kempinski Hotel Shanghai. "Kempinski Hotels Opens Grand Kempinski Hotel Shanghai," Kempinski Hotels Press Release, May 14, 2013.

55 "Sherritt May Sell Its Stakes in Cuban Hotels, Agribusiness," CubaNews, December 1, 2003.

56 Rob Sequin, "Real Estate in Cuba—Still No 'Groundbreaking' Events," Havana Journal, December 2, 2010.

57 "Proyectan Inversiones para Ampliar Oferta Turística en Cuba," Prensa Latina, July 19, 2022.

58 Roc Hotels (hotels Arenas Doradas and Barlovento in Varadero and hotel Presidente in Havana) and Warwick Hotels (hotel Naviti Varadero) had signed management contracts that took effect on January 1, 2015. SuperClubs and Warwick (which also began to manage the Cayo Santa Maria Resort in mid-2015) have left Cuba. MINTUR data obtained by the authors in November 2016.

59 Meliá Hotels International, "Informe Annual 2015," https://www.meliahotels international.com/es/accionistas-e-inversores/informacion-financiera/informe -integrado.

60 "La Hotelera Sol Meliá Tuvo Ingresos de 3.664 Millones de Dólares en Cuba," EFE, December 22, 2010.

61 Calculations of the authors from ONEI 2015a, 2009a, 2003, 1998.

62 Anita Snow, "Cuba's Castro Oversees Resort Opening," Associated Press, January 21, 2003.

63 "Hotel Revolution in Cuba," Bouygues, October 6, 2017, https://www.bouygues .com/en/publications-at-a-glance/magazine/hotel-revolution-in-cuba/.

64 "Sherritt International Sells Stake in Cubacel," *The Globe and Mail*, September 6, 2003; "Telecom Italia Sells ETECSA Stake for $706 Million," Reuters, January 31, 2011.

65 The full text of the Helms-Burton law, also known as Cuban Liberty and Democratic Solidarity (Libertad) Act of 1996, can be found at: https://home .treasury.gov/system/files/126/libertad.pdf

66 Doreen Hemlock, "SuperClubs Closing Two Resorts," *South Florida Sun Sentinel*, June 18, 2004.

67 For the list of Blue Diamond's hotel properties in the Caribbean, see https://www .bluediamondresorts.com/.

68 Fidel Castro's speech delivered in the city of Santiago de Cuba on July 26, 1993, http://www.cuba.cu/gobierno/discursos/1993/esp/f260793e.html.

69 Fidel Castro's speech delivered in Havana before the Cuban Communist Party's Assembly of Balance Sheet, Renovation, and Ratification of Mandates, November 7, 1993, http://www.fidelcastro.cu/es/discursos/discurso-pronunciado-en-la -clausura-de-la-asamblea-de-balance-del-trabajo-renovacion-y.

70 Ciara Nugent, "How Doctors Became Cuba's Biggest Export," TIME, November 30, 2018; Sarah Marsh and John Zodzi, "Cuba Punches Above Weight with 'White Coat Army' During Pandemic," Reuters, September 14, 2020.

71 In 2002, gross revenues from tourism activities accounted for a remarkable 72.2 percent of Cuba's total exports of services (ONEI 2003).

72 Apart from medical and other professional services, only exports of biotechnology-derived pharmaceuticals and those of oil products (including re-exports of crude oil and fuels supplied by Venezuela) expanded in relative terms between 2004 and 2014.

73 Marc Frank, "Global Crisis, Storms Hit Cuba Finances," Reuters, December 17, 2008; Gerardo Arreola, "Cuba Tiene Congeladas Desde Hace 6 Meses las Cuentas Bancarias de Empresas Extranjeras," *La Jornada*, May 24, 2009.

74 "Ban Lifted on Cubans Staying in Tourist Hotels," Associated Press, March 31, 2008.

75 MINTUR data obtained by the authors in May 2015.

76 The Tourism Dependency Index (TDI) recently launched by the Inter-American Development Bank (IADB) will be examined in Chapter 4.

77 Marc Frank, "Cubans Cast Aside Coronavirus Fears to Search for Scarcer Food," Reuters, April 8, 2020.

78 Between 1959 and 1989, during which the vast majority of exports and imports went to and originated from the Soviet Union, Cuba recorded a positive merchandise trade balance in only two years (small surpluses in 1960 and 1974) and ran up a huge cumulative deficit (Mesa-Lago 1993, 138).

79 Calculations of the authors from ONEI 2015a, 1998.

80 Quintana et al. (2004, 186–7) claimed that tourism revenues might have financed nearly 30 percent of Cuba's total imports of goods and covered around 64 percent of the country's merchandise trade deficit between 1990 and 2003.

81 The Tourism Satellite Account has also been developed by the World Tourism Organization (UNWTO), the Organization for Economic Cooperation and Development (OECD), and the Statistical Office of the European Communities (Eurostat).

82 The Tourism Satellite Account contrasts data from the demand side (goods and services purchased by tourists) with data from the supply-side of the economy (the value of goods and services produced by industries to meet tourist demand) (United Nations 2010a, 1).

83 The official exchange rate of one Cuban peso for one US dollar applied to business entities and served exclusively for compiling national income, fiscal, and enterprise accounting aggregates.

84 Figueras Pérez (2002, 116) claimed that, for every 100 Cuban workers in the tourism sector, fifty-three jobs were created in the manufacturing industry, thirty-six in construction, fourteen in agriculture and silviculture, twenty-nine in transportation, three in communal services, and three in communications.

85 The *current account* on the balance of payments records the net flow of money arising from the trade of goods and services, factor income like dividends paid to foreign investors, and other cash transfers like remittances. The *capital and financial accounts*, which gauge the net change in foreign assets, including foreign investment, loans, and reserves, are two additional key components of a country's BOP, but Cuba provides little information about them (Pérez-López 2011b, 44–6).

86 For a thorough examination of the unofficial economic activities that mushroomed in Cuba in the first half of the 1990s, see Pérez-López 1995.

87 Fidel Castro's speech at the evening session of the National Assembly of the People's Government, ANPP, at the Havana Convention Center on July 11, 1992, http://lanic.utexas.edu/project/castro/db/1992/19920712.html.

88 Cuban citizens receive their wages and pensions in regular CUPs. Yet due to generalized shortages of goods available, they are compelled to buy dollar-pegged convertible pesos, or CUCs (or US dollars until they were taken out of circulation in late 2004), in order to purchase the products they need in expensive retail stores that deal only with the latter currency. At the end of 1993, the nominal average monthly salary in Cuba was 182 CUPs, and the unofficial exchange rate was 100 CUPs per dollar (that rate soared to 130:1 in mid-1994). The current rate is 25CUP:1CUC.

89 Dalia Acosta, "Cuba: Pimps and Prostitutes Expelled from 'Blue Paradise,'" Inter Press Service, June 11, 1996.

90 As a way to stifle internal migration from the provinces to Havana, the Cuban government enacted a new law in 1997 that imposed considerable bureaucratic hurdles on people who wanted to move to and reside permanently in the capital city. "Decreto 217: Regulaciones Migratorias Internas para la Ciudad de la Habana y Sus Contravenciones," April 22, 1997, https://www.gacetaoficial.gob.cu/html/regulacionesmigratoriasparaC.H.html.

91 "Havana Tamer After Castro Cracks Down on Illegal Sex," CNN, February 3, 1999.

92 "Cuba Unveils New Economic Freedoms for Businesses," Associated Press, October 26, 2010.

93 Ricardo Torres Pérez, "A Crucial Moment for Cuba's Tourism Industry," *Progreso Weekly*, September 6, 2016.

94 Surveys of inbound visitors are conducted either on an ongoing basis or at specific moments during the high and low seasons (United Nations 2010b, 36).

95 "Ley No. 113," *Gaceta Oficial*, November 21, 2012. Approved by the National Assembly in July 2012, Cuba's new tax code took effect in January 2013.

96 Marc Frank, "In Communist Cuba, the Tax Man Cometh," Reuters, November 28, 2012.

97 Fabio Valdés García, "Cuba 3.0: Tiendas en MLC, un Nuevo Impulso para la Economía Cubana," *Granma*, November 3, 2019.

98 Carlos Lage Dávila, remarks, Noticiero Nacional de la Televisión Cubana, March 18, 2001.

99 "Cuba: Economía y Turismo," Economics Press Service, May 31, 2001.

100 Gilda Fariñas and Susana Tesoro, "Turismo: Monedas al Aire," *Bohemia,* May 16, 2003.

101 Gross hard currency revenues from tourism activities in Havana increased from $571.7 million in 2003 to $642.1 million in 2008 and plummeted to $480.5 million in 2013 (ONEI 2015d, 2009b).

102 On June 4, 2002, then-tourism minister Ibrahim Ferradaz said on Cuban national television that 67 percent of goods for the country's tourism sector were produced in Cuba.

103 The 8 percent reevaluation of the Cuban convertible peso occurred in April 2005 and lasted until March 2011 when the value of the CUC returned to be pegged at par with the US dollar (Spadoni 2014, 53).

104 According to the US-Cuba Trade and Economic Council, US food sales to Cuba increased from only $4.3 million in 2001 to $392 million in 2004 and, after a couple of years of slight downward trends, peaked at $710.1 million in 2008 (USCTEC 2009).

105 Cuba's total imports of goods from China rose from less than $900 million in 2005 to about $1.5 billion in 2008, remained well above $1 billion annually (and above $2 billion in 2015 and 2016) until 2019, and stood at nearly $1 billion in 2020–1 (ONEI 2022a, 2021, 2020a, 2019b, 2013, 2011a).

106 Marc Frank, "Cubans Allowed to Stay at Tourist Hotels," Reuters, March 31, 2008.

107 Omar Everleny Pérez Villanueva, "La Inversión Extranjera Directa en Cuba: Un Aporte al Desarrollo Nacional," OnCuba, November 8, 2021.

3

International Tourism in Cuba, 2015–2019

For better or for worse, the performance of the Cuban tourism sector during the 2015–19 period was significantly affected by two major events, one that occurred just a few days before the beginning of this period and the other that took place on the second Tuesday of November 2016. On December 17, 2014 (17D), the US president, Barack Obama, and the Cuban president, Raúl Castro, announced in simultaneous televised speeches the beginning of a historic process toward the normalization of relations between the United States and Cuba after more than a half-century of unremitting hostility. The US and Cuban presidents unveiled the decision to restore full diplomatic relations between their countries with the reopening of embassies and hailed a landmark exchange of prisoners accused and convicted of spying. Obama also announced a series of measures aimed at facilitating travel to the island from the United States and fostering economic linkages to Cuba and urged Congress to start working to end the long-standing US embargo against Cuba, whose main provisions are codified under the Torricelli law of 1992, the Helms-Burton law of 1996, and the Trade Sanctions Reform and Export Enhancement Act (TSRA) of 2000. In particular, TSRA codified a ban on tourist travel to Cuba for US citizens.

Progress in strengthening US-Cuba ties after the 17D was substantial. Along with notable diplomatic achievements, Obama's measures fueled a sizable growth in US visitors to Cuba, producing tangible economic benefits for the island's tourism sector (including a major boost for Cuba's private entrepreneurs) and creating business opportunities for US companies, even if relatively few commercial deals were completed. At any rate, embargo restrictions that are codified into law remained a powerful stumbling block to greater economic interaction between the United States and Cuba. For its part, Cuba did agree to business deals with US firms that brought more revenues to the government, but it did little to enhance the United States' ability to engage with the Cuban private sector. Moreover, by the time Obama left the White House in January 2017, the market-oriented economic reform process in Cuba launched by Raúl Castro a decade earlier after he replaced his brother Fidel at the helm of the Cuban state,[1] had come to a virtual standstill as Cuban authorities were preparing the terrain for the post-Castro era (Raúl stepped down from the presidency in February 2018) while facing economic problems amplified by a profound crisis in the island's key benefactor, Venezuela.[2] And to further complicate things, the surprising victory of Donald Trump in the US presidential election of November 8, 2016, ushered in a more aggressive US stance on Cuba. Trump rolled back the Havana-Washington rapprochement of the Obama era and reinforced considerably the economic sanctions against the island by imposing tighter restrictions on US travel to and business dealings with Cuba.

US and Cuban Policy Changes

This section delves into the amendments to the Cuba sanctions regulations introduced during the Obama and Trump administrations. However, it also reviews the main policy and institutional changes in the tourism sector that the Cuban government implemented during the 2015–19 period while facing two very different situations, at first a sharp increase of international visitors to Cuba, and especially of visitors from the United States amid improved relations with the US government, and later tougher US economic sanctions that took a toll on visitor arrivals.

As part of his policy of engagement with Cuba, President Obama issued six sets of amendments to sanctions regulations after the 17D (in January and September 2015, in January, March, and October 2016, and one set of amendments in June-July 2015 linked to the rescission of Cuba's designation as a state sponsor of terrorism), easing restrictions on travel, remittances, trade, investment, banking, and other commercial dealings with the island.[3] Changes in the area of travel were the most consequential. Although tourist travel to Cuba by persons subject to US jurisdiction remained prohibited, all specific licenses in the twelve existing categories of authorized travel were converted into general licenses which do not require prior approval from the US Treasury Department.[4] Americans were allowed to use US-issued credit and debit cards (and spend as much as they want) in Cuba, open bank accounts there, and bring back unlimited amounts of Cuban rum and cigars for personal use. A key change that Trump ultimately reversed, new rules in 2016 established that Americans could visit Cuba independently for people-to-people non-academic educational purposes without having to go in organized group tours. Furthermore, the Obama administration authorized cruises and ferry services from the United States to Cuba and reached an agreement with Havana's officials to resume direct commercial flights between the two countries. Transportation by cruise ships of authorized US travelers was permitted by general license, and certain related lodging services aboard vessels used for such travel were permitted as well.[5]

Additional amendments to the Cuba sanctions program facilitated US-based travel to the island, created better conditions for the Cuban tourism sector as a whole, and for Cuban private entrepreneurs within it, to prosper, and helped US firms pursue business operations in the Cuban market despite a continued prohibition (except in the telecommunications sector and a few additional areas) on direct investments. In May 2015, the United States removed Cuba from its list of state sponsors of terrorism. Washington's decision represented a crucial step in the process of rapprochement with Havana. On July 20, 2015, the United States and Cuba officially restored full diplomatic relations with the reopening of embassies in each other's capitals. And on the eve of President Obama's historic visit to Havana in March 2016, the United States removed Cuba from its list of countries deemed to have insufficient security (anti-terrorism) measures in their ports, thus allowing easier access to US ports to cargo vessels, cruise ships, and possibly ferries arriving from the island.[6] Moreover, limits on donative remittances to Cuban nationals other than certain prohibited Cuban government and Communist Party officials, as well as on money transfers to support the development of private businesses in Cuba were removed entirely. Limits on remittances that authorized US travelers and Cuban nationals could carry to Cuba were also eliminated completely. Unlimited remittances (and visits) by Cuban Americans to family members on the island had been allowed since 2009. Finally, US firms were permitted to provide services to improve Cuba's outdated telecommunications and internet infrastructures, and even to sign joint ventures with Cuba's telecommunications monopoly ETECSA. Rules allowing US companies to set up fiber-optic cables and satellite links and enter into roaming agreements with Cuba had been in effect since September 2009.[7]

In mid-2015, as Obama's new rules made it easier to travel and send money to Cuba from the United States, and prospects for the Cuban tourism industry looked bright, Havana's authorities reorganized the structure and functions of the Cuban Ministry of Tourism. Decree Law 330 and the Council of Ministers Agreement 570, both enacted in May 2015, established that MINTUR is responsible for leading and controlling the application of government policies on tourism.[8] More specifically, MINTUR should play a prominent role in tourist accommodation activities, travel agencies, various attractions, and tourism products offered by Cuba like golf tourism and real estate development, tourist transportation, and tourism marketing and promotion. It is also responsible for identifying and supervising areas of high importance for tourism, approving and controlling hotel investment projects and hotel administration contracts, promoting private (the actual term used is *no estatal* or non-state) sector tourism, and recruiting workers. To carry out these functions, MINTUR was restructured into seven large business groups known as Higher Business Management Bodies (*Organizaciones Superiores de Dirección Empresarial*, OSDEs) that, in theory, should operate independently as corporate bodies of an enterprise system rather than administrative bodies (Salinas, Mundet, and Salinas 2018, 226). In other words, the idea was to separate MINTUR's functions from those of the state, reducing bureaucratic hurdles and moving away from top-down decision-making arrangements. Still the same today, MINTUR's new structure consisted of four accommodation OSDEs (Cubanacán, Gran Caribe, Islazul, and Campismo Popular),[9] one for supplementary hotel services (Cubasol), one for travel agencies (Viajes Cuba), and one for tourism services (Servitur) (MINTUR 2015b). Separate from MINTUR, the Gaviota group also operates in the Cuban tourism sector.

The positive shock caused by relaxed US travel rules, coupled with the negative effects of dwindling Venezuelan aid and reduced soft credits from China and a few other commercial partners, forced Cuban authorities not only to step up the promotion of international tourism to boost the island's foreign exchange earnings but also to reconsider their approach toward certain aspects of the tourism industry (Feinberg and Newfarmer 2016, 9). Besides betting on the economic benefits of cruise tourism that took off after US cruise companies entered the Cuban market in 2016, different modalities were pursued in seeking hotel investment projects with foreign partners. Between 2016 and 2018, Cubanacán, Gran Caribe, and Islazul signed at least nine hotel management contracts with foreign chains in which the latter contributed financing for the renovation of hotels. The capital involved was substantial in some hotels of Gran Caribe and Cubanacán, among them the Iberostar Riviera (over $30 million worth of financing) and Meliá Tryp Habana Libre ($25 million) in Havana, Barceló Solymar-Occidental Arenas Blancas ($10 million) in Varadero, and Valentin Tropicoco ($8 million, today known as Valentin Marazul) on the beach of Santa María del Mar, just twelve miles east of Havana. Even the smaller, budget friendly Islazul secured foreign capital to finance the refurbishment of some two- and three-star hotels in Havana and Camagüey.[10]

Moreover, Cubanacán and Gran Caribe began once again to actively pursue joint ventures with foreign companies for the construction of new hotels. Published on an annual basis, MINTUR's portfolio of business opportunities for foreign investors listed about twenty or more hotel joint venture projects each year between 2015 and 2018, mainly to build hotels in Havana, Cienfuegos, and various beach areas on the northern coast of the provinces of Camagüey, Las Tunas, and Holguín. But almost all of these projects had yet to materialize by the end of 2018. In early 2015, there were twenty-six joint ventures formed with overseas investors in Cuba (including one operating in China), of which thirteen had executed investments in fifteen hotels with a combined 5,906 rooms. About four years later, twenty-seven established joint ventures existed, of which fourteen had executed investments in fifteen hotels with a combined 4,995 rooms (MINTUR 2018, 2017, 2016, 2015a). The traditional focus on hotel management contracts also continued during this period, but neither Cubanacán nor Gran Caribe was able to significantly increase their hotel portfolio.

In stark contrast to MINTUR's hotel companies, the Gaviota group has been the biggest contributor to Cuba's expanding hotel room capacity over the past two decades, and even more so since 2015 (Henthorne 2018, 75). Controlled by Cuba's military conglomerate Grupo de Administración Empresarial S. A. (GAESA), which is involved in all sectors of the Cuban economy, Gaviota has reserved for itself most tourism development projects in Havana and other choice locations across the island, and in 2016 took over control of Habaguanex, the consortium that runs a number of boutique hotels, stores, and restaurants in colonial Old Havana.[11] Gaviota also operates marinas, like the Marina Gaviota Varadero, rental car firms, taxi and tour bus businesses, convenience stores, the Cuban Export-Import Corporation CIMEX (Cuba's largest commercial company), and Almacenes Universales, the logistics and warehousing company with activities at the port of Mariel and all other Cuban ports.

Gaviota, generally, allows foreign participation only through hotel management contracts rather than joint ventures involving external financing. After the 17D, probably betting on a sustained growth of international tourism in Cuba following the détente with the United States, Gaviota went on a hotel building spree that far outpaced projects under the control of Cubanacán and Gran Caribe. In an attempt to cater to a higher-income segment of the tourism market, Gaviota also focused increasingly on luxury hotels (Cintra Roque 2017, 54). All Gaviota's properties built with Cuban capital, major five-star plus establishments opened in Havana like the Gran Hotel Manzana Kempinski in May 2017, the Iberostar Grand Packard in September 2018, and the Accor Sofitel Paseo del Prado hotel (later renamed Royalton Habana) in November 2019. And more luxury hotels have been built since then. The hotel projects of Gaviota are executed by GAESA's real estate company, Almest, along with the French firm Bouygues Construction. Furthermore, in December 2016, Gaviota's subsidiary AT-Comercial S.A. signed its first joint venture with Iberostar for the production of hotel supplies in the Mariel Special Development Zone.[12] As for MINTUR's companies, a notable achievement was the official opening of the 946-room luxury beachfront five-star resort Meliá Internacional Varadero, owned by Gran Caribe, in September 2019.[13]

The post-17D boom in American visitors to Cuba did not last very long, though. After little more than two years of warming ties during Obama's tenure, US-Cuba relations soured under the administration of Donald Trump. As it was the case with all US presidents in the post-Cold War era, from George H. W. Bush to Bill Clinton, George W. Bush, and even Barack Obama when he first ran for the presidency in 2008 (Spadoni 2010, 45–56),[14] domestic political considerations linked to the need to obtain support from the influential Cuban American community in South Florida clearly influenced Trump's stance on Cuba. In an interview with the Daily Caller in September 2015, then-candidate Trump said that, even if he would have made a better deal, the concept of opening with Cuba was "fine."[15] Nevertheless, a year later, at a rally in Miami before a crowd of conservative Cuban Americans, he embraced hardline embargo politics by pledging to reverse Obama's "concessions" to the Castro regime unless the latter met US demands for a free Cuba.[16] Trump began to fulfill his campaign pledge on June 15, 2017, when he traveled again to Miami to announce new restrictive measures on Cuba. He was accompanied by Senator Marco Rubio and Representative Mario Diaz-Balart, both Cuban Americans from Florida, who had supported him for president. Arguing that Obama had negotiated a "terrible and misguided deal" with Havana,[17] Trump ordered tighter restrictions on travel to and commercial dealings with Cuba, mainly to deprive the Cuban military of US dollars and empower ordinary Cubans by funneling those dollars to their small-scale private activities. As a White House fact sheet released the next day stated, "The new policy channels economic activities away from the Cuban military monopoly . . . while allowing American individuals and entities to develop economic ties to the private, small business sector in Cuba."[18]

But it was only in November 2017 that Trump officially started to roll back parts of Obama's Cuba policy by implementing the policy changes he had announced earlier in June.

Trump banned individual people-to-people trips to Cuba and reintroduced the requirement for Americans traveling under this category to be part of licensed groups. He also prohibited US firms and citizens from doing business with about 180 entities owned or controlled by the Cuban military's holding company GAESA (the so-called Cuba Restricted List has been updated and expanded several times since then), whose tentacles reach deep into the Cuban economy.[19] In addition to the loss of significant business opportunities for US entities, this meant that American visitors to Cuba were now barred from patronizing numerous military-owned hotels (83 establishments at that time, of which 27 were in Havana and Old Havana, and 116 today), restaurants, stores, and other enterprises. Furthermore, the regulatory changes came at a complicated time for prospective travelers to Cuba as the State Department had warned US citizens in late September not to travel to the island because of mysterious acoustic attacks against the personnel of the US embassy in Havana.[20]

Except for the addition in November of more than a dozen hotels and a few other properties to the list of Cuban entities that Americans are prohibited from patronizing because they are owned by the Cuban military, no new US restrictions on travel and business with Cuba were enacted in 2018. In August, the State Department softened its travel warning for Cuba from Level 3 ("reconsider travel") to Level 2 ("exercise increased caution").[21] However, traveling to Cuba from the United States would soon become much more difficult. In early June 2019, making good on his promise to reverse Obama's Cuba policy, Trump ended the general license authorizing people-to-people educational activities in Cuba and banned US-based cruises to the island effective immediately. Moreover, he suspended commercial flights between the United States and Cuban destinations other than Havana in October and did the same for all US charter flights three months later.[22] To intensify economic pressure on Cuba, the Trump administration also restricted Cuba's access to the US financial system by denying authorization for so-called U-turn transactions,[23] and capped Cuban American family remittances to $1,000 per quarter. Trump even activated Title III of the Helms-Burton law of 1996. After being suspended every six months for twenty-three years, Title III went into effect on May 2, 2019, giving US citizens, including naturalized Cuban Americans, the right to sue so-called "traffickers" in properties expropriated from them by the Cuban government, and potentially affecting a wide swath of Cuba's foreign business partners that used land and facilities claimed by plaintiffs.[24]

Put simply, the Trump presidency dashed Cuba's hopes of a long-term increase in American visitors that could have been a veritable game-changer for the island's tourism industry. In July 2019, speaking to a permanent commission of the country's National Assembly, the Cuban tourism minister, Manuel Marrero Cruz, acknowledged the negative impact of harsher US sanctions on visitor arrivals and outlined Cuba's strategy to restore tourism growth. In order to stimulate not only the flow of foreign visitors but also vacations by Cuban residents and Cubans living abroad, Marrero said that a number of measures would be implemented. They included a better use of the internet and social media for promotion purposes and an increased focus on online sales, an exhaustive market analysis to identify shortcomings and areas of improvement, a greater use of local products for the tourism industry "not as an option but as an urgent need" to foment sustainability and productive linkages, and the construction of new hotels even if the private sector was "a necessary ally for the development of tourism in the country."[25]

Meanwhile, it is worth noting that Cuba's political landscape was changing dramatically. Miguel Díaz-Canel had succeeded Raúl Castro as president of the Council of State in April 2018. That office evolved into president of the Republic in December 2019 after a new Cuban Constitution sought to decentralize the job of former leaders Fidel and Raúl Castro by establishing a division of functions among the president of the Republic, the prime minister of the government, and the first secretary of the Central Committee of the Communist Party (Rojas 2020).[26] Marrero Cruz, who served as a colonel in the Cuban armed forces and previously worked at Gaviota, was named Cuba's prime minister in December 2019. The government reshuffle was completed in April 2021, when Díaz-Canel also took the reins of the Communist Party from Raul Castro.[27]

Trends in Overnight and Cruise Visitors

A considerable increase in travelers to Cuba from the United States, especially US citizens of non-Cuban origin, was one of the most visible direct effects of Obama's regulatory changes to the Cuba sanctions program. The Obama-era engagement with Havana demonstrated how improved relations with its powerful northern neighbor could open up new, beneficial economic opportunities for Cuba, even more so at a time when shock waves from Venezuela's economic malaise had finally reached the island. Relaxed embargo rules also created some significant business opportunities for US companies in an effort to strengthen US economic interests in Cuba and make the opening to America's former Cold War foe "irreversible."[28]

As shown in Table 3.1, international visitors to Cuba increased by more than 50 percent between 2014 and 2017. During this period, Cuba attracted each year about half a million more visitors from abroad than the previous year. Approximately three million in 2014, international arrivals broke the four million mark in 2016 and reached 4.65 million in 2017. The post-17D growth of US-based travel to Cuba was impressive and made by far the biggest contribution. In fact, the United States alone accounted for over 40 percent of the growth in visitor arrivals to Cuba in 2014–17. As noted in Chapter 2, Cuban official statistics make a distinction between Cuban Americans who travel to Cuba with a Cuban passport and American citizens traveling there with a US passport. Between 2014 and 2016, annual trips to Cuba by US citizens of non-Cuban descent tripled from 91,254 to 284,552, and Cuban American trips grew nearly 30 percent to reach 329,430. The flow of US travelers to Cuba received further stimulus in 2016 from the launch of non-stop daily commercial flights between the United States and Cuba, and from the resumption of cruise services between the two countries.[29] But when it comes to stimulating overnight trips to Cuba by US citizens, no other factor was more consequential than Obama's new rules of March 2016 allowing Americans to travel to the island on their own under the people-to-people category.[30] More specifically, travelers could plan their own itineraries as long as they kept records for five years showing they had engaged in a full-time schedule of loosely defined educational activities.[31] Besides reducing travel costs, this rule change greatly diluted the US ban on American tourism to Cuba since US officials, despite some monitoring, practically used the honor system to regulate authorized travel to Cuba.

Practically, from March 2016 until November 2017, when Trump ended individual people-to-people Cuba travel, US citizens traveling to Cuba not for family visits could simply book a flight and a hotel (except beach resorts) or a private accommodation online, and self-certify in an affidavit that they had complied with US regulations. Six US air carriers, namely Delta, United, JetBlue, American Airlines, Southwest, and Alaska Airlines, offered commercial flights

Table 3.1 International Visitors by Origin, 2014–19

	2014	2015	2016	2017	2018	2019
Canada	1,175,077	1,300,405	1,205,809	1,133,824	1,109,630	1,120,077
Cuban residents abroad	361,228	390,111	427,747	517,561	600,306	623,972
Cuban Americans	258,837	292,692	329,430	432,642	521,059	552,895
United States	91,254	162,972	284,552	618,346	638,365	498,538
Russia	69,237	44,208	65,386	105,258	137,440	177,977
Germany	139,138	175,507	242,355	243,172	208,506	174,956
France	103,475	138,972	187,468	209,239	197,521	169,394
Mexico	82,820	105,767	131,353	141,540	171,945	166,995
Spain	77,099	107,903	153,340	168,949	167,969	146,339
Italy	112,076	137,970	191,585	227,829	177,852	133,436
UK	123,910	156,052	194,805	205,562	189,963	123,745
Argentina	68,849	85,367	94,727	99,435	97,484	95,578
Chile	38,500	49,378	51,860	44,041	49,889	57,998
Others	559,812	685,563	778,182	938,803	965,040	786,583
Total	3,002,475	3,540,175	4,009,169	4,653,559	4,711,910	4,275,588

Sources: ONEI 2022a, 2021a, 2020a, 2016a; Data of Cuban Ministry of Tourism (MINTUR) obtained by the authors in May 2022.

to Cuba. In addition to the online home-rental service Airbnb, which entered the Cuban market in 2015, the online hotel reservation services Booking.com (owned by the Priceline Group), TripAdvisor, and Expedia all reached agreements with Cuban hotels to join their booking system.[32] As for the size of the business and the specific economic benefits to US companies, the Havana Consulting Group calculated that air travel to Cuba from the United States generated roughly $1.5 billion in revenues in 2016 through the sale of plane tickets, tourist packages, and health insurance, and through baggage fees. The sale of visas raised an additional $150 million, and a portion of that money was pocketed by US firms through processing fees (THCG 2017). In a clear sign that the US prohibition on travel to Cuba for tourism purposes had become less threatening, statistics of the Cuban Ministry of Tourism reveal that only one in ten Americans who visited Cuba in 2017 did so through a third country, as opposed to more than one in three in 2015 and in the first four months of 2016.[33] Not surprisingly, together with 432,644 trips by Cuban Americans, trips to Cuba by US citizens of non-Cuban origin more than doubled in 2017 to 618,346 to bring the total number of arrivals from the United States to over one million. Underscoring the key role of the United States in fueling the post-17D tourism boom in Cuba, the share of US-based travelers in the total number of international visitors to Cuba shot up from 11.7 percent in 2014 to 22.4 percent in 2017 (Perelló Cabrera 2019; ONEI 2018a).

Actually, with the notable exception of the largest source market, Canada, international arrivals to Cuba from virtually all countries increased between 2014 and 2017. The case of Canada, which exhibits the most repetitive and seasonal tourism in Cuba among all major source markets, is quite significant. From 1994 until 2014, Canadian tourist arrivals to Cuba grew every single year except in 2002, right after the September 11, 2001, terrorist attacks on the United States. They even grew 10.7 percent in 2015 to a peak of 1.3 million tourists before suffering a 12.8 percent drop in 2015–17. Some Cuban scholars argued that a combination of factors might have caused the aforementioned fall in the number of Canadian tourists to Cuba, namely higher hotel prices due to increased demand, a depreciation of the Canadian dollar vis-à-vis the US dollar, and a growing presence of US visitors (Carrazana Amador and Herrera Cárdenas 2022). In reality, it is unlikely that hotel prices played a major role in the decline of Canadian tourism in Cuba. Increased demand surely boosted accommodation prices in Havana and in sightseeing destinations that Americans were visiting for educational purposes, but not so much the prices of Cuban beach resorts that attract most Canadian tourists. The number of British tourists, who, like Canadians, tend to visit Cuba for beach vacations, nearly doubled in 2014–17 even if the British pound suffered a considerable devaluation during that period. Arrivals from major European source markets like Germany, France, Italy, and Spain, that send tourists to Havana and various beach destinations in Cuba, grew markedly as well. In short, it is not a secret that many Canadian and European tourists like to go to Cuba because there are relatively few Americans. What seems to have happened in the post-17D is that both Canadians and Europeans, to a certain extent, stepped up their visits to Cuba to anticipate a looming avalanche of US tourists. The spike in Canadian trips to the island simply faded faster than that in European trips.

In 2018, Cuba welcomed a record-breaking 4.71 million international visitors (ONEI 2019b). However, a closer look at these travelers that differentiates the more lucrative stay-over or overnight tourism from cruise tourism indicates that the latter's growth was the main reason why foreign arrivals in Cuba peaked in 2018. As highlighted in Table 3.2, overnight visitors to Cuba rose from 2.98 million in 2014 to 4.08 million in 2017, which was the best year for the Cuban tourism industry in terms of revenues, but fell around 6 percent in 2018 to 3.83 million and remained practically stagnant in 2019. The extensive damage from the passage of Hurricane Irma in September 2017,[34] and the US travel advisory a few weeks later to reconsider traveling to Cuba because of mysterious sonic attacks, negatively affected stay-over visits to the island in 2018. Yet the main contributing factor was Donald Trump's decision to end individual people-to-people trips to Cuba from the United States in November 2017. Forcing people-to-people trips to be taken with an authorized group stifled US-based travel to Cuba by making those trips more expensive and less appealing to independent travelers who prefer to design their own itineraries (Spadoni 2021, 246). Less impacting was Trump's decision to place dozens of Cuban hotels off-limits to American visitors, even though the inclusion of several hotels in Havana and

Table 3.2 International Visitors to Cuba by Concept, 2014–19

Concept	2014	2015	2016	2017	2018	2019
Overnight visitors	2,983,172	3,511,192	3,789,185	4,076,116	3,834,415	3,827,935
Cruise visitors	8,085	9,922	174,063	433,339	625,909	305,116
Others[a]	11,218	19,061	45,921	144,104	251,586	142,537

Sources: MINTUR data obtained by the authors in May 2020; ONEI 2016a.
Note:
[a] Includes cruise crew members, yachtsmen, other visitors.

Old Havana (twenty-six establishments) in the first list of restricted Cuban entities of November 2017 might have complicated the accommodation plans of some US organized groups.

Over 60 percent of the fall in overnight visitor arrivals to Cuba in 2018 was accounted for by a drop in US arrivals. Interestingly, US stay-over visits to Cuba rose more than 20 percent in the first four months of 2019 and managed to end 2019 at roughly the same level as the previous year, even if the Trump administration had added several Cuban hotels to the list of entities banned for Americans, and eliminated people-to-people travel (and cruise travel) to Cuba in early June.[35] Between November 2018 and November 2019, Trump blacklisted twenty-nine additional Cuban hotels owned by the Cuban military, including the Iberostar Grand Packard, the Sofitel Paseo del Prado, and even the Gran Hotel Bristol Kempinski that had yet to open its doors. The other ultra-luxury hotel of the Swiss chain Kempinski in Havana, the Gran Hotel Manzana, was already on the US list of restricted Cuban entities.[36] What definitely did not hurt US-based travel to Cuba was the US Department of State's decision in August 2018 to soften its travel warning for Cuba from Level 3 ("reconsider travel") to Level 2 ("exercise increased caution").[37]

Cuba received 4.27 million international visitors in 2019, a 9.2 percent drop from the previous year, essentially due to the US ban on cruises to the island. In the meantime, visitor arrivals to Cuba, mostly overnight travelers, from European countries had fallen significantly. Between 2017 and 2019, annual arrivals of tourists from Italy plummeted by 41.4 percent, and so did tourists from Germany (−28.1 percent), France (−19 percent), and Spain (−13.4 percent). British tourists also plunged by 39.8 percent, but this drop was largely fueled by the collapse, in September 2019, of the UK travel company Thomas Cook that sent around 95,000 visitors to Cuba annually.[38] At any rate, an increase in Cuban American arrivals and in visitors from Russia, Mexico, and a few other countries helped offset the decline in European arrivals. After fluctuating up and down during the 2014–16 period, the annual flows of Canadian tourists to Cuba varied little in 2017–19. Canada remained Cuba's top visitor source market in 2019 with 1.12 million visitors, followed by the United States (552,895 Cuban Americans and 498,538 US citizens of non-Cuban descent), Russia (177,977), Germany (174,956), France (169,394), Mexico (166,995), Spain (146,339), Italy (133,436), and the UK (123,745) (ONEI 2022a).

Among the trends in visitor arrivals to Cuba during the 2015–19 period, the rise and fall of cruise tourism was undoubtedly the most remarkable development.

From 1990 until the 17D, despite an uptick in cruise passenger arrivals in the second half of the 1990s when Italy-based Costa Crociere operated its Costa Playa cruise ship and managed a passenger ship facility at the port of Havana, and in the mid-2000s when Spain-based Pullmantur Cruises operated its Holiday Dream from Havana, cruise tourism in Cuba achieved very modest results and never really blossomed.[39] In 2006, to further complicate things, the US cruise lines Carnival and Royal Caribbean purchased all the European companies that used Havana as home port, which led to the immediate suspension of Cuba from their itineraries and to nearly a decade of hardly any cruise arrivals to the island (Perelló Cabrera and Betancourt 2018, 65–9).[40] In December 2015, Switzerland-based MSC Cruises became the first major cruise line to homeport in Cuba, sailing out of Havana to Jamaica, Cayman Islands, and Cozumel with its 2,150-guest MSC Opera. However, it was the US giant Carnival Corporation's historic cruise voyage to Havana from Miami under the banner of its new Fathom brand five months later that marked the beginning of a new era for the Cuban cruise industry, albeit a short one. In May 2016, two months after receiving the green light from the Cuban government, Carnival's 704-guest Fathom Adonia cruise ship became the first US cruise vessel to dock in Cuba in almost four decades. The Cuban government was so eager to relaunch US-Cuba cruises that it lifted a controversial ban on Cuban Americans sailing to their native country that it had sparked protests from exiles.[41] The Adonia sailed to Cuba from the United States for about a year, offering weeklong voyages with calls in Havana, Cienfuegos, and Santiago de Cuba, and, from the end of 2016, adding a call in Santiago de Cuba to its cruises to the Dominican Republic.[42]

Carnival's Fathom cruise line stopped sailing to Cuba in June 2017 to join the company's P&O Cruises fleet in the UK. Yet Carnival quickly launched Cuban itineraries from Tampa, Florida, aboard its 2,052-guest Carnival Paradise cruise ship. One of the largest cruise ships ever to call at Havana, Carnival Paradise offered four- and five-day cruises to Cuba with a daytime and overnight call in Havana and with a stop in either Cozumel, Mexico, or Key West, Florida.[43] Other major US cruise lines like Norwegian and Royal Caribbean, and a number of additional ship operators (Pearl Seas, Holland America, Seabourn, Regent Seven Seas Cruises, and Virgin Voyages, among others) started Cuba trips as well (Henthorne 2018, 110). Norwegian Cruise Line offered cruises to Cuba from Miami on the 2,004-passenger Norwegian Sky ship and from Port Canaveral, Florida, on the 1,936-passenger Norwegian Sun ship. Royal Caribbean International used a refurbished 1,602-passenger vessel, the Empress of the Seas, to sail to Cuba from Miami. It also used its 2,350-guest Majesty of the Seas ship to sail to three Cuban ports from either Tampa or Fort Lauderdale, with stops to Cozumel, Key West, or Nassau, Bahamas. Many of Norwegian's and Royal Caribbean's itineraries included overnight stays in Havana.[44]

Thanks to the easing of US sanctions rules, Cuba was now a highly coveted cruise destination. Less than 10,000 per year in 2014–15 (ONEI 2016a), cruise passenger arrivals in Cuba reached 174,063 in 2016 and skyrocketed to 433,339 in 2017. That year, twenty-three cruise lines made 706 calls in nine Cuban destinations. Of those calls, 304 were in Havana, 145 in Cienfuegos, and 104 in Santiago de Cuba (Perelló Cabrera 2019). As cruise lines sailing to the island expanded their offerings and more cruise ships entered the Cuban market, Cuba welcomed a record 625,909 cruise passengers in 2018. To shed more light on the key role of the United States in the revival of the Cuban cruise market, Figure 3.1 (which excludes Cuban Americans) provides official Cuban data on US visitors to Cuba by mode of travel between 2014 and 2019. The slight increase in arrivals to Cuba from the United States in 2018 was due entirely to a spectacular growth in cruise visitors whose Cuba-bound voyages had to eschew sun-and-beach activities in favor of cultural and educational exchange programs to avoid the ban on tourism. The number of US overnight visitors traveling to Cuba by air fell more than 30 percent from 448,475 in 2017 to 297,360 in 2018. Arrivals of US cruise passengers, on the other hand, grew from 171,302 to 341,005. As

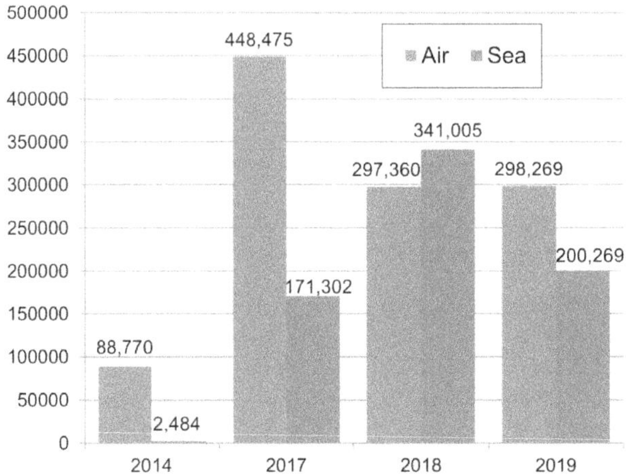

Figure 3.1 US visitors to Cuba by mode of travel, 2014–19. MINTUR data obtained by the authors in February 2020.

2018 neared its end, the demand for Cuba cruises remained high and the island seemed poised to grow further as a cruise hotspot. Stressing that no port in Cuba was operating at full capacity and taking into account consumer demand in the medium term, an analysis by the Economist Intelligence Unit (EIU) in March 2018 forecasted that the number of cruise passengers to Cuba was likely to rise to 1.5 million by 2022 even if the country's investment projects for upgrading its maritime infrastructure, especially the planned (yet never completed) expansion of the cruise terminal of the port of Havana, were to be delayed.[45]

As displayed in Figure 3.2, more than half (54.5 percent) of international cruise passengers to Cuba in 2018 were Americans. Other significant, yet much smaller, cruise source markets were Germany (6.8 percent of total passengers), Italy (6.7 percent), the UK (5.4 percent), Spain (4.5 percent), France (4.1 percent), Mexico (3.9 percent), and Canada (2.3 percent). Traveling by cruise had become the easiest way for US citizens of non-Cuban descent to visit Cuba as cruise lines were unaffected by US restrictions on Americans staying at certain hotels on the island and supposedly were able to ensure compliance with US law and its ban on tourism in Cuba. Although it seemed a bit odd, the notion that a US cruise traveler on a multi-destination trip was a tourist when disembarking in the Bahamas or the Dominican Republic but not when visiting Cuba, the latter was set up well for educational and cultural tours since its main cruise ports in Havana, Cienfuegos, and Santiago de Cuba were all in areas recognized by the

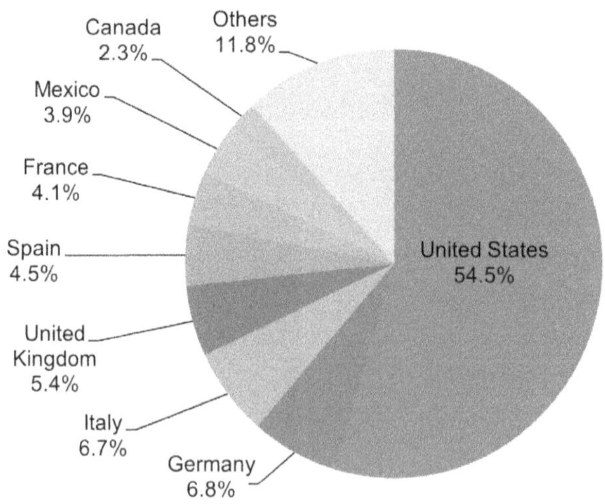

Figure 3.2 Cruise visitors by country of origin in 2018 (percentage distribution). MINTUR data obtained by the authors in February 2019.

United Nations Educational, Scientific, and Cultural Organization (UNESCO) as World Heritage Sites. In the first four months of 2019, compared with the same period in 2018, arrivals of US cruise travelers to Cuba more than tripled, and so did total international cruise arrivals.[46] However, cruise tourism in Cuba virtually came to a halt in June 2019 when the Trump administration ended the permission for US cruise ships to carry licensed travelers to Cuba. Carnival, Norwegian, Royal Caribbean, and other cruise lines serving the Cuban market hurried to replace their calls in Cuban ports with other destinations and offered new itineraries and options to 800,000 passengers who had booked cruises to Cuba.[47] As for the negative impact on the Cuban cruise sector, Havana's officials claimed that Cuba aimed to welcome one million cruise passengers in 2019 and that Trump's move meant for Cuba the loss of 560,000 cruise visitors in the second half of the year.[48]

More Hotels, More Private Rooms

In the post-17D era, at least until 2017, tourism was the one bright spot in Cuba's ailing economy, which was struggling with plummeting merchandise exports and reduced economic ties with Venezuela. With annual international overnight visitor arrivals growing by more than one million in 2014–17, or an average of about 360,000 per year, Cuba scrambled to expand its hotel capacity, building several four- and especially five-star establishments in popular tourist regions across the island. Besides handing over the management of most of these properties to international chains, the Cuban government courted foreign hotel operators to start developing the untapped high-end tourism market, as some ultra-luxury five-star plus hotel were built, primarily in Havana. In May 2017, just six months before Trump's move to ban individual people-to-people travel to Cuba, then Cuban tourism minister Manuel Marrero Cruz said that the country was adding 2,000 hotel rooms a year to meet growing tourism demand.[49] Sun-and-beach vacations in all-inclusive resorts in Varadero, Jardines del Rey, Cayo Santa María near the town of Caibarién in the province of Villa Clara, Playa Guardalavaca in the province of Holguín, and other locations on Cuba's northern coast, continued to be the flagship product of Cuba's tourism sector even though a substantial number of new hotels were built in urban areas like Havana that offer a heterogeneous range of cultural, architectural, and social experiences for leisure. Mainly triggered by soaring American travelers to Cuba and concentrated in Havana and other popular urban locations, another important development was the sizable growth of private homestays as many Cuban property owners, to make extra cash, obtained a government license to legally convert rooms within their primary residence to accommodate international visitors.

To underscore the construction efforts of the principal Cuban hotel chains and the overall availability of rooms for international tourism in Cuba with the contribution of private accommodations, Table 3.3 provides information on Cuba's lodging infrastructure by sector (state versus private) and by entity in 2014 and 2019. Two things stand out. First, the state-owned military-run Gaviota hotel group was, and still is today, the most dynamic tourism group in Cuba (Cintra Roque 2017, 35–6).[50] In an interview with the US magazine Hotels in early 2016, Gaviota's business vice president, Raniel Leyva Medina, confirmed that the company's model of business was to avoid granting shares in its properties and pursue only management contracts with international hotel chains. Stressing that his company had 26,000 hotel rooms in the pipeline for 2025, Leyva Medina also revealed that Gaviota was "moving forward into high-class tourism, luxury hotels," and that it was "basically emphasizing Havana as the number one Cuban plaza and continuing with Varadero (Matanzas Province), Cayo Santa Maria (Santa Clara Province), Cayo Coco (Ciego de Ávila Province), Cayo Cruz (Camagüey Province) and Holguín (Holguín Province)."[51] Cuba's hotel room inventory rose from 62,090 rooms in 2014 to 74,211 rooms in 2019, an increase of 12,121 rooms. Gaviota accounted for more than 90 percent of that increase since it added 11,182 rooms to its stock

Table 3.3 Lodging Capacity by Sector and Entity in 2014 and 2019 (year-end, existing rooms)

Sector (Entity)	2014	2019	2014–19 Growth
State sector (of which)	62,090	74,211	12,121
Gaviota	22,903	34,085	11,182
Cubanacán	15,368	15,937	569
Gran Caribe	12,250	13,859	1,609
Islazul	9,983	9,316	−667
Other entities	1,586	1,014	−572
Private sector	18,742	26,224[a]	7,482
Total Cuba	80,832	100,435	19,603

Source: MINTUR data obtained by the authors in May 2022; Perelló Cabrera 2015a.
Note:
[a] June 2019.

during the same period. At the end of 2019, Gaviota had 34,085 rooms (mostly in five-star hotels), or 45.9 percent of Cuba's total hotel rooms. Struggling with a lack of financial resources, Cubanacán and Gran Caribe, with 15,937 rooms (21.5 percent of the total) and 13,859 rooms (18.7 percent), respectively, in 2019, had experienced little growth since 2014. In Havana, Gaviota's hotel rooms surged from approximately 1,300 in 2014 to nearly 2,500 at the end of 2018.[52]

The second major development linked to the expansion of Cuba's lodging infrastructure after the 17D was the proliferation of privately run bed and breakfasts, or *casas particulares*, in various districts of Havana, from Miramar to Vedado, Plaza de la Revolución, Central Havana, and Old Havana, as well as in places like Trinidad, Cienfuegos, Viñales in the province of Pinar del Rio, and Baracoa, all crucial tourist areas providing good opportunities for earning hard currency (Bye 2020, 90). These private accommodations also exist in various beach areas, but in smaller numbers than in urban sites. Accompanied by a rapid growth of private restaurants, or *paladares*, and additional private businesses catering to foreign tourists, the number of rooms in Cuba's private sector increased from 18,742 in 2014 to 26,224 in June 2019. By then, the private sector accounted for about 25 percent of total available rooms for international tourism on the island and trailed only Gaviota in overall capacity. Concentrated in Havana, US citizens of non-Cuban descent who traveled to Cuba in 2015–17 made a key contribution to the development of the burgeoning private tourism sector on the island. Havana ended 2017 with 11,552 *casas particulares*, 2,959 more than the previous year. In 2017, according to the Cuban Ministry of Tourism, a staggering 80 percent (357,737) of overnight US visitors to Cuba (448,475) stayed in private accommodations.[53] For instance, the US online lodging service Airbnb, which entered the Cuban market in mid-2015, reported that it processed room reservations in *casas particulares* for over 12 percent (some 35,000 people) of all US travelers to Cuba in 2016. Initially allowed to deal only with Americans and later authorized to offer its listings to all foreign travelers to the island, Airbnb also revealed that more than 560,000 guests stayed in Cuban homes listed on Airbnb between May 2015 and May 2017, and that $40 million was paid to Cuban individuals for sharing their homes (Airbnb 2017).

American media reported that overnight US visitors to Cuba gravitated to private bed and breakfasts as they could not stay in resorts because of the US embargo. They also suggested that Cuban private renters would suffer little from Trump's tighter rules of November 2017 since Airbnb successfully lobbied the

White House to secure language in those rules to allow American individual travelers to stay in Cuban private homes, like those listed on Airbnb, if they visited the island under the "support for the Cuban people" general travel license.[54] Both claims are incorrect. Until November 2017, American legal travelers to Cuba could stay in any Cuban hotel except in beach establishments where stays could expose them to possible fines for violation of the tourism ban. Many US visitors rented rooms in private homes either because the capacity of top-end foreign-run hotels in Havana and other urban areas was being pushed to the limit, or, quite frequently, because they preferred private accommodations that offered a more authentic and more pleasant experience, and at more affordable prices. An internal survey of international visitors to Havana conducted by Cuban tourism officials in 2016 revealed that *casas particulares* ranked significantly higher than hotels in quality/price satisfaction and general customer satisfaction.[55] They also exhibited higher occupancy rates than Cuban hotels (Kozak 2016, 106). It is worth noting that after the 17D, Cuban travel agencies signed thousands of contracts with local private entrepreneurs, principally in the transportation, lodging, and food services sectors, to meet growing tourism demand. By the end of 2017, these contracts had become so popular that the Cuban government placed limits on their values. Earlier in August, the government had temporarily suspended the issuance of new licenses for *casas particulares*, *paladares*, and other private occupations, though existing licensees were allowed to continue to operate (Henken 2018, 232; LeoGrande 2017).

Furthermore, it was disingenuous to believe that travel under the "support for the Cuban people" license could simply replace the individual people-to-people travel license. While the wording describing the general requirements for the two licenses was similar, it was clear that the days of little or no policing of the comings and goings of those making individual people-to-people trips to Cuba were gone.[56] As a matter of fact, intensified scrutiny of potential violations of the tourism ban and worsening US-Cuba relations discouraged a sizable number of prospective US travelers from going to Cuba. Many Americans knew little about Cuba and thus believed that Trump's combative tone and regulations meant the island was again off-limits. In essence, the end of self-directed people-to-people tours (which Trump charged they were being used to circumvent the tourism ban) hurt the Cuban private sector and its cash revenues from US visitors because organized groups tend to rely on state-run services for logistical reasons. Since it is difficult to accommodate large groups in *casas particulares*, quite expensive packaged tours lodge visitors in state-owned hotels and oftentimes arrange lunches and dinners at designated state-owned restaurants. To sum up, many Cuban private renters had been struggling since November 2017 due to falling arrivals of US overnight visitors. Trump's move to ban group people-to-people educational travel to Cuba in June 2019 delivered an even bigger blow to Cuba's private tourism businesses.

Meanwhile, in the state sector, the Gaviota-led hotel-building effort significantly boosted Cuba's availability of luxury accommodations for international visitors. At the end of 2019, as shown in Table 3.4, more than 70 percent of Cuba's total hotel rooms were in four- and five-star hotels. However, it was the top category accommodation segment that had experienced the highest growth in the previous five years. Between 2014 and 2019, the share of four-star hotel rooms in Cuba's total hotel rooms fell from 37.6 percent to 30.3 percent as the country added just about 2,300 rooms in this category. During the same period, Cuba's hotel capacity incorporated some thirty new five-star hotels with a combined 11,000 rooms, raising the stock of five-star hotel rooms to 30,785, accounting for 41.5 percent of total hotel rooms on the island. For the first time, Cuba's hotel portfolio included three five-star plus hotels, all in Havana, with a combined 817 rooms. And yet, according to Cuban official sources, notwithstanding considerable luxury hotel construction to attract international business travelers and higher-income tourists, Havana still had only 2,615 five-star hotel rooms at the end of 2018, representing 20.7 percent of the city's total hotel rooms (12,608). Besides 3,100 four-star hotel rooms (24.6 percent of the total), 6,893 hotel rooms, or 54.7 percent of the total, were in the three-star category or below, raising the need for more upscale development.[57] There were also a large number of rooms out of service. Cuba's official newspaper

Table 3.4 Hotel Capacity by Category in 2019 (year-end)

Category	Hotels	% of Total	Rooms	% of Total
Five-star plus	3	0.7	817	1.1
Five-star	59	14.0	30,785	41.5
Four-star	110	26.1	22,519	30.3
Three-star	108	25.6	11,087	14.9
Two-star	78	18.5	6,336	8.6
One-star	17	4.0	930	1.3
Unrated	47	11.1	1,737	2.3
Total	422	100.0	74,211	100.0

Source: MINTUR data obtained by the authors in May 2022.

Granma reported at the end of 2018 that, ahead of the 500th anniversary of its founding in November 2019, Havana was set to open twelve new hotels and incorporate 1,121 rooms that were currently out of service.[58] Just to give a sense of the magnitude of the problem, at some point in 2017, one-fourth of Havana's total hotel rooms were out of service. This helps to explain why in 2017, despite record arrivals of US overnight visitors, Havana's hotel occupancy rate was 70.5 percent against available rooms and only 53.9 percent against existing rooms.[59]

To offer additional evidence of the kind of tourism that Cuba is promoting, Table 3.5 provides data on Cuba's hotel capacity by tourism modality and by region in 2019. Cuba had 422 hotels at the end of 2019, of which 188 (44.5 percent of the total) were for sun-and-beach tourism, 178 (42.2 percent) for urban or city tourism, and the rest for nature, nautical, health, and other tourism modalities. A look at hotel rooms, though, reveals how Cuba's main tourism product was by far sun-and-beach tourism. The latter accounted for 55,444 rooms or 74.7 percent of the country's 74,211 hotel rooms in 2019, followed by urban tourism with 15,392 rooms (20.7 percent of the total) and the remaining modalities with a combined 3,375 rooms (4.6 percent). Among beach locations, Varadero had added fifteen new resorts and nearly 5,000 rooms to its hotel capacity since 2014. Varadero boasted sixty-six hotels with a combined 21,863 rooms, or approximately 30 percent of Cuba's total rooms, in 2019, followed by the keys of Caibarién (Cayo Santa Maria, the largest of all, Cayo Ensenachos, and Cayo Las Brujas) on the north coast of the province of Villa Clara with 12,220 rooms (16.4 percent), Jardines del Rey with 9,851 rooms (13.3 percent), and the beaches north of Holguín like Playa Guardalavaca, Playa Esmeralda, Playa Pesquero, and others with 6,012 rooms (8.1 percent). As for urban locations, the primacy of Havana was beyond question. Havana had incorporated some twenty-four new hotels with a combined 3,700 rooms into its hotel supply since 2014 to reach 13,348 rooms in 2019, and its share of Cuba's total hotel rooms that year was virtually unchanged from 2014 (18.0 percent versus 18.4 percent) despite a sizable expansion of Cuba's total hotel room stock.

Regarding Havana, it should be mentioned that there are a number of high-end hotels that are managed by Cuban state companies rather than by foreign chains. Using a dataset based on customer comments from the American TripAdvisor platform about 107 Cuban hotels (state- and foreign-run) in Havana and Varadero, a 2014 study by two Spanish scholars found that Cuban hotel companies were ranked lower than foreign chains in all categories of customer service. Among the top twenty companies for customer satisfaction,

Table 3.5 Hotel Capacity by Tourism Modality and by Region in 2019 (year-end)

Modality	Hotels	%	Rooms	%	Region	Hotels	%	Rooms	%
Sun and beach	188	44.5	55,444	74.7	Varadero	66	15.6	21,863	29.5
Urban	178	42.2	15,392	20.7	Havana	96	22.7	13,348	18.0
Nature	33	7.8	1,359	1.8	Caibarién	33	7.8	12,220	16.4
Nautical	8	1.9	951	1.3	Jardines del Rey	34	8.1	9,851	13.3
Health	8	1.9	937	1.3	North of Holguín	29	6.9	6,012	8.1
Others	7	1.7	128	0.2	Others	164	38.9	10,917	14.7
Total	422	100.0	74,211	100.0	Total	422	100.0	74,211	100.0

Source: MINTUR data obtained by the authors in May 2022.

Cubanacán, Gran Caribe, and Islazul were at the bottom of the ranking, Gaviota was ranked fifteenth, and Habaguanex, which manages hotels in Old Havana and was taken over by Gaviota in 2016, was the only Cuban chain to break into the top ten in the eighth position. Seven Spanish chains such as Blau, Iberostar, Be Live, Meliá Hotels International, H10 Hotels, Barceló, and BlueBay Resorts, were in the top ten. The aforementioned study, though, also found that in Havana, unlike in Varadero, state-run hotels beat foreign-managed hotels when rated on their locations while lagging behind in the quality-price relationship. Eight Cuban-managed hotels located in Havana, namely six Habaguanex-run establishments such as the hotels Terral, Saratoga, Palacio del Marqués de San Felipe y Santiago de Bejucal, Los Frailes, Conde de Villanueva, and El Comendador, the famous Hotel Nacional managed by Gran Caribe, and the Gaviota-run Hotel Quinta Avenida, were in the top twenty list of individual hotels ranked by customer satisfaction (Cerviño and Fraiz 2014).[60] Yet most premium Cuban-managed hotels in Havana, in particular those of Habaguanex, are historic facilities that necessitate frequent and expensive renovations, hence Gaviota's plans to promote high-end tourism require not only the opening of new upscale hotels but also significant investments in the improvement of existing hotels.

Official data compiled in Table 3.6 indicate that, at the end of 2018, foreign chains managed twenty-three hotels in Havana with a combined 5,319 rooms, or 42.2 percent of the city's total hotel rooms. This share was significantly lower than the share (60.1 percent) of Cuba's total hotel rooms run by international chains that year. Havana had eighty-eight hotels in 2018, thus only about one-fourth of them were foreign-run. There were ten five-star or five-star plus hotels, of which foreign chains managed six, specifically the Meliá Habana, Meliá Cohiba, Iberostar Parque Central, Grand Packard, Gran Hotel Manzana Kempinski, and the Hotel Quinta Avenida that was now Four Points by Sheraton following a multimillion-dollar deal in March 2016 with the US Starwood Hotels & Resorts Worldwide (a few months later acquired by Marriott International).[61] The remaining five-star hotels, all Cuban-managed, were the Hotel Nacional and the Gaviota-run Palacio del Marqués de San Felipe y Santiago de Bejucal, Santa Isabel, and Saratoga that was partially destroyed in a suspected gas explosion in May 2022.[62] Moreover, of the thirty-two four-star hotels that existed in Havana at the end of 2018, foreign chains managed only the Riviera, Memories Miramar, Capri, Victoria, Mercure Sevilla, Presidente, and Habana Panorama. The remaining four-star hotels included the Gran Caribe-run Plaza and

Table 3.6 Hotels Managed by Foreign Chains in Havana in 2018 (year-end)

	Hotel	Company	Category	Rooms
1	Meliá Habana	Meliá Hotels International	5	397
2	Meliá Cohiba	Meliá Hotels International	5	462
3	Tryp Habana Libre	Meliá Hotels International	3	572
4	Iberostar Parque Central	Iberostar Hotels & Resorts	5	427
5	Grand Packard	Iberostar Hotels & Resorts	5+	321
6	Riviera	Iberostar Hotels & Resorts	4	352
7	Memories Miramar	Blue Diamond Hotels & Resorts	4	427
8	Aparthotel Montehabana	Blue Diamond Hotels & Resorts	2	164
9	Capri	NH Hotels	4	220
10	Victoria	NH Hotels	4	31
11	Gran Hotel Manzana Kempinski	Kempinski Hotels S.A.	5+	246
12	Hotel Quinta Avenida	Marriott	5	186
13	Mercure Sevilla	Accor Hotels	4	178
14	Presidente	Roc Hotels	4	158
15	Habana Panorama	H10 Hoteles	4	320
16	Copacabana	Be Live Hotels—Globalia	3	168
17	Vedado	Elite Club Vacanze	3	203
18	St. Johns	Elite Club Vacanze	3	88
19	Paseo-Habana	Sercotel Hoteles	3	30
20	San Alejandro	Sercotel Hoteles	3	78
21	Caribbean	Sercotel Hoteles	2	38
22	Lido	Sercotel Hoteles	2	65
23	Tropicoco	Valentín Hotels	2	188
	Total			5,319

Source: MINTUR data obtained by the authors in June 2019.

Inglaterra, the Cubanacán-run Comodoro and Club Arenal, and the Gaviota-run Telegrafo, Park View, Ambos Mundos, Los Frailes, Raquel, Terral, Palacio San Miguel, Conde de Villanueva, Habana 612, El Comendador, Armadores de Santander, Palacio O'Farill, Tejedillo, Mesón de la Flota, Beltrán de Santa Cruz, and Marques de Prado Ameno. As for the property of the Havana hotels, only the Meliá Habana and Iberostar Parque Central were joint ventures with foreign companies. The hotels of Gran Caribe and Cubanacán are owned by the state tourism real estate firm Inmobiliaria del Turismo (Inmotur). Gaviota's hotels are owned by the military-controlled construction firm Almest.

As explained in Chapter 2, international hotel chains running Cuban hotels brought crucial benefits to Cuba, enhancing quality standards and the level of service, transferring knowledge and skills, establishing ties to major tour operators and foreign airlines, and helping the Cuban government obtain international credits for tourism development. And, of course, foreign chains benefited as well (Figueras Pérez 2020). During a period of strong growth of international visitor arrivals, Cuba's hotel rooms under management contracts with foreign companies rose from 32,415 in 2014 to 42,593 in 2018, and their share of Cuba's total hotel rooms increased from 52.2 percent to 60.1 percent. Nevertheless, information presented in Table 3.7 reveals that the number of hotel rooms in Cuba that were administered by foreign companies fell considerably in 2019 to 37,940 or 51.1 percent of the island's total hotel rooms, most likely because several expiring management contracts were not renewed amid worsening conditions of the Cuban tourism sector. At the end of 2019, fourteen international chains operated 83 hotels in Cuba, less than one-fifth of the country's 422 hotels. Spain's Meliá remained the leading foreign hotel operator in Cuba with contracts to manage thirty-four hotels with a combined 14,511 rooms, followed by Canada's Blue Diamond (14 hotels with 8,462 rooms), and Spain's Iberostar (13 hotels with 6,086 rooms). These three companies administered over 75 percent of Cuba's total hotel rooms under foreign management. The remaining eleven foreign chains, with much smaller portfolios, included seven Spanish firms, India's MGM Muthu Hotels, France's Accor, Switzerland's Kempinski, and the US Marriott.[63] Since 2019, many new hotel contracts with overseas companies have been signed but, as we will see in Chapter 5, at the end of 2022, the share

Table 3.7 Foreign Hotel Operators in Cuba in 2019 (year-end)

	Company	Country of Origin	Number of Hotels	Number of Rooms	% of Total Rooms
1	Meliá Hotels International	Spain	34	14,511	38.25
2	Blue Diamond Hotels & Resorts	Canada	14	8,462	22.30
3	Iberostar Hotels & Resorts	Spain	13	6,086	16.04
4	H10 Hoteles	Spain	4	2,041	5.38
5	Barceló Hotels & Resorts	Spain	3	1,353	3.57
6	Valentín Hotels	Spain	2	1,208	3.18
7	MGM Muthu Hotels	India	2	1,000	2.64
8	Be Live Hotels—Globalia	Spain	3	781	2.06
9	Accor Hotels	France	2	744	1.96
10	Roc Hotels	Spain	2	603	1.59
11	Blau Hotels & Resorts	Spain	1	395	1.04
12	Sercotel Hoteles	Spain	1	324	0.85
13	Kempinski Hotels S.A.	Switzerland	1	246	0.65
14	Marriott	United States	1	186	0.49
	Total		83	37,940	100.00

Source: MINTUR data obtained by the authors in May 2022.

of Cuban hotel rooms under foreign management was still lower than its level in 2018.

Finally, in April 2016, the Cuban government unveiled an ambitious plan (2030 Vision Plan) to add 108,000 hotel rooms of three-star quality or better to Cuba's room stock by 2030 to accommodate ten million international visitors (not including an additional five million projected cruise visitors) and triple the country's tourism revenues (Rutty and Richardson 2019, 2; Salinas, Mundet, and Salinas 2018, 227–8).[64] Approximately 70 percent of the new rooms were expected to be built with Cuban capital and the rest with foreign capital. Setting impressive construction goals that, in the estimation of some scholars, would require a total of $33 billion in new investments over a fifteen-year period to be met (Feinberg and Newfarmer 2016, 33), the Cuban plan envisioned the island's hotel capacity to reach about 89,000 hotel rooms in 2020, 132,000 rooms in 2025, and some 166,000 rooms in 2030. This would mean building an average of more than 7,000 rooms per year, which is nearly three times higher than the annual average number of rooms (2,424) that were added to Cuba's stock in 2014–19. As for the principal locations of the new hotel rooms, around 30,000 were supposed to be built in the province of Camagüey (Camagüey city, Cayo Cruz, Cayo Sabinal), 15,700 in the province of Havana (Monte Barreto, Vedado, Playas del Este, Old/Central Havana), 13,600 in the province of Holguín (Pesquero, Guardalavaca, Península del Ramón), 12,800 in the province of Ciego de Ávila (Cayo Coco, Cayo Guillermo, Cayo Paredón Grande), 8,000 in the province of Matanzas (Varadero), and 6,500 in the province of Sancti Spiritus (Trinidad, Península Ancón) (Sierra 2017). However, Cuba's tourism goals have been scaled back since the global Covid-19 pandemic started to wreak havoc on the island's leisure industry in early 2020. Cuba's current plan is to increase its hotel capacity to 95,000 rooms by 2030 and receive annually over six million international visitors.[65]

Main Tourism Results

Along with overseas remittances that were fueled by growing arrivals of Cuban residents abroad, primarily Cuban Americans, international tourism was virtually the only bright spot of the Cuban economy after the 17D, at least until the end of 2017 when Donald Trump started to reverse Barack Obama's rapprochement with Havana's government. Cuba's gross hard currency revenues from tourism activities increased considerably after 2014 and reached record levels of more than $3 billion per year in 2016–17 before experiencing a major decline in 2018–19. Remittances, according to unofficial estimates, more than doubled during this period to around $3.7 billion per year and possibly emerged as Cuba's second-largest source of hard currency revenues after exports of professional services (Morales 2020, 2019).[66] Nevertheless, Cuba's annual GDP averaged only 2 percent in 2014–18 and shrank 0.2 percent in 2019 (ONEI 2020a, 2018). After 2017, in particular, Cuba's financial conditions deteriorated markedly as the island faced an acute balance of payments crisis involving, among other things, severe foreign exchange shortages, falling international reserves, and repeated failures to respect debt payments even if Cuba had been able to obtain substantial debt relief from a number of key creditors (Spadoni 2021, 230–1). Cuba's foreign exchange earnings plummeted in tandem with the economic woes of its ally Venezuela and a tightening of the decades-old US trade embargo, including new restrictions on US travel. Unofficial sources also revealed that the international reserves of Cuba rose noticeably from $7.4 billion in 2011 to $12 billion in 2016. However, by the end of 2019, these reserves reportedly had fallen to less than $10 billion as Cuba felt the blow of intensified US economic sanctions and once again struggled to pay its debt to foreign firms and creditor countries (EIU 2020, 2016).

Table 3.8 highlights the relative contribution of different sectors of the Cuban economy to the country's total hard currency revenues from exports of goods and services between 2014 and 2019. Cuban exports of goods and services dropped by nearly 30 percent in 2014–19 from $17.8 billion to $12.6 billion, but Cuba's annual overall trade balance continued to record surpluses (albeit shrinking ones) as import levels, in particular Venezuelan supplies of oil and refined products to the island, fell as well (Hernández-Catá 2019, 32). During this period, exports of goods plummeted by over half from $4.9 billion to just $2.1 billion, and exports of services decreased by nearly 20 percent from $13 billion to $10.6 billion. Among Cuba's top export goods, only tobacco products witnessed a small increase in earnings from international sales while exports of sugar, nickel, and pharmaceuticals all declined. The share of Cuba's total exports of goods and services accounted for by nickel, though, rose slightly (ONEI 2022a, 2021, 2020a).

Regarding services, until 2018, Cuba never published official disaggregated statistics on its service trade, with the only exception being tourism activities. Even if they increased somewhat in relative terms, hard currency revenues

Table 3.8 Exports of Goods and Services by Sector, 2014–19 (percentage distribution)

Year	Sugar	Nickel	Tobacco	Pharmaceuticals	Other Goods	Tourism	Other Services[a]
2014	2.3	4.2	1.3	3.1	16.4	14.3	58.4
2015	2.9	3.5	1.4	4.1	10.5	18.9	58.7
2016	2.6	3.4	1.7	3.2	6.0	22.4	60.7
2017	3.4	4.4	1.7	2.3	5.2	23.4	59.6
2018	1.3	5.1	1.8	2.8	5.3	19.1	64.6
2019	1.7	4.9	2.1	2.1	5.5	20.9	62.8

Sources: ONEI 2022a, 2021a, 2020a.
Note:
[a] Mainly medical and other professional services.

generated by non-tourism services plunged from $10.4 billion in 2014 to $7.6 billion in 2019. Cuba's exports of services other than tourism practically refer to revenues from exports of medical and other professional services and, to a lower extent, telecommunications services, given that official data on gross tourism earnings already include earnings from international transportation services. Mesa-Lago and Vidal Alejandro (2019, 9) estimated that sales of Cuban professional services abroad (mainly in Venezuela) fell from $9.8 billion in 2014 to $7.8 billion in 2017. At any rate, since 2018, Cuba's Office of National Statistics and Information (*Oficina Nacional de Estadísticas e Información*, ONEI) has published critical details of the island's external service trade, offering evidence that the sale of medical services overseas (worth over $6 billion in 2018) is Cuba's single-largest source of hard currency revenues. ONEI also revealed that revenues from these services have experienced a considerable drop in recent years. This is due to a reduced presence of Cuban healthcare providers in Venezuela and the end of Cuban medical missions to Brazil and a few other Latin American countries where Cuba-allied leftist presidents lost power.[67] On the other hand, the share of Cuba's total exports of goods and services accounted for by tourism activities shot up between 2014 and 2017 from 14.3 percent to 23.4 percent as record numbers of international overnight visitor arrivals proved to be strong revenue boosters for the Cuban tourism industry. Nevertheless, a fall in stay-over visitor arrivals in 2018–19 cut down Cuba's tourism income and reduced the relative contribution of tourism (19.1 percent in 2018 and 20.9 percent in 2019) to Cuba's total earnings from merchandise and service exports (ONEI 2022a, 2019b).

Table 3.9 collects key indicators of international tourism in Cuba to shed additional light on the performance of the Cuban tourism industry during the 2014–19 period. To begin with, the average length of stay (ALOS) of overnight visitors to Cuba fell considerably from 8.6 days in 2014 to 7.5 days in 2018 and to 7.1 days in 2019.[68] A critical parameter reflecting the number of guest nights relative to arrival numbers, the length of stay influences service offers, transportation and other infrastructure requirements, energy use, and, most importantly, revenue generation. Generally, the longer the period of time tourists spend in a destination, the higher the overall expenditures, which in turn affects job creation and income (Gössling, Scott, and Hall 2018, 2088). In the case of Cuba, the ALOS of foreign visitors grew slightly in 2014–16 in tandem with soaring international arrivals to the country. Yet, this indicator had already started to decline in 2017 despite a record number of overnight visitors.

Table 3.9 Selected Indicators of International Tourism, 2014–19

	2014	**2015**	**2016**	**2017**	**2018**	**2019**
International overnight visitors	2,983,172	3,511,192	3,789,185	4,076,116	3,834,415	3,827,935
Average length of stay (days)	8.6	8.5	8.8	8.1	7.5	7.1
Hotel rooms	62,090	64,231	66,547	67,665	70,884	74,211
Average hotel occupancy rates (%)	58.6	58.7	61.5	52.1	49.1	49.8
Gross revenues (US millions)	2,546.1	2,818.6	3,068.6	3,301.7	2,782.2	2,645.1
Gross revenues per tourist (US dollars)	848	796	765	709	590	619
Gross revenues per room (US dollars)	41,006	43,882	46,112	48,795	39,250	35,643
Import coverage (%)	19.5	24.1	29.9	32.5	24.2	26.7
Trade deficit coverage (%)	3[illegible].1	33.7	38.6	42.5	30.5	33.7
Total contribution to GDP (%)	8.8	10.1	10.1	10.7	10.6	9.9
Total employment (thousands)	413.8	475.0	488.8	513.1	497.6	538.0

Sources: ONEI 2022, 2021a, 2020a, 2016a; MINTUR data obtained by the authors in May 2022; WTTC 2022a, 2019, 2018.

To a significant degree, this trend had to do with the boom in American visitors following the thaw in relations between Washington and Havana. The ALOS of US citizens of non-Cuban descent who visited Cuba in 2017 was 6.7 days, even shorter than that of Canadian tourists (7.7 days) who tend to stay in Cuba for shorter periods than many other travelers. The ALOS of American visitors, many of whom stayed in *casas particulares* (which tend to exhibit shorter ALOS than other types of accommodations) in Havana (which exhibits shorter ALOS than beach locations), was the shortest among the ten largest groups of foreign travelers to Cuba except for Mexicans (6.5 days). The ALOS of Cuban residents abroad was 11.5 days while that of European tourists, namely Germans, Italians, British, Spanish, and French, ranged from 14.3 days to 11.5 days.[69] In essence, the stunning growth of US overnight visitors to Cuba in 2017 pushed down the overall average length of stay of international visitors to the island. The ALOS in Cuba decreased further in 2018–19 mainly because of a sizable drop in European arrivals.

Low hotel occupancy is another issue of concern for Cuba. While the island's stock of five-star and four-star hotel rooms kept expanding at a rather fast pace, annual average hotel occupancy rates in Cuba grew little in 2014–16 from 58.6 percent to 61.5 percent, fell to 52.1 percent in 2017, and plunged to less than 50 percent in 2018–19 when hotel construction actually picked up speed (ONEI 2020a). Still, a recurrent strain on luxury hotels, particularly during the high season from November to March, and more intensely in Havana, could not be avoided. For instance, foreign media reported that the post-17D surge of international visitors, especially US travelers, to Cuba overwhelmed the island's tourism infrastructure, from airport services to restaurants and above all hotel availability. Havana was the place under the most pressure. With several high-end hotels completely booked, many visitors had no other option but to rent rooms in private homes or seek lodging at lower-end state-run facilities.[70] And yet the annual hotel occupancy in Havana showed little variation in 2014–17, averaging just over 50 percent. Thus, the Cuban tourism industry's difficulties in meeting tourist demand after the 17D were due not so much to an insufficient number of hotel rooms as to broader critical problems. Cuba's emphasis on the construction of new luxury hotels instead of renovations and upgrades of deteriorated or obsolete properties has resulted in a massive number of rooms out of service. At the end of 2018, Cuba had more than 12,000 hotel rooms out of service, representing about 18 percent of the country's hotel room capacity. More than 4,000 of those rooms belonged to Cubanacán, some 3,800 to Gaviota, and nearly 2,000 each to Gran Caribe and Islazul. Havana had the largest stock of hotel rooms not available for use (3,700), but Varadero (2,500) and provinces with popular beach areas like Ciego de Ávila (1,600), Villa Clara (1,500), and Holguín (500) had many hotel rooms out of service as well.[71]

To be fair, the Cuban tourism industry underwent substantial expansion after the 17D, and a number of its key indicators were positive. Between 2014 and 2017, international visitor arrivals to Cuba increased by an annual average of 15.7 percent, and the country's gross hard currency revenues from tourism activities rose by an annual average of 9 percent, from $2.54 billion to a record $3.3 billion. Also positive during this period, gross revenues per room grew from about $41,000 to nearly $49,000, which was the highest level since the mid-2000s when the government of Fidel Castro reverted some of the timid market-oriented economic reforms of the 1990s and launched an aggressive policy of recentralization and militarization of the tourism sector (Espino 2008, 131). Moreover, the ratio of Cuba's tourism earnings to the total value of Cuban imports of goods increased from 19.5 percent in 2014 to 32.5 percent in 2017, and so did the ratio of tourism revenues to the country's merchandise trade balance, which soared from 31.1 percent to 42.5 percent (ONEI 2020a). Both the import coverage and trade deficit coverage of tourism revenues would have grown even if Cuba's merchandise imports and its trade deficit had not shrunk, as they did, during this period. In addition, according to data of the World Travel & Tourism Council, the total contribution of travel and tourism to Cuba's GDP was 10.7 percent in 2017, the highest level since the mid-2000s. That year, the number of jobs supported directly and indirectly by the Cuban tourism industry reached 513,100 (direct jobs were 123,500), or approximately 10 percent of total employment in Cuba (WTTC 2018). Hence it is safe to argue that, to a certain extent, the Cuban tourism industry performed relatively well in the post-17D period, at least until 2017, and that it was a spike in US visitors to the island that had the greatest impact on this performance.

After 2017, though, the conditions of Cuba's tourism sector worsened appreciably, and this happened before the travel shock caused by the coronavirus (Covid-19) pandemic wreaked havoc on Cuban tourism. Annual inflows of foreign visitors to Cuba decreased by 8.1 percent between 2017 and 2019, and Trump's economic sanctions and their negative impact on US-based travel to the island were only part of the reason. Arrivals to Cuba from top European source markets also suffered a decline that continued in the early months of 2020 before Cuba closed its borders to international tourists and non-resident visitors due to the Covid outbreak.[72] A nearly 20 percent plunge in gross hard currency revenues from tourism activities in 2017–19 (gross earnings amounted to $2.64 billion in 2019) was accompanied by sizable drops in tourism income's import and trade deficit coverage, its total contribution to Cuba's GDP, and gross revenues per room. The latter stood at just $35,643 in 2019, their lowest level since 1994.

But the most worrying trend, which signaled major efficiency problems for the Cuban tourism industry and the limitations of Cuba's massive sun-and-beach tourism model, was the substantial decline of gross revenues per tourist that plummeted from $848 in 2014 to $709 in 2017 amid growing arrivals of overnight visitors to the country.[73] The level of gross revenues per tourist in 2017, a year that broke the record of overnight arrivals and tourism income, was the lowest since Cuba's opening to international tourism in the early 1990s. Not surprisingly, this crucial indicator deteriorated further in 2018–19 amid thriving cruise tourism that brings far less economic benefits to destinations than overnight tourism. Based on port fees and an average of $140 per cruise passenger in expenses while in Cuba, the US-Cuba Trade and Economic Council estimated that US cruises could generate $83 million in revenues for Cuba in 2017–19, or just about $28 million per year (USCTEC 2017).[74] However, Cuba's gross revenues per tourist experienced a downward trend in the post-17D period even if cruise visitors are excluded from the calculations. Taking into account only overnight travelers, gross revenues per tourist decreased from $853 in 2014 to $810 in 2017, and they fell to just $691 in 2019.[75] In short, international visitors to Cuba are spending on average progressively less, and the Cuban government's efforts to build luxury hotels around the island, especially in Havana, to tap into high-end tourism that in theory should boost foreign exchange earnings, have yet to produce meaningful positive results.[76]

Another important trend in the Cuban tourism sector in the wake of Obama's overture to Cuba was the emergence of Cuban private businesses, mainly in basic services like lodging and restaurants, as stiff competitors to state entities and as major hard currency earners. Concentrated outside of beach resort areas in locations like Havana and other popular tourist places more suitable for some sort of individual tourism, and often relying on remittances from abroad to finance their expansion, Cuban private entrepreneurs connected to the tourism industry have established quite successful businesses that, in many cases are being run more efficiently than their state competitors (Bye 2020, 89–90; Feinberg and Newfarmer 2016, 12). During a speech at the presentation of the annual review of the Ministry of Interior Commerce's operations in February 2019, Cuban president Miguel Díaz-Canel candidly admitted that state-owned restaurants, shops, and other service companies should imitate practices of the private sector to effectively address their endemic problems of inefficiency, poor customer service, or theft of resources. He asked himself, "why don't we do what the non-state sector does?" and added that "state [services] cannot be the ugliest."[77] Indeed, after the 17D, the dramatic surge in US visitors to Cuba basically supported Cuba's incipient private sector by pouring hundreds of millions of dollars into privately owned bed-and-breakfasts, restaurants, taxi services, and other activities. Put simply, American travelers to Cuba made strong connections with ordinary Cubans and put money directly into their hands.

As evidence of the growing importance of Cuban private tourism entrepreneurs during the years of the historic US-Cuba thaw and of the decline in their business after Trump reverted Obama's Cuba policy, Table 3.10 provides official ONEI data on visitor days and gross tourism revenues by sector in Cuba during the 2014–19 period. A combination of number of visitors and length of stay, visitor days by sector for 2014–17 show that the total number of days that international visitors to Cuba stayed in private accommodations (*casas particulares*) grew markedly and at a much faster pace than the same indicator for state-owned installations, which actually decreased in 2017. Slightly more than three to one (3.07) in 2014, the ratio of visitor days in the state sector to visitor days in the private sector was about two to one (2.01) in 2017. Data from the Cuban Ministry of Tourism for the same period show a similar trend in that ratio from 3.07 to 1.81. This means that with approximately one-third of the rooms for tourists available in the state sector, *casas particulares* recorded in 2017 around half as many overnight visits as Cuban hotels and other state accommodation facilities. It should also be underscored that tourists in all-inclusive beach resorts spend little outside of the hotel while guests of *casas particulares* make use of private and state extra-hotel services like restaurants (*paladares*) and cafeterias, rental cars, taxis, excursions, and other activities (Betancourt 2020, 37–8). As for the impact on the Cuban economy, the state/private sector ratio of gross tourism revenues shrank from 5.24 in 2014 to 3.50 in 2017. During this period, gross tourism revenues of the state sector increased by 24.7 percent from $1.99 billion to $2.48 billion while gross earnings of the private sector grew by 86 percent from $378.9 million to $706.6 million (ONEI 2018a).

Table 3.10 Visitors Days and Gross Tourism Revenues by Sector, 2014–19

	Visitor Days			Gross Tourism Revenues ($US millions)		
	State Sector	Private Sector	Index State/Private	State Sector	Private Sector	Index State/Private
2014	19,424,438	6,315,095	3.07	1,988.4	378.9	5.24
2015	21,830,066	8,086,702	2.69	2,115.6	485.2	4.36
2016	24,248,358	8,954,100	2.70	2,369.8	537.3	4.41
2017	22,113,174	10,972,649	2.01	2,479.3	706.6	3.50
2018	20,942,391	7,897,851	2.65	2,228.7	473.9	4.70
2019	21,019,796	6,217,894	3.38	2,223.1	373.1	5.95

Source: ONEI 2020a.

At any rate, as it lost its principal customers, the US visitors, the Cuban private sector suffered disproportionately from Trump's tougher measures against Cuba that, ironically, were enacted with the stated goal of supporting the island's private entrepreneurs (Franco Rodriguez, Cisneros Mustelier, and Echarri Chávez 2021, 3; Honey and Betancourt 2020, 8; Leogrande and Newfarmer 2017). As a senior White House official unequivocally said about Trump's new restrictions on travel and business with Cuba that were announced in June 2017, "The new policy will empower the Cuban people. . . . It does not target the Cuban people but the measures are designed to restrict the flow of money to oppressive elements of the Cuban regime."[78] In reality, Cuba's private tourism entrepreneurs were hurt badly. During the first full year under Trump's new policy (the end of self-directed people-to-people tours) that went into effect in November 2017, occupancy of private bed-and-breakfasts in Havana plummeted to 44 percent after years at close to full capacity. The elimination of group people-to-people travel and the ban on cruises to Cuba in June 2019 wiped out more business of the Cuban private sector, including the rather small yet helpful economic contribution of cruise visitors on land (especially in Havana) as they booked taxi drivers and tour guides and spent at private restaurants, bars, and shops.[79] President Díaz-Canel himself told Cuba's National Assembly in December 2019, "The suspension of cruise ship arrivals and flights to the provinces, cuts in remittances, the closure of consular offices and limitations on travel licenses, among others, hit the non-state sector of the economy particularly hard."[80] In effect, the visitor days in the private sector plunged by 43 percent in 2018–19 and the private sector's gross tourism earnings fell by 47 percent to $373.1 million. The state/private sector ratios of visitor days and gross tourism revenues in 2019 were higher than their level in 2014. The strong negative impact of Trump's policy on Cuba's private business sector is beyond question.

Unresolved Challenges

While the outbreak of the global Covid-19 pandemic in early 2020 delivered a heavy blow to international tourism in Cuba and around the world, whose full implications have yet to be seen, the post-17D era exposed a number of serious challenges for the Cuban tourism industry that remain unresolved. First of all, notwithstanding sustained growth in international visitor arrivals, expenditures per visitor have declined steadily since the mid-1990s, with low repeat visits or return rates. This stands in contrast with other popular Caribbean and Latin American tourist destinations and raises the need for the Cuban tourism sector to boost its competitiveness by somehow distinguishing itself from the sun-and-beach tourism model that remains dominant in the region (Rutty and Richardson 2019, 2). The removal of the US ban on tourism in Cuba would certainly give a major boost to the flow of American travelers to the island. After all, the United States can fairly be considered Cuba's "natural market" since historically, different phases of development of international tourism in Cuba have been deeply related to the status of the country's diplomatic and economic relations with the United States (Perelló Cabrera 2019). But it is hard to deny that, under the current conditions, the traditional intensive model of sun-and-beach tourism has limited room for further growth. Visitors from the United States accounted for more than 50 percent of the increase in total international arrivals to Cuba between 2014 and 2019 and, with the exception of Cuban Americans and tourists from Russia and Mexico, all of Cuba's top twenty-five source tourism markets experienced either a drop (especially European markets) or virtually no growth in 2017–19 (ONEI 2020a). And if we consider that Cuba has been catering to the low-income end of those markets with the promotion of inexpensive vacation deals, which resulted in progressively less spending by foreign visitors, it is clear that the country should favor a more comprehensive tourism model of diverse offerings that satisfy new demands.

Low hotel occupancy rates and a declining average length of stay (ALOS) of foreign visitors to Cuba are two additional challenges that have not been adequately addressed, let alone overcome. If anything, Cuba's recent hotel construction fever, fueled almost entirely by projects of the military-run corporation Gaviota, has been increasingly difficult to justify given the disproportionate tourism investment in relation to the modest hotel occupancy and to plummeting visitor arrivals and spending (Monreal 2020; Triana Cordoví 2019). Aside from the billions of dollars in hotel investments whose recovery is by no means guaranteed, the building spree seems even more ill-advised given the fact that Cuba's underutilization of hotel capacity is amplified by a huge number of hotel rooms out of service.[81] Consistently low occupancy rates, which, to a large degree, negatively affect efficiency and hotel profitability, are the result of excessive construction and limited hotel maintenance work that every year during the high season keeps between 10 percent and 15 percent of existing rooms out of order while international standards set a maximum level of 3 percent (Figueras Pérez and Pajón Espina 2021, 117).

Regarding the shrinking average length of stay of international visitors to Cuba, the problem is twofold. First, visitors from Canada, by far Cuba's top source market, exhibit one of the shortest ALOS among the main groups of foreign travelers to the island (the same can be said for US citizens of non-Cuban descent), much shorter than that of Cuban Americans and of the principal European markets. Second, the ALOS of visitors to Havana (around 3–3.5 days), where there are no low-priced all-inclusive holiday packages like in beach areas and more opportunities are available for tourist spending in restaurants, bars and cafés, night clubs, boutiques and craft markets, museums, and art galleries, among other venues, is about half the national average. Havana also has lower hotel occupancy rates than Cuba as a whole (Perelló Cabrera 2015a; ONEI 2020a). Moreover, the ALOS of tourists in *casas particulares* that are concentrated in Havana and other urban areas, is only approximately two days.[82] An unpublished 2019 study by specialists of the Cuban Ministry of Tourism, besides calling for the expansion of hotel management contracts with international chains that would help increase the average occupancy rate (about 70 percent in 2016–18) of flights to Cuba, argued that a joint business strategy with foreign chains was needed to boost the ALOS and tourist-days because "the tourist offer in Cuba is growing faster than the demand."[83] The promotion of diaspora tourism to smooth out the strong seasonality of international tourism in Cuba and raise the utilization of hotel capacity (Pérez-López and Díaz-Bríquets 2011; Espino 2013), the revival of European tourist arrivals, and effective plans to stimulate urban and cultural tourism in Havana and other places would surely make the aforementioned goals more achievable.

On the subjects of efficiency and hotel profitability, Table 3.11 offers a rare view of Cuba's top twenty highest-revenue hotels in 2017, which was, in the words of a high-level MINTUR official, "historically the best year for [tourism in] Cuba."[84] The information presented in this table is extrapolated from a MINTUR database of key performance indicators covering the operations of 144 Cuban hotels in 2017. That year, Gaviota's hotels accounted for 42.2 percent of total hard currency earnings of Cuba's hotel sector (about 1.4 billion Cuban convertible pesos, or CUCs—or US dollars), followed by Cubanacán (30.1 percent), Gran Caribe (21.5 percent), and Islazul (6.2 percent). Among the top twenty hotels in terms of revenues (between 34 million CUCs and 21 million CUCs), there were ten hotels of Gaviota, eight of Cubanacán, and two of Gran Caribe. Most notably, underscoring the importance of foreign participation, the Hotel Nacional was the only hotel managed by a Cuban chain, Gran Caribe. The

Table 3.11 Cuba's Top Twenty Highest-Revenue Hotels in 2017

	Hotel	**Cuban Chain**	**Management Company**	**Revenues (CUC 000')**	**RevPAR[a] (CUCs)**
1	Parque Central	Cubanacán	Iberostar	34,248	219.7
2	Hotel Playa Pesquero	Gaviota	Gaviota	34,183	93.7
3	Hotel Paradisus Varadero	Gaviota	Meliá	33,971	117.2
4	Hotel Grand Memories Varadero	Gaviota	Blue Diamond	33,699	83.2
5	Hotel Cohiba	Cubanacán	Meliá	31,979	190.9
6	Iberostar Varadero	Cubanacán	Iberostar	31,469	222.2
7	Hotel Nacional	Gran Caribe	Gran Caribe	31,227	191.0
8	Complejo Hotelero Solymar-AB	Gran Caribe	Barceló	31,017	96.0
9	Hotel Paradisus Princesa del Mar	Gaviota	Meliá	30,859	134.2
10	Royalton Hicacos (former Sandals)	Cubanacán	Blue Diamond	30,663	207.8
11	Hotel Iberostar Laguna Azul	Gaviota	Iberostar	30,095	101.3
12	Meliá Habana	Cubanacán	Meliá	28,594	198.3
13	Sol Palmeras	Cubanacán	Meliá	27,617	125.6
14	Hotel Memories Paraíso y Azul	Gaviota	Blue Diamond	24,946	45.3
15	Meliá Varadero	Cubanacán	Meliá	24,399	136.4
16	Hotel Memories Las Salinas	Gaviota	Blue Diamond	23,855	63.1
17	Hotel Meliá Jardines del Rey	Gaviota	Meliá	22,562	52.6
18	Meliá Las Americas	Cubanacán	Meliá	21,875	178.9
19	Hotel Ocean Vista Azul	Gaviota	H10	21,407	124.8
20	Hotel Meliá Marinas	Gaviota	Meliá	21,111	75.0

Source: MINTUR data obtained by the authors in November 2018.
Note:
[a] RevPAR (Revenue per Available Room).

other nineteen hotels were all managed by international chains, and six of them (Iberostar Parque Central, Iberostar Varadero, Meliá Habana, Sol Palmeras, Meliá Varadero, and Meliá Las Americas, all Cubanacán's hotels) were joint ventures. Furthermore, sixteen were all-inclusive beach resorts and only four were city hotels, all located in Havana.

In terms of Revenue per Available Room or RevPAR, which represents the revenue generated per available room, whether occupied or not, the top twenty rankings in 2017 included ten hotels of Cubanacán (eight of them joint ventures), five of Gran Caribe, and five of Gaviota. While RevPAR does not take into account operational costs and other expenses and thus does not truly reflect profitability, it is commonly used to measure a hotel's financial performance, and in some ways its overall success, by assessing the hotel's ability to fill its available rooms at an average rate. If a property's RevPAR increases, that generally means either its average room rate, or its occupancy rate, or both are improving. In Cuba, the RevPAR ranking uncovered one critical difference from the ranking based on revenues. Twelve of Cuba's top twenty hotels and nearly half of the top fifty, almost all located in Havana (among them the hotels Iberostar Parque Central, Meliá Habana, Gran Hotel Kempinski, Nacional, Cohiba, Sevilla, Capri-Victoria, Inglaterra, and Four Points by Sheraton) and including various properties in Old Havana managed by Gaviota and transferred from Habaguanex, were not all-inclusive beach resorts.[85] As a matter of fact, some of the all-inclusive establishments in the top-twenty ranking based on revenues were not even in the top-fifty RevPAR ranking. In short, Havana had thousands of hotel rooms out of service that pushed down the city's hotel occupancy rate, and the ALOS of its visitors was lower than that of tourists in beach areas, but many of its hotels showed a better revenue-generating performance than all-inclusive resorts.

Even more striking, the top twenty ranking of Cuba's hotels in 2017 in terms of Net Revenue per Available Room, or NRevPAR, which gives a more accurate assessment of a hotel's overall financial performance (and arguably of efficiency levels and profitability) by factoring in distribution costs and other expenses, included only two all-inclusive beach resorts. Hotels in Havana dominated the top ranking positions, with several properties exhibiting net results (NRevPAR as a share of RevPAR) of 60 percent or above. But a disturbing development also emerged from the MINTUR dataset. Approximately one-fifth of the hotels surveyed, principally Gaviota's hotels, operated at a loss in 2017. Virtually all of these properties were all-inclusive beach resorts located in Varadero, Cayo Coco, Cayo Santa Maria, and other popular beach locations in Cuba. While the MINTUR database collected information for a single year and several factors might have contributed to the dismal business performance of many Cuban beach resorts in 2017, the overall picture casts legitimate doubts on the economic benefits of Cuba's still dominant sun-and-beach tourism model centered on relatively cheap vacation travel packages sold by tour operators that have traditionally marketed Cuban tourism.

Another questionable aspect of Cuba's massive investments in the construction of new hotels is that the largest share of hard currency earnings of the Cuban tourism sector comes from gastronomic services, which need to be improved, rather than lodging. Between 2014 and 2019, as shown in Table 3.12,

Table 3.12 Gross Tourism Revenues by Spending Category, 2014–19 (percentage distribution)

Category	2014	2015	2016[a]	2017	2018	2019
Gastronomy	39.0	38.9	37.2	38.8	39.1	41.1
Lodging	29.9	28.8	32.0	30.4	28.4	26.1
Retail commerce	9.7	9.5	8.4	10.7	12.4	12.6
Transportation	16.5	18.3	18.1	12.2	8.5	8.1
Recreation	1.0	1.0	1.3	1.5	1.8	1.8
Others	3.9	3.5	3.0	6.4	9.8	10.3

Sources: Calculations of the authors from ONEI 2020b, 2019a, 2018b, 2017, 2016b, 2015b.
Note:
[a] January-June 2016.

goods and services offered by restaurants, cafeterias, snack bars, and other food outlets generated, on average, close to 40 percent of Cuba's gross revenues from tourism activities. During the same period, hotels, beach resorts, and other accommodation businesses generated, on average, less than 30 percent of total tourism income. The relative average contributions of retail commerce and transportation were slightly over 10 percent, and the contribution of recreation activities was just about 1.5 percent. It would therefore make sense for Cuba to step up efforts to raise the quantity and quality of its gastronomic offers in tourist areas and across the country, and especially to further encourage the expansion of private restaurants to generate forward and backward value chains and reduce imports of foodstuffs and raw materials.[86] The number of private restaurants in Cuba increased considerably during Raúl Castro's presidency to reach 1,700 in 2016, but hardly any growth took place in 2017–19 as Havana's authorities more than once suspended the issuing of new licenses for *paladares* citing concerns over tax evasion, labor code violations, illegal purchases of supplies in the black market, and additional violations (Bye 2020, 90).[87] Cuban private restaurants not only established very successful operations but also reinforced tourism-agricultural linkages through farm-to-table arrangements with Cuba's agroecological farmers and by their demand for high-quality produce in free farmers' markets largely supplied by the island's small private farmers (Deere and Royce 2019). Yet private businesses have very limited access to wholesale markets and must rely on state intermediaries to import the products they need.[88]

Indeed, one of the biggest problems of the Cuban tourism sector as a foreign exchange earner is the hefty amount of importing that it has to do, from foodstuffs to bedsheets and many other goods. The sector's reliance on purchases from abroad has actually increased in recent years (Figueras Pérez 2019). Speaking on national television in June 2019, Cuba's minister of economy and planning, Alejandro Gil Fernández, said that for every dollar of income generated by tourism, more than 60 cents worth of imports were spent, that is, net income was less than 40 percent of gross income. The Cuban minister added that the island's authorities were working on "the creation of incentives and conditions for domestic production to competitively substitute those purchases from abroad."[89] As a matter of fact, around the same time, foreign media revealed that Cuba was trying to revive businesses, above all in the agricultural and textile sectors, by turning them into part of the supply chain for the tourism industry.[90] This is a very challenging task. As we will see in detail in Chapter 5, today's Cuban economy is mainly a service economy with widespread inefficiencies and a weak production base exemplified by a decapitalized, poorly performing manufacturing industry along with an agricultural sector that is producing well below its potential despite significant reforms (Mesa-Lago 2019, 4–6). Gil Fernández also pointed out that tourism is a strategic sector that "has the virtue of depending essentially on external demand," and it can grow regardless of domestic demand "to boost other sectors of the economy."[91] However, to do so, tourism needs a more dynamic national industry and an agricultural sector that is capable of moving enough produce to satisfy visitors. Havana's authorities keep referring to Cuban tourism as the locomotive of the economy,[92] but as it stands right now, tourism is more like a locomotive with mostly empty wagons.

As for domestic demand, based on its purchasing power, national tourism cannot be a real alternative to international tourism as an engine of growth. Converted into CUCs, the sum of all salaries paid in Cuba in 2019 was 1.6 million CUCs (equivalent to 38.5 million Cuban pesos), well below the $2.6 billion in revenues from international tourism. Furthermore, total overnight stays of national tourists in all of Cuba's accommodation facilities in 2019 were just about 30 percent of overnight stays of international visitors, and only one-fourth of stays of national tourists were in five- and four-star hotels (the same indicator for foreign tourists was more than 60 percent) as they tend to concentrate on homes and cabañas for camping, villas, and motels (ONEI 2020a). The very different profiles of national and foreign tourists make it difficult to channel the former toward high-category hotels that were designed for the latter. At any rate, certainly fueled by income originating outside of Cuba, in particular family remittances, the relative share of national tourists' stays in four- and five-star Cuban hotels more than doubled in 2014–19, and the hard currency revenues from national tourism activities increased from 84.2 million CUCs to 426.8 million CUCs (ONEI 2020c, 2015c). Since domestic tourism demand lacks sufficient scale, is stymied by prices and standards that, for the most part, do not respond to the average profile of the national tourist, and largely depends on external income supported by family ties, an effective strategy would be to build on those ties by targeting Cuban residents abroad as key drivers of Cuba's tourism sector. Besides exhibiting a longer average length of stay and being far less seasonal than other top source markets, the Cuban diaspora is a source of visitors with considerable scale, strong motivations to travel to Cuba beyond leisure purposes, and a marked demand for food services, that is, the biggest component of tourism spending (Monreal 2020).

Lastly, with respect to the decline of overnight visitors to Cuba in 2018–19, both external and internal factors played a role. On the one hand, the sizable drop of US visitors to the island was due, without a doubt, to Trump's travel restrictions and to other measures of his administration aimed to squeeze Cuba's hard currency revenues from tourism. Aside from limiting, and then ending, the heavily used people-to-people educational category of travel that functioned as an exemption to the overall ban on US tourism, Washington's government fined a number of American companies for selling Cuba travel tours and packages in violation of US embargo rules and forced major online platforms to remove Cuban hotels from their offerings. Most notably, in mid-2019, the Seattle-based digital travel industry conglomerate Expedia Group, agreed to pay $325,406 for breaching US sanctions with respect to Cuba. Similarly, Hotelbeds USA, the Florida-based subsidiary of Spain's Hotelbeds Group, agreed to pay $222,705 for providing Cuba-related travel services to non-US persons in violation of US law, and California-based Cubasphere was fined $40,320 for engaging in unauthorized travel-related transactions involving the category of professional services (Betancourt and Rugh 2020, 6).[93] In any case, as damaging as they could be, Trump's measures, which included a ban on US cruises to Cuba and the activation of Title III of the Helms-Burton law, reflected another shift (possibly short-term) in US policy that was beyond Cuba's control and that it should not strongly affect Cuba's plans for the long-term development of its tourism industry.

On the other hand, the fall of non-US overnight visitors to Cuba, especially European tourists, in 2018–19 and the early months of 2020 before Covid became an issue, was a concerning trend mainly caused by internal factors that highlighted additional serious challenges for the Cuban tourism industry. Even if it mirrored to some extent a deceleration in the flows of European vacationers to the Caribbean region (Perelló Cabrera 2020), the decline of non-US tourists to Cuba was triggered by a number of shortcomings of the Cuban leisure industry that made the island's tourism market less competitive than other popular Caribbean destinations like the Dominican Republic, Jamaica, and the Bahamas. These shortcomings included relatively low hotel quality standards (especially of Cuban-managed establishments) in terms of comfort, food quality and variety, and customer service; several obsolete hotels in need of refurbishment and modernization; insufficient extra-hotel amenities; major increases in hotel and restaurant prices; expensive car rentals; exceptionally slow and spotty internet connectivity; and poor infrastructure conditions (Mesa-Lago and Svejnar 2020, 21).

Regarding the infrastructure, since the opening to international tourism in the early 1990s, Cuba has made major investment efforts to renovate seven international airports, with the largest investments being reserved for the Varadero and Havana airports, and build six new marinas, especially a huge facility (Gaviota Varadero Marina) located at the tip of the Hicacos peninsula near the beach of Varadero with 1,200 berths for small boats and mega yachts (Figueras Pérez and Pajón Espina 2021, 116). And yet, along with poorly maintained roads, Cuba's current airports remain, for the most part, outdated and need to be greatly improved to be able to cope with larger flows of travelers, including the main terminals (terminal two for flights coming from the United States, and terminal three for other international flights) of Havana's José Martí International Airport, which receives more than half of the total international passengers flying every year to Cuba.[94] Furthermore, even if the outlook for cruise tourism is bleak at the moment due to US restrictions, to become a leading cruise destination in the Caribbean region, Cuba must address some key infrastructure deficiencies, in particular those limiting the size and number of ships that can berth in the port of Havana. Unlike popular cruise destinations in the Caribbean, Havana's harbor cannot accommodate multiple large ships per day. To do so, it would need to significantly expand pier space that is dedicated to cruise ship dockage. Currently, Havana's cruise terminal can only handle a single 2,000-guest (approximately) ship and a second smaller vessel of 500–700 passengers (Neiberg et al. 2021).[95] In short, Cuba's budget should not be devoted so heavily to expand the country's supply of hotel rooms but rather to conduct extensive maintenance and restoration of existing establishments and, most importantly, to upgrade other types of infrastructure that also make up the traveler's experience.

Conclusion

For better or worse, the performance of the Cuban tourism industry during the 2015–19 period was strongly affected by external factors linked to Cuba's relationship with the United States. The historic rapprochement between Washington and Havana that began toward the end of 2014 during the Obama administration brought a relaxation of US travel restrictions with respect to Cuba that gave a strong boost to the influx of American visitors to the island, while US companies explored potential business opportunities in the Cuban market. Although trips for tourism purposes remained prohibited, travel to Cuba from the United States was made easier by a licensing process for loosely defined individual educational trips that involved just a simple declaration, and by the resumption of direct flights and cruise services between the two countries. Also triggered by a rather surprising surge in arrivals from traditional European source markets and some other countries, possibly to anticipate a complete lifting of the embargo that would unleash a flood of US tourists, international visitor arrivals to Cuba and the country's foreign exchange earnings from tourism activities grew considerably in 2014–17. As a matter of fact, 2017 was the best year for the Cuban tourism industry in terms of overnight visitor arrivals and gross hard currency income, with a notable share of that income benefiting the emerging private sector in Cuba, in particular private room rentals, restaurants, and taxi drivers in Havana. Foreign visitors to Cuba increased somewhat in 2018 thanks to booming American cruise travelers, but by then the thaw in US-Cuba relations was coming to an end as the Trump administration adopted a more hard-edged approach that resulted in new US restrictions on business with Cuba's military-owned enterprises and on remittances, a ban on educational and cultural exchanges, and a ban on US cruises to Cuba. Trump's measures had a significant negative impact on the flow of US visitors to Cuba, and especially on the Cuban private sector.

During the 2015–19 period, Cuba built numerous hotels in its principal tourist areas, signed new hotel management contracts with foreign chains, and continued to promote all-inclusive sun-and-beach tourism as the top product of the country's tourism sector. There were some key differences from previous years, though. The vast majority of new hotels belonged to Gaviota and construction gained speed, there was an increased focus on luxury establishments, particularly in Havana, and private accommodations witnessed a major expansion. Meanwhile, annual inflows of European visitors to Cuba fell considerably in 2018–19, and a number of shortcomings of the Cuban tourism industry, mostly old problems never effectively addressed, became increasingly evident. These problems included a shrinking average length of stay of international visitors, low hotel occupancy rates and a very large number of hotel rooms out of service, declining gross revenues per tourist, excessive import levels, insufficient hotel quality standards, and lack of proper infrastructure, amenities, and internet connectivity. The fact that all-inclusive beach resorts, at least based on data for 2017, exhibit worse efficiency results than other types of hotels also shows the limitations of Cuba's tourism model and lends support to arguments in favor of diversifying the country's tourism products. In sum, Cuba's tourism sector was showing clear signs of exhaustion prior to the emergence of the global Covid-19 pandemic, whose impact will be analyzed in Chapter 5. But before that, the next chapter compares the tourism industry of Cuba with that of competing Caribbean nations to identify critical structural problems that Cuba must address to raise competitiveness vis-à-vis its neighbors.

Notes

1 Announced in July 2007 and formally launched in 2008, Raúl Castro's reform process was arguably the "strongest pragmatist cycle" of economic reforms in Cuba since 1959 (Mesa-Lago and Pérez-López 2013, 3).

2 Along with the shortcomings of Cuba's socialist economic system, Havana's authorities were receiving backlash over low state wages, soaring consumer prices, and growing income inequality. Marc Frank and John Paul Rathbone, "Cuba's communists dig in as Castro's reform drive grinds to a halt," *Financial Times*, March 28, 2017.

3 "Cuba: Revisions to License Exceptions," *Federal Register*, Vol.81, No.200, October 17, 2016, 71365–67; "Cuban Assets Control regulations," *Federal Register*, Vol.81, No.200, October 17, 2016, 71372–78; "Cuba: Revisions to License Exceptions and Licensing Policy," *Federal Register*, Vol.81, No.51, March 16, 2016, 13972–74; "Cuban Assets Control Regulations," *Federal Register*, Vol.81, No.51, March 16, 2016, 13989–94; "Cuba Licensing Policy revisions," *Federal Register*, Vol.81, No.17, January 27, 2016, 4580–83; "Cuban Assets Control Regulations," *Federal Register*, Vol.81, No.17, January 27, 2016, 4583–86; "Enhancing Support for the Cuban People," *Federal Register*, Vol.80, No.182, September 21, 2015, 56898–904; "Cuban Assets Control Regulations," *Federal Register*, Vol.80, No.182, September 21, 2015, 56915–26;

"Cuba: Implementing Rescission of State Sponsor of Terrorism Designation," *Federal Register*, Vol. 80, No. 140, July 22, 2015, 43314–20 (see also "Cuban Assets Control Regulations: Terrorism List Governments Sanctions, Regulations," *Federal Register*, Vol. 80, No. 114, June 15, 2015, 34053–54); "Cuba: Providing Support for the Cuban People," *Federal Register*, Vol.80, No.11, January 16, 2015, 2286–91; "Cuban Assets Control Regulations, "*Federal Register*, Vol.80, No.11, January 16, 2015, 2291–302.

4 The twelve allowed travel categories are: family visits; government business; people-to-people non-academic educational activities; religious activities; journalism; professional research and meetings; public performances, clinics, workshops, competitions, and exhibitions; humanitarian projects; support for the Cuban people; private foundation work and research; exportation, importation, or transmission of information or informational materials; and certain authorized export transactions.

5 For a comprehensive review of Obama's amendments to the Cuba sanctions program, see Spadoni 2017b, 236–9.

6 Josh Lederman, "U.S. eases security for ships visiting Cuba," Associated Press, March 17, 2016.

7 "Cuban Assets Control Regulations," *Federal Register*, Vol. 74, No. 172, September 8, 2009, 46000–07.

8 Decree Law 330 was enacted on May 5, 2015, and it was published in Cuba's Official Gazette about a month later. "Decreto-Ley No. 330," *Gaceta Oficial*, June 4, 2015. "Acuerdo No. 7742," Consejo de Ministros, May 30, 2015.

9 The Campismo Popular (Popular Camping) business group runs over one hundred facilities in Cuba linked to nature tourism, four of them for international tourists and the rest for national tourists. Ana Ibis Falcón Lamoth, "Cuba's Tourist Camping Offers Are Promoted in Italy," Prensa Latina, July 4, 2022.

10 MINTUR data obtained by the authors in November 2018.

11 Andrea Rodriguez, "Cuban Economic Boom a Boon to the Military," Associated Press, September 10, 2016.

12 Eric Caraballoso, "Havana's Future Hotels," OnCuba, May 18, 2019; Katheryn Felipe Gonzalez, "En Primera Persona: de Hoteleras y Empresarios," *Granma*, May 26, 2017.

13 Ventura de Jesús, "Inauguración del Hotel Internacional Confirma Vitalidad del Desarrollo del Turismo a Pesar del Bloqueo," *Granma*, September 15, 2019.

14 While in 2004 Obama said that he supported the end of the embargo that had "utterly failed in the effort to overthrow Castro," at a meeting in Miami in August 2007 he called for the preservation of such a tool as "an important inducement for change." Aaron Wiener, "Obama Suggests Cuba Policy Reform," *Washington Independent*, December 31, 2008.

15 Jamie Weinstein, "Donald Trump on His Nuclear Doctrine, Democracy Promotion and Why He Refuses to Use Term 'Supreme Leader,'" *Daily Caller*, September 7, 2015.

16 Marc Caputo, "Trump Morphs Back into a Cuban Hardliners," Politico, September 16, 2016.

17 Steve Holland, "Trump Rolls Back Parts of What He Calls 'Terrible' Obama Cuba Policy," Reuters, June 16, 2017.

18 "Fact Sheet on Cuba Policy," The White House, June 16, 2017, https://trumpwhitehouse.archives.gov/articles/fact-sheet-cuba-policy/.

19 "Amendments to Implement United States Policy Toward Cuba," *Federal Register*, Vol.82, No.216, November 9, 2017, 51983–86; "Cuban Assets Control Regulations," *Federal Register*, Vol.82, No.216, November 9, 2017, 51998–2004; "Treasury, Commerce, and State Implement Changes to the Cuba Sanctions Rules," US Department of the Treasury, November 8, 2017.

20 Arshad Mohammed and Sarah Marsh, "U.S. Cuts Staff in Cuba Over Mysterious Injuries, Warns Travelers," Reuters, September 29, 2017.

21 Courtney McBride, "U.S. Eases Security Warning for Travel to Cuba," *Wall Street Journal*, August 23, 2018.

22 "Restricting the Temporary Sojourn of Aircraft and Vessels to Cuba," *Federal Register*, Vol.84, No.108, June 5, 2019, 25986–88; "Cuban Assets Control Regulations," *Federal Register*, Vol.84, No.108, June 5, 2019, 25992–93; "United States Restricts Scheduled Air Service to Cuban Airports," US Department of State, October 25, 2019; "United States Further Restricts Air Travel to Cuba," US Department of States, January 10, 2020.

23 U turn transactions refer to the transfer of funds from a foreign bank that is cleared by a US financial institution and then transferred out to another foreign bank.

24 Gisela Salomon and Michael Weissenstein, "Under Trump Change, Cuba Business Partners Can Now Be Sued," Associated Press, May 2, 2019.

25 Declarations of Manuel Marrero Cruz are reported in: Oscar Figueredo Reinaldo, Abel Padrón Padilla, Lissett Izquierdo Ferrer, and Dianet Doimeadios Guerrero, "Asamblea Nacional: Ministerios de Turismo e Industrias Brindaron Información a los Diputados, Asistió el Presidente Cubano," Cubadebate, July 11, 2019, http://www.cubadebate.cu/noticias/2019/07/11/asamblea-nacional-ministerios-brindan-informacion-a-los-diputados-asiste-el-presidente-cubano/.

26 Rafael Rojas (2020, 3) argued that the new Cuban Constitution of February 2019, rather than decentralizing power, shifted to a differentiation of responsibilities among three leadership positions (president, prime minister, and first secretary of the party), with the differences in roles ultimately determined by the Cuban

Communist Party. The latter has the highest ideological and political authority, which is the reason why the president should also be the head of the party.

27 Nelson Acosta and Sarah Marsh, "Raul Castro Retires but Cuban Communist Party Emphasizes Continuity," Reuters, April 19, 2011.

28 "Statement by the President on the Presidential Policy Directive on Cuba," Office of the Press Secretary, The White House, October 14, 2016, https://obamawhitehouse.archives.gov/the-press-office/2016/10/14/statement-president-presidential-policy-directive-cuba.

29 Alan Gomez, "First U.S. Commercial Flight in 5 Decades Lands in Cuba," *USA TODAY*, August 31, 2016; "Cuba Gives Green Light to US Cruises," BBC News, April 22, 2016.

30 Julie Hirschfeld Davies, "U.S. Eases Restrictions on Travel to Cuba and Bank Transactions," *New York Times*, March 15, 2016.

31 The US Department of the Treasury defined (and still defines) compliance with people-to-people regulations as "a full-time schedule of educational exchange activities intended to enhance contact with the Cuban people, support civil society in Cuba, or promote the Cuban people's independence from Cuban authorities and that will result in a meaningful interaction between the traveler and individuals in Cuba." "Treasury and Commerce Announce Significant Amendments to the Cuba Sanctions Regulations ahead of President Obama's Historic Trip to Cuba," Press Release, US Department of the Treasury, March 15, 2016.

32 Mimi Dwyer, "Priceline Unit Booking.com Strikes Cuba Deal for Americans," Reuters, March 21, 2016; Melanie Lieberman, "TripAdvisor Adds Booking for Trips to Cuba," Fortune, October 29, 2016; Sarah Marsh, "Expedia Begins Offering Online Booking for Hotels in Cuba," Reuters, May 23, 2017.

33 MINTUR data obtained by the authors in February 2017.

34 Nick Miroff, "Hurricane Irma Pounds Cuba's North Coast," *Washington Post*, September 9, 2017.

35 MINTUR data obtained by the authors in February and May 2020. Sarah Marsh and Marc Frank, "Trump Administration Tightens Restrictions on Cuba Travel," Reuters, June 4, 2019.

36 "List of Restricted Entities and Subentities Associated With Cuba Effective as of November 19, 2019," Bureau of Economic and Business Affairs, US Department of State, November 19, 2019.

37 Courtney McBride, "U.S. Eases Security Warning for Travel to Cuba," *Wall Street Journal*, August 23, 2018.

38 "Cuba Acelera Inversiones en Turismo," Inter Press Service, December 11, 2019.

39 Katherine Hutt Scott, "Cruises Quietly Dock in Havana," *South Florida Sun-Sentinel*, July 22, 1999; "Report: Cuban Economy Hurt by Pull-Out of Pullmantur," Travel Weekly, September 27, 2007.

40 Carnival had already purchased a 50 percent interest in Costa Crociere in 1997 and the remaining 50 percent in 2000.
41 Marc Frank, "Emotional Return As First US Cruise in Decades Reaches Cuba," Reuters, May 2, 2016.
42 Tom Stieghorst, "Fathom's Dominican Republic Cruises to Add Cuba Call," Travel Weekly, December 22, 2016.
43 Gene Sloan, "Carnival Cruise Line Kicks Off Cuba Voyages," *USA TODAY*, June 30, 2017.
44 Gene Sloan, "Norwegian Cruise Line Adds Cuba Cruises From Port Canaveral," *USA TODAY*, July 10, 2017; "Royal Caribbean Expands Sailings to Cuba in 2018," Travel Agent Central, December 1, 2017.
45 "Cruise Tourism Continues to Boom Despite US Restrictions," Economist Intelligence Unit (EIU), March 22, 2018; Julio Martínez Molina, "Crucerismo, una Puerta Más a un Destino Seguro," *Granma*, June 11, 2018.
46 MINTUR data obtained by the authors in November 2019.
47 Adam Leposa, "Cuba Cancelled: How the Cruise Lines Are Reacting," Travel Agent Central, June 5, 2019.
48 "Cuba Prevé Perder un 10 percent de Turistas Este Año por las Sanciones de EEUU," EFE, July 11, 2019.
49 Sarah Marsh, "Luxury Hotels Arrive in Cuba Despite Fears Trump May Hurt Tourism Boom," Reuters, June 7, 2017.
50 Yisel Martínez García, "Continua en Crecimiento la Planta Hotelera del País," *Granma*, July 19, 2018.
51 Jeff Weinstein, "Hotels Exclusive: Insights on Cuba From Gaviota's Medina," *Hotels*, February 9, 2016.
52 MINTUR data obtained by the authors in June 2019; Minerva Hernández Basso, "Fortalece Su Presencia en La Habana," *Opciones*, May 8, 2014. Gaviota had already about 1,000 hotel rooms in Havana in the early 2000s (Colantonio 2006a, 205), thus its room inventory in Havana grew very little until 2014.
53 MINTUR data obtained by the authors in February 2018 and June 2018.
54 Carmen Sesin, "Sharp Decline in US Travel to Cuba Spurs Overall Drop in Tourism for the Island," NBC News, April 25, 2018; Melanie Zanona, "Airbnb's Cuba Lobbying Blitz Pays Off," The Hill, November 10, 2017.
55 MINTUR data obtained by the authors in May 2017.
56 For additional information on the various categories of authorized travel to Cuba for US citizens, see the Frequently Asked Questions page of the US Department of the Treasury at: https://home.treasury.gov/policy-issues/financial-sanctions/faqs/topic/1541.
57 MINTUR data obtained by the authors in June 2019.

58 "Potenciarán Inversiones en el Turismo por Medio Milenio de La Habana," *Granma*, December 31, 2018.
59 MINTUR data obtained by the authors in February 2018.
60 For additional information on the 2014 list of hotel chains and Cuba's top hotels ranked by customer satisfaction on TripAdvisor, see Johannes Werner, "Researchers Rank Cuban Hotels," Cuban Standard, August 6, 2014, https://www.cubastandard.com/researchers-rank-cuban-hotels/.
61 Marc Frank, "Starwood Signs First U.S.-Cuba Hotel Deal Since 1959 Revolution," Reuters, March 19, 2016.
62 "Death Toll From Explosion at Havana Hotel Rises to 40," Associated Press, May 9, 2022. The Hotel Saratoga is run by the Cuban joint venture company (without a foreign operator) Hotel Saratoga S. A., which belongs to Gaviota.
63 MINTUR data obtained by the authors in May 2022.
64 Nancy San Martin, "Cuba Plans to Build 108,000 New Rooms by 2030 to Accommodate Tourists," *Miami Herald*, April 18, 2016.
65 "Cuba Aspira Llegar a Seis Millones de Visitantes para 2030," Agencia Cubana de Noticias, January 18, 2022.
66 Besides cash remittances, Morales (2019) estimated that Cubans received $2,975 million worth of in-kind remittances (made up primarily of clothing, medicines, and electronics products) in 2018.
67 The number of Cuban doctors and nurses abroad plunged from more than 50,000 in 2015 to around 28,000 in 2020, mainly because they had to leave countries like Bolivia, El Salvador, Ecuador, and especially Brazil (8,300 left after Bolsonaro won the presidency in October 2018). "Mercy and Money; Cuban Doctors," *The Economist*, April 4, 2020.
68 Calculations of the authors from ONEI 2020a.
69 MINTUR data obtained by the authors in November 2018.
70 Jaime Hamre, "Surge of Americans tests limits of Cuba's tourism industry," Reuters, January 26, 2016.
71 MINTUR data obtained by the authors in February 2019.
72 José Luis Perelló Cabrera, "Turismo en Cuba Ante el Deterioro Económico Global: Nuevos Retos y Estrategias," Excelencias Cuba, October 23, 2019.
73 Calculations of the authors from ONEI 2018a, 2015a.
74 More specifically, the US-Cuba Trade and Economic Council estimated for the 2017–19 period that US cruises could generate $19 million in port fees to Cuba and that passengers would spend approximately $64 million while on the island. It also estimated that cruises could generate a staggering $623 million in gross revenues to the US cruise companies (USCTEC 2017).

75 Calculations of the authors from ONEI 2020a and from MINTUR data obtained in May 2020.

76 Katell Abiven, "Cuba Taps into High-End Luxury Tourist Market," Agence France-Presse, March 10, 2019; "Cuba Hosts Luxury Hotels as It Preps To Be Major Travel Hub," Xinhua, May 17, 2017.

77 "Díaz-Canel Pide lo Imposible a los Comercios Estatales: Imitar a los Privados," EFE, February 19, 2019.

78 Julian Borger, "Donald Trump to Announce New Restrictions on Cuba Trade and Travel," *The Guardian*, June 16, 2017.

79 Andrea Rodriguez, "Trump's Cuba Policy Hurts Private Sector, New Figure Says," Associated Press, January 15, 2019; Sarah Marsh, "Cuba's Private-Sector Workers Say Trump's Travel Curbs Will Hurt Them Too," Reuters, June 5, 2019.

80 Speech by Cuba's President, Miguel Díaz-Canel, before the Nation Assembly, December 23, 2019, https://en.granma.cu/speeches-diaz-canel/2019-12-23/diaz-canel-to-the-people-of-cuba-united-we-have-won-united-we-will-win.

81 "El 15% de las Habitaciones Hoteleras de Cuba Está Fuera de Servicio," ADN Cuba, December 19, 2019.

82 In the first half of 2019, the average length of stay of tourists in Cuba's *casas particulares* was 1.97 days. Oscar Figueredo Reinaldo, Abel Padrón Padilla, Lissett Izquierdo Ferrer, and Dianet Doimeadios Guerrero, "Asamblea Nacional: Ministerios de Turismo e Industrias Brindaron Información a los Diputados, Asistió el Presidente Cubano," Cubadebate, July 11, 2019.

83 Unpublished study of the Cuban Ministry of Tourism, May 2019.

84 Carmen Sesin, supra note 54.

85 The top fifty RevPAR ranking of Cuban hotels in 2017 also included the Complejo Las Terrazas (with a non-hotel activity component) in the Artemisa province, the Iberostar Trinidad, and the hotels H10 Panorama, Presidente, and Habana Libre in Havana.

86 Sayli Sosa Barceló, "Turismo Cubano: Equilibrios y Desbalances," *Invasor*, April 3, 2023.

87 Andreas Knobloch, "Cuba Restrictions Hold Back Private Sector," Deutsche Welle (DW), July 18, 2018; Marc Frank, "Havana Suspends New Licenses for Private Restaurants, Owners Fret," Reuters, October 17, 2016.

88 Yudy Castro Morales, "¿Cómo Accederán las Formas No Estatales de Gestión a los Servicios del Comercio Exterior?," *Granma*, August 3, 2020.

89 Remarks by Alejandro Gil Fernández on Cuban national television's Mesa Redonda program, June 25, 2019, https://www.mep.gob.cu/es/noticia/intervencion-de-alejandro-gil-ministro-de-economia-en-el-espacio-mesa-redonda.

90 Andrea Rodriguez, "Cuba Looks to Tourism to Ease Economic Crisis," Associated Press, May 18, 2019.
91 Alejandro Gil Fernández, supra note 89.
92 René Tamayo, "El Turismo Cubano Continuará Siendo la Locomotora de la Economía," Cubadebate, March 29, 2021.
93 "OFAC Publishes New Sanctions Against Companies for Services to Cuba," OnCuba, June 14, 2019.
94 "Compañías Francesas Administrarán Aeropuerto de La Habana," Xinhua, August 4, 2016.
95 Gene Sloan, "Analyst: New Cuba cruises could be approved within days," *USA TODAY*, March 21, 2016.

4

International Tourism in Cuba and the Caribbean, 1990–2019

A Comparison

This chapter compares the tourism industry of Cuba with that of competing Caribbean nations. The timeframe covers the post-Cold War period up until 2019, the last full year before the outbreak of the global coronavirus (Covid-19) pandemic hit tourism-dependent economies hard, shattering borders, canceling holidays and business trips, greatly curbing flight numbers, and wreaking havoc on the world's travel and tourism industry. An analysis of how the Cuban tourism industry coped with the Covid-19 crisis and how it fared compared to other Caribbean countries in its recovery will be the subject of the next chapter.

Cuba has become a major force in Latin American and Caribbean tourism. Whereas in 1990 Cuba ranked twenty-third among the twenty-five most visited countries by international tourists in Latin America, by the end of 2019 it occupied seventh place. In 2019, among the leading Caribbean destinations, only the Dominican Republic and Cancún, Mexico received more stopover international visitors than Cuba (UNWTO 2021a).[1] And this was achieved despite the US economic embargo and a ban on tourist travel from the United States to the island. Before Covid-19, Cuba was the second-largest destination for Canadian travelers to Latin America after Mexico,[2] and the second-most popular Caribbean destination for European travelers (Spadoni 2014, 55). Yet Cuba's tourism industry fared poorly in terms of efficiency, especially if compared to other Caribbean countries offering all-inclusive arrangements. Gross revenues per tourist and gross revenues per room in Cuba were much lower than in the Bahamas, Jamaica, Puerto Rico, and the Dominican Republic. To some degree, this was due to the socioeconomic profile of leisure tourists to Cuba, who tend to book low- to mid-price package tours. But the shortcomings of the Cuban tourism sector identified in Chapter 3, in particular the low-quality standards of several hotels, the poor state of infrastructure, and the lack of adequate complementary services, were major contributing factors. Moreover, the average year-round hotel occupancy in Cuba has been traditionally low since many rooms are either out of service or in very precarious conditions. Put simply, the Cuban tourism industry must address critical structural problems in order to raise competitiveness vis-à-vis its Caribbean neighbors.

Importance of Tourism for the Caribbean

Not surprisingly, with its many compelling tourism attributes like clear azure waters, pristine white-sand beaches, and other spectacular natural wonders, mild temperatures year-round, and a multicultural heritage reflected in its different languages, cuisine, rums, music, and customs, the Caribbean region is highly dependent on tourism. If we consider the economic benefits of tourism (above all as a driver of economic growth and a source of hard currency income) to the Caribbean, the latter is widely seen as the most tourism-intensive and tourism-dependent region in the world (Cannonier and Burke 2019, 86). In 2019, according to rankings from the World Travel & Tourism Council based on the economic contribution of tourism as a share of a country's GDP, eight out of the ten most tourism-dependent countries in the world were small states in the Caribbean region. They were, namely, Antigua and Barbuda, by far the most tourism-dependent state (tourism accounted for over 80 percent of that country's GDP), Aruba, St. Lucia, Anguilla, the British Virgin Islands, St. Kitts and Nevis, Grenada, and the Bahamas. The Dominican Republic instead had the largest tourism sector in the region, followed by Cuba, Puerto Rico, the Bahamas, and Jamaica (WTTC 2022b).

A team of analysts at the Inter-American Development Bank (IDB) has developed a more comprehensive way to assess a country's dependence on tourism by focusing not only on the impact of tourism on economic output but also on its effects on employment and export earnings. Based on available data from 2015 to 2019 covering over 160 countries, the IDB's Tourism Dependency Index (TDI) is measured using five-year averages for the total contribution of tourism to exports of goods and services, real gross domestic product, and national employment. The TDI, which ranges from zero to 100, with 100 representing total dependence on tourism, is calculated for thirty-five countries in Latin America and the Caribbean. All the top ten most tourism-dependent nations in the Americas were located in the Caribbean region even though the sector was fairly significant for some of the largest countries in Latin America as well, in particular for Mexico, Chile, Peru, Argentina, and Brazil.[3] As for the global ranking, six of the ten most tourism-dependent countries in the world and nearly a dozen of the top twenty were Caribbean nations (Gómez García et al. 2021, 21; Mooney and Zegarra 2020, 4–5). The importance of tourism for the economy of many Caribbean countries is beyond question.

Table 4.1 collects data on the recently launched Tourism Dependency Index for the fifteen most tourism-dependent countries in the Caribbean region. More specifically, it reports the TDI scores for the 2015–19 period with the average contributions of tourism to export earnings, GDP, and employment. It also offers statistics on the annual average international tourist arrivals (overnight visitors) to these Caribbean countries and on their population and land area. Calculated by the authors using official Cuban data, the same indicators of tourism dependence are provided for Cuba, which is not included in the TDI ranking of the Inter-American Development Bank.

Based on the three critical variables that made up the TDI, Aruba was the most tourism-dependent country in the world (79.8 index score) out of 166 countries surveyed. Fueled by an average of more than one million international tourist arrivals per year, export earnings from tourism represented almost 90 percent of Aruba's total export revenues, and the sector accounted for about 70 percent of the country's GDP and 80 percent of total employment. Other small Caribbean nations like Grenada (58.5 TDI score, no. 4 ranking), Antigua and Barbuda (55.1, no. 5), the Bahamas (54.5, no. 6), St. Lucia (53.4, no. 7), and Dominica (42.0, no. 9) rounded out the top ten list of the most tourism-dependent economies globally. Underscoring a lack of economic diversification with respect to exports and a strong vulnerability to external shocks affecting global demand for travel and tourism, a large share of total export receipts (generally over 50 percent) was linked to tourism in all these countries, with peaks of 87.8 percent in Grenada and 73.8 percent in the Bahamas. Similarly, tourism accounted for one-third or more of total GDP and of total employment in all these countries, with peaks of 42.8 percent in Grenada and 40.6 percent in the Bahamas with respect to GDP, and of 83.5 percent in Antigua and Barbuda and 71.8 percent in St. Lucia with respect to employment. Barbados (41.4 TDI score, no. 11 ranking), Jamaica (38.7, no. 13), Belize (37.9, no. 15), St. Kitts and Nevis (35.9, no. 18), and St. Vincent and Grenadines (35.9, no. 19) completed

Table 4.1 Indicators of Tourism Dependence for Caribbean Countries

	Population (2019)	Land Area (000 km^2)	Tourism Dependency Index[a] (2019)	Percentage of Total Exports (2015–19)	Percentage of Total GDP (2015–19)	Percentage of Total Employment (2015–19)	Annual Average Tourist Arrivals (2015–19)
Aruba	106,442	180	79.8	88.0	70.9	80.6	1,110,300
Grenada	122,724	345	58.5	87.8	42.8	44.8	162,800
Antigua and Barbuda	92,117	442	55.1	42.4	39.4	83.5	256,000
Bahamas	404,557	13,940	54.5	73.8	40.6	49.0	1,504,600
St. Lucia	178,583	539	53.4	50.7	37.7	71.8	362,400
Dominica	71,428	750	42.0	51.0	36.5	38.5	73,900
Barbados	280,180	431	41.4	65.3	28.5	30.6	617,800
Jamaica	2,813,773	10,990	38.7	57.3	28.6	30.2	2,242,200
Belize	389,095	22,965	37.9	43.7	33.9	36.0	392,800
St. Kitts and Nevis	47,712	261	35.9	27.9	25.8	54.0	117,400
St. Vincent and Grenadines	104,924	389	35.9	39.8	26.3	41.5	76,200
Cayman Islands	66,134	264	27.0	25.0	24.4	31.5	406,800
Dominican Republic	10,881,882	48,671	24.1	38.8	16.3	17.2	5,891,540
Haiti	11,160,438	27,750	16.4	31.1	9.1	9.2	468,040
Cuba	11,316,697	109,884	12.6	20.5	12.0	5.2	3,557,400
Trinidad and Tobago	1,519,955	5,127	7.7	5.6	7.7	9.9	406,200

Sources: Gómez García et al. 2021; Mooney and Zegarra 2020; United Nations data (UNdata), "Popular Statistical Tables, Country (Area) and Regional Profiles," data.un.org; Calculations of the authors from ONEI 2020a and MINTUR data obtained in June 2020.

Note:

[a] The Tourism Dependency Index (TDI) is calculated using the total contribution of tourism (five-year averages for 2015–19) to total exports of goods and services, GDP, and employment for each country. The TDI's range is from 0 to 100, with 100 meaning total dependence on the sector.

the top twenty list of the most tourism-dependent states in the world (Gómez García et al. 2021, 21).

Overall, the TDI ranking for 2015–19 confirmed that the Caribbean is heavily dependent on tourism and that the region's small countries, in particular, are the most tourism-dependent economies in the world. Among the twelve countries classified by the International Monetary Fund (IMF) as Caribbean Small Developing States or SDS,[4] only Trinidad and Tobago, Guyana, and Suriname are commodity producers (oil and gas exporters) with rather low tourism dependency.[5] The others, all island states except Belize with a combined population of just 1.7 million and a total area of about 40,000 km^2, are the Caribbean countries with the highest TDIs along with Aruba (a small autonomous country within the Kingdom of the Netherlands) and Jamaica. While the IMF list of SDS includes only countries with populations under 1.5 million, Jamaica is in the World Bank's Small States Forum (SSF) list, which includes some countries with populations over 1.5 million but with characteristics similar to those of countries under the threshold (IEO 2022, 2–3). Caribbean SDS are classified by the World Bank as either high- or upper-middle-income countries.[6] They are characterized by relatively high levels of development and institutional capacity, but they exhibit critical weaknesses arising from public debt overhangs, limited access to concessional finance, fairly high pricing and cost structures that undermine competitiveness in the global market for tourism, and few possibilities to diversify their economies. They are also subject to shocks emanating from their principal tourism markets and extremely vulnerable to natural disasters (the Caribbean is one of the world's most disaster-prone regions)[7] that frequently force expensive repairs and reconstruction work, with significant negative impacts on economic expansion and fiscal stability. Between 2010 and 2019, the real GDP growth of Caribbean SDS averaged a meager 1 percent and real incomes per head rose cumulatively by just 3.5 percent (Da Costa and Rustomjee 2022).

Unlike most SDS, the two Caribbean countries with the largest tourism sectors, the Dominican Republic and, at least until 2019, Cuba, maintain economies with a somewhat deeper mix of industries and sectors, including agriculture, manufacturing, and various services. They are both classified as upper-middle-income countries by the World Bank. The Dominican Republic has a mixed economy with extensive private ownership that has exhibited strong dynamism and resilience since the early 1990s, while Cuba has a primarily planned economy still dominated by state-run enterprises (alongside a rapidly emerging private sector) that has been mired in stagnation for several years. Yet trade and investment remain insufficiently diversified in both countries, especially in Cuba, and tourism continues to be a key pillar of both economies.

The Dominican Republic has been the fastest-growing economy in Latin America and the Caribbean during the 2010s, with its GDP expanding by 5.8 percent annually, well above the region's annual growth rate of 2 percent. It also experienced an average annual GDP growth rate of more than 5 percent during the twenty-five years before the Covid-19 outbreak (UNWTO 2022a; OECD 2020a, 32). With an average of nearly 6 million international tourist arrivals per year, the Dominican Republic recorded a TDI score of 24.1 for the 2015–19 period, ranking thirtieth globally among the countries most dependent on tourism. More specifically, the Dominican tourism sector generated approximately 16 percent of the country's total GDP, 17 percent of national employment, and nearly 40 percent of total revenues from exports of goods and services. Tourism receipts were the Dominican Republic's largest source of foreign exchange, followed by remittances from abroad and free trade zone exports (medical devices, tobacco products, electrical machinery, textile and apparel, and other industrial products).[8] Helped without a doubt by a general consensus among Dominican political leaders about the need to support tourism, the latter is widely considered to have played a critical role in boosting the Dominican Republic's economic expansion over the past three decades. It was not, though, the only factor. The country's remarkable growth was also the result of a series of market-oriented economic reforms implemented in the 1990s and 2000s, in particular the liberalization of foreign exchange transactions, the improvement of monetary control, the removal of certain barriers to international trade like import tariffs and quotas, the signing of new trade agreements, and the elimination of restrictions on foreign direct investment in nearly every sector coupled with a series of laws offering generous tax incentives to overseas investors. High growth was accompanied by a significant reduction of poverty levels and inequality, especially since the mid-2000s (Meyer 2020; Parra, Pienknagura, and Sayed 2018).

As for Cuba, which received an average of roughly 3.5 million international overnight visitors per year in 2015–19, the authors calculated that tourism accounted for 12 percent of GDP, around 5 percent of national employment, and 20.5 percent of total exports during that period, resulting in a TDI score of 12.6 that would place Cuba at around the sixtieth position in the IDB's TDI ranking. Nevertheless, the importance of tourism for the Cuban economy is certainly higher than the TDI score would suggest. Between 35 percent and 45 percent during the 1990s and in the first half of the 2000s, the share of Cuba's total exports of goods and services accounted for by tourism has decreased notably since the mid-2000s as exports of professional (mainly medical) services became the island's top hard currency earner. Even so, Cuba has continued to depend heavily on tourism revenues, along with exports of some key products (nickel, sugar, cigars, and pharmaceuticals) and remittances from abroad that are captured in state-owned hard currency stores, to alleviate its chronic foreign exchange shortages because the services provided by Cuban professionals in Venezuela and other countries are generally part of agreements involving Cuban purchases of energy products and additional goods from these countries. Between 2015 and 2019, Cuba's international tourism receipts trailed remittances from abroad (Morales 2023),[9] but they were higher than the country's total foreign exchange earnings from merchandise exports. Moreover, tourism is one of Cuba's few export sectors whose activities have considerable linkages with other productive sectors (Spadoni 2023, 654). In Cuba, the tourism sector remains not only a key revenue generator but also the principal agent of dynamism in the economy and a crucial driving force behind production chains that, despite considerable room for improvement, ensure import substitution while helping to reduce tourism costs. Needless to say, with its strict state controls that stifle market opportunities, Cuba has a less attractive business environment than the Dominican Republic.[10]

In sum, the variables that make up the TDI reflect the three principal dimensions through which tourism activities affect a destination's economy, namely their contributions to foreign exchange earnings, national output, and employment. In the Caribbean region, tourism is a leading source of hard currency revenues that represent a sizable share of total exports of many countries. It is also a major component of many countries' GDP and a major employer, providing countless jobs for low-skilled workers together with jobs for a more skilled workforce. During the 2010s, the Caribbean tourism sector grew at an annual average of 3 percent, outpacing the overall expansion of the region's economy (1.3 percent) but failing to keep pace with the average growth rate of tourism worldwide (4.2 percent). The World Travel and Tourism Council saw this as evidence that the Caribbean has not fully exploited its potential for tourism growth, which it attributed mainly to obstacles such as limited air connectivity and high aviation fees, inadequate investments in digital and physical infrastructure, low diversification of tourism products and services, and insufficient efforts to support preparedness for future crisis (WTTC

2022b). Besides, tourism can produce significant cross-sectoral spillover effects within a country's economy, promoting backward linkages to transportation, construction, telecommunications, agriculture, and retail trade. For instance, among its many contributions, tourism growth stimulates the construction of hotels, airports, roads, ports, and other infrastructure, and enhances the integration with domestic agriculture by boosting local food production for tourists. Yet these linkages between tourism and the domestic economy have not been fully developed in the Caribbean (Acevedo, Laframboise, and Wong 2017a, 40). Despite the economic importance of international tourism for the Caribbean, the region must address serious challenges in order to reap the full potential benefits of tourism.

Inbound Tourism in the Caribbean

Following two notable cycles in the 1920s and in the 1950s, and encouraged by major political, economic, and technological changes worldwide in the context of a rapid intensification of the processes of globalization, the third and longest cycle of tourism development in the Caribbean region began in the 1980s and gained steam in the 1990s after the end of the Cold War. The rising international integration of capital and product markets, largely fueled by liberalizing economic policies and by reductions in transportation and communication costs, brought to light a growing specialization in services in many countries around the world. There were, though, two different forms of structural transformation. Whereas developed economies transferred productive resources mainly from the industrial sector to services, developing economies like those in the Caribbean experienced a shift in the sectoral composition of output from the agricultural sector to services (UNCTAD 2017). In fact, most Caribbean countries simply moved directly from primary-sector-led development to tertiary-sector-led development, with travel and tourism emerging as the most significant service exports. To facilitate this shift and promote the competitiveness (especially vis-à-vis other destinations in the region) and sustainability of the tourism sector, Caribbean governments granted significant tax concessions and other fiscal incentives to both existing and potential investors, in particular overly generous corporate tax holidays and import-related tax exemptions that positively affected foreign direct investment. Nevertheless, besides a biased distribution of benefits in favor of large operators that produced economic distortions and sub-optimal results, tax incentives imposed considerable costs on host economies, resulting in sizable revenue losses and aggravating the already vulnerable fiscal conditions of most countries (McIntyre 2017, 121; CEPAL 2003a).[11]

As a result of liberalization policies and measures (no Caribbean government resorted to protectionist measures to boost its tourism industry), tourism became, and still is, one of the most open and unregulated sectors in the Caribbean region. Furthermore, the 1980s and especially the 1990s saw the consolidation of tourism trends that had been taking shape in the previous two decades, namely the prioritization of the sun-and-beach modality and leisure tourism, the promotion of all-inclusive packages, and the development of cruise tourism (González Núñez 2022, 105). During the 1990s, the Caribbean experienced a spectacular growth in arrivals of both overnight and cruise visitors. According to the standard definitions of the World Tourism Organization, an overnight or stay-over visitor is a traveler taking a trip to a particular country for any main purpose (business, leisure, or other personal purpose) and staying there at least one night but less than a year. A cruise tourist instead is a same-day visitor, also known as an excursionist.[12] Starting from a base of approximately four million international overnight visitors in 1970, the Caribbean already received over seven million visitors per year by the mid-1980s and over ten million by the end of that decade as air travel became more economically accessible to middle-class North Americans and Europeans. During the next decade, the region's tourism sector expanded further, with annual stay-over arrivals increasing from 11.6 million in 1990 to around 16.1 million in 1999 (UNWTO 2000). Similarly, as sea cruises became more and more the vacation of choice (and more affordable due to increased competition) with the launch of plenty of new ships and the addition of new destinations, travel lengths, themes, ship design concepts, onboard activities, and shore excursions, cruise arrivals to the Caribbean grew noticeably from 7.7 million passengers in 1990 to 12.1 million passengers in 1999.[13]

The rapid expansion of inbound tourism volumes in the Caribbean decelerated markedly in the 2000s as the region's tourism industry was negatively affected by the terrorist attacks against the United States on September 11, 2001, the stock market downturn of 2001–2, and especially the global financial and economic crisis of 2008–9. Annual stay-over visitors to the region had reached nearly 19 million by 2009, but the average growth per year in the 2000s was less than half its level in the 1990s. The same can be said for cruise tourist arrivals that amounted to 19.1 million in 2007 yet dropped to 17.1 million in 2009 as they suffered greatly from the global financial crisis of the late 2000s.[14] Since then, both tourism segments have continued to expand even if their performance has been uneven. In 2019, the Caribbean welcomed about 26.3 million international overnight visitors and 30.2 million cruise tourists (UNWTO 2021a;, CTO 2020; Perelló Cabrera 2017). Most importantly, though, the Caribbean's share of the global tourism market has declined steadily since the early 2000s. After staying fairly constant during the 1990s at roughly 2.7 percent, the region's share of international tourist arrivals worldwide (overnight visitors) had fallen to approximately 2 percent by 2010 and to just 1.8 percent by 2019. Along with the considerable impact of global political and economic issues, this trend was primarily due to the surge of new destinations in various regions across the world, the frequency and intensity of extreme weather and natural disasters, and the lack of real innovation in the tourist products being offered (Perelló Cabrera 2020).

Regarding the international tourist flows to specific Caribbean countries, and to highlight the principal trends, Table 4.2 offers data from the United Nations World Tourism Organization (and from Cuban official sources) on annual overnight visitor arrivals to the Caribbean region by destination country from 1990 until 2014. During the 1990s, the most visible trend was the CARICOM countries' loss of market share to the Hispanic Caribbean, which comprises Puerto Rico, the Dominican Republic, Cuba, and Mexico (viz.Cancún and Cozumel that are not included in the UNWTO statistics for the Caribbean). This was mainly the result of the declining popularity of English-speaking Caribbean destinations among European and Canadian tourists in favor of Spanish-speaking destinations. Consisting of fifteen small countries,[15] many of which cater to the high-end segment of the tourism market, yet including sizable tourist recipients like the Bahamas and Jamaica, the Caribbean Community or CARICOM lost regional market share to newly emerging, lower-cost tourism destinations in the Hispanic Caribbean. It also suffered, reportedly, from a deterioration in the non-price competitiveness vis-à-vis other Caribbean destinations, lagging behind in non-price factors such as quality of services, product design, technological advances, distribution channels, marketing and advertising, and air access (Randall 2006, 294). By the early 2000s, the Hispanic Caribbean accounted for more than half (52.6 percent) of total tourists to the region while CARICOM countries accounted for just 25.4 percent, down from around 34 percent in 1990 (WTTC 2004; CEPAL 2003a, 2003b).[16]

Table 4.2 Overnight Visitor Arrivals to Caribbean Countries, 1990–2014 (thousands)

	1990	1995	2000	2005	2010	2014
Anguilla	31	38	44	62	62	71
Antigua & Barbuda	206	220	237	245	300	249
Aruba	433	619	721	732	824	1,072
Bahamas	1,562	1,598	1,544	1,608	1,370	1,422
Barbados	432	442	545	547	532	520
Bermuda	433	387	332	270	232	224
British Virgin Islands	177	219	281	337	330	386
Cayman Islands	253	361	354	168	288	383
Cuba	327	742	1,741	2,261	2,507	2,983
Curacao	208	224	191	222	342	454
Dominica	45	60	70	79	76	81
Dominican Republic	1,530	1,776	2,973	3,691	4,124	5,141
Grenada	82	108	129	98	110	133
Guadeloupe	288	640	807	863	392	486
Haiti	120	145	140	112	255	465
Jamaica	841	1,147	1,323	1,479	1,922	2,080
Martinique	281	457	526	484	478	490
Montserrat	13	18	10	10	6	9
Puerto Rico	2,560	3,140	3,466	3,702	3,103	3,246
St. Kitts and Nevis	76	79	73	127	98	105
St. Lucia	138	232	270	318	306	338
St. Maarten	565	449	432	468	443	500
St. Vincent and Grenadines	54	60	73	95	72	71
Trinidad and Tobago	194	260	399	463	388	412
US Virgin Islands	462	454	546	594	590	602
Total	11,638	14,209	17,617	19,035	19,539	22,282

Sources: ONEI 2015a, 2011a, 2006, 1998; UNWTO 2016, 2015, 2001, 1997; United Nations World Tourism Organization, "145 Key Tourism Statistics," https://www.unwto.org/tourism-statistics/key-tourism-statistics.

Among the countries with the largest tourism sectors (excluding Mexico), Puerto Rico was the Caribbean's top recipient of tourists throughout the 1990s and until the mid-2000s, when annual arrivals reached 3.7 million. As arrivals declined over the next decade, Puerto Rico was surpassed by the Dominican Republic. Annual international visitors to the Dominican Republic nearly doubled during the 1990s from 1.5 million to almost 3 million, and by 2014 had topped 5 million. Cuba, instead, was the Caribbean's fastest-growing tourist destination in the 1990s, and it continued its rise as a leading tourist recipient in the region afterward. While in 1990 Cuba attracted only 327,000 international visitors and ranked tenth among the most popular countries for tourists in the Caribbean, in 2014 it was visited by some 3 million foreign travelers and ranked third after the Dominican Republic and Puerto Rico. By 2014, if we also consider Cozumel (205,000 stay-over tourists received) and especially Cancún (4.3 million tourists), which is a large enough destination for sun-and-beach tourism within Mexico that competes directly with the Caribbean (particularly in the US market), the Hispanic Caribbean accounted for 59.3 percent of total tourists to the region while CARICOM's share was 24.8 percent (UNWTO 2016; SECTUR 2015). At any rate, among CARICOM countries, Jamaica saw its annual tourist arrivals more than double between 1990 and 2014 from 841,000 to some 2.1 million. As a result, Jamaica, remarkably, increased its share of total international visitors to the Caribbean, and in the second half of the 2000s it surpassed the Bahamas as CARICOM's largest tourist destination. Aside from the tiny Montserrat, and to a great extent because of the maturity of its market, the Bahamas was the only Caribbean country whose tourism sector shrank in terms of arrivals during 1990–2014, from nearly 1.6 million to 1.4 million (Acevedo, Laframboise, and Wong 2017a, 41).

Despite a small drop in 2018 mainly fueled by the extensive damages that hurricanes Irma and Maria of September 2017 caused in certain countries like Anguilla, British Virgin Islands, Dominica, St. Maarten, US Virgin Islands, and Puerto Rico (UNDP 2017), overnight visitors to the Caribbean grew at an annual average of 2.3 percent during the 2015–19 period from 24.1 million to 26.3 million (Table 4.3). Considering visitor arrivals to the Mexican Caribbean (Cancún alone received more than 6 million foreign tourists), the Hispanic Caribbean's share of total tourists to the region reached 61.3 percent in 2019, whereas CARICOM's share decreased further to 24.0 percent.[17] The Dominican Republic consolidated itself as the most popular tourist destination in the Caribbean. Stopover tourists to the Dominican Republic soared from 5.6 million in 2015 to about 6.6 million in 2018 and were still close to 6.5 million in 2019, even if a string of suspicious deaths in the summer of that year involving eleven US citizens raised serious safety concerns among would-be visitors and convinced many Americans to rethink plans to vacation in the Dominican Republic.[18] In fact, after suffering a 10 percent drop in total visitor arrivals (almost entirely due to a 30.7 percent reduction in the flow of US tourists) between July and October 2019 as compared to the same period in 2018, and following the announcement by Dominican officials in September that new safety measures were put in place to reassure travelers,[19] inbound tourism in the Dominican Republic fell at a slower pace in November and December and closed the year less than 2 percent below its level in 2018. So, it would seem that

Table 4.3 Overnight Visitor Arrivals to Caribbean Countries, 2015–19 (thousands)

	2015	2016	2017	2018	2019
Anguilla	73	79	68	55	95
Antigua & Barbuda	250	265	247	269	301
Aruba	1,225	1,102	1,070	1,082	1,119
Bahamas	1,496	1,499	1,452	1,633	1,804
Barbados	592	632	664	676	713
Bermuda	220	244	270	282	269
British Virgin Islands	393	408	335	192	302
Cayman Islands	385	385	418	463	503
Cuba	3,511	3,789	4,076	3,834	3,828
Curacao	468	441	399	432	464
Dominica	75	78	72	63	90
Dominican Republic	5,600	5,959	6,188	6,569	6,446
Grenada	155	156	168	186	188
Guadeloupe	512	581	659	735	—
Haiti	516	445	467	447	286
Jamaica	2,123	2,182	2,353	2,473	2,681
Martinique	487	519	536	537	556
Montserrat	9	9	9	10	10
Puerto Rico	3,542	3,736	3,513	3,068	3,180
St. Kitts and Nevis	118	116	115	125	120
St. Lucia	345	348	386	395	424
St. Maarten	505	528	402	178	320
St. Vincent and Grenadines	75	79	76	80	86
Trinidad and Tobago	440	409	395	375	389
US Virgin Islands	642	667	535	381	514
Total	24,091	25,235	26,046	25,800	26,341

Sources: ONEI 2020a; MINTUR data obtained by the authors in May 2020; UNWTO 2023a, 2021b, 2019, 2018, 2017; United Nations World Tourism Organization, "145 Key Tourism Statistics," https://www.unwto.org/tourism-statistics/key-tourism-statistics.

the situation had improved. In reality, this was the result of increased arrivals of Dominican residents abroad since US tourist arrivals were down 31.2 percent in November and December 2019. They were also down 36.7 percent in the first two months of 2020 before Covid-19 became an issue, suggesting that US travelers' anxieties about the lack of safety in the Dominican Republic were still high at that time.[20]

As for other countries, annual tourist arrivals in Puerto Rico fell noticeably in 2017–18 and rebounded timidly in 2019 to 3.2 million as the US territory struggled to recover from hurricanes Irma and Maria. Meanwhile, Cuba emerged as the second-most visited tourist destination in the Caribbean (excluding Cancún). Stimulated by Barack Obama's lifting of some restrictions on US-based travel to the island, overnight visitors to Cuba jumped from 3.5 million in 2015 to about 4.1 million in 2017, but plunged to 3.8 million in 2019 after Donald Trump reversed key pieces of the Obama-era engagement with Havana. Regarding CARICOM's largest tourist recipients, visitor arrivals in the Bahamas increased from 1.5 million in 2015 to around 1.8 million in 2019. Yet, the Bahamas was devastated by the category five hurricane Dorian in September 2019 that pummeled Abaco and Grand Bahama Islands for two days, leaving behind massive destruction and causing an estimated $3.4 billion worth of damage, losses, and additional costs, an amount equivalent to a quarter of the country's GDP (IDB 2020).[21] Finally, visitors to Jamaica rose from 2.1 million in 2015 to some 2.7 million in 2019, a record-breaking year for tourism on the English-speaking Caribbean island. In 2015–19, Jamaica's inbound tourism recorded the highest annual average growth rate (6 percent) among the Caribbean's leading tourist destinations, followed by the Bahamas (5.1 percent), the Dominican Republic (3.6 percent), Cuba (2.4 percent), and Puerto Rico (-2.4 percent). It should also be stressed that US tourist arrivals in Jamaica, unlike US tourists to the Dominican Republic, grew 12.9 percent in 2019 (JTB 2020). Thus, albeit facing challenges similar to other countries in the region, when Covid-19 began to wreak havoc on travel and tourism worldwide in March 2020, the tourism sector of Jamaica was arguably in better conditions than that of its main Caribbean competitors, which were coping with pressing issues involving safety and security, travel restrictions, and extreme weather events.

Leaving aside Cuba's cruise tourism boom from mid-2016 until mid-2019 during the short-lived rapprochement with the United States, the Caribbean's leading recipients of stopover tourists (Bahamas, Puerto Rico, Jamaica, and the Dominican Republic) are also popular destinations for cruise tourists along with Cozumel, the Cayman Islands, Belize, Turks and Caicos, and a number of small islands in the Lesser Antilles. The Caribbean's warm and generally dry climate during the winter, its proximity to the United States, and its appeal to visitors from North America, Europe, South America, and other parts of the world have played a key role in stimulating the growth of the cruise industry in the region. The Caribbean's geographic structure allows short-distance navigation to visit multiple island states and ports with diverse customs, languages, and colonial roots. Caribbean cruises, therefore, offer a broad array of experiences, excursions, and cultural tours that can hardly be matched anywhere else in the world.

The cruise industry as we know it today dates back to the early 1970s, when the creation of the North American cruise industry following the founding of major lines such as Princess Cruises (1965),[22] Norwegian Cruise Line (1966), Royal Caribbean (1968), MSC Cruises (1970), and Carnival (1972) transformed the industry from a trans-ocean carrier service mostly reserved for rich people into an affordable alternative to a typical land vacation. Since then, both the number and the size of cruise ships operating worldwide have increased considerably. About two dozen new ships were built in the 1970s, almost forty in the 1980s, another eighty in the 1990s, and many more afterward. At the end of 2021, there were 323 cruise ships in service across the globe with a combined capacity of 581,200 passengers.[23] Another sixteen ships reportedly debuted in 2022, and forty-four new ships with a combined capacity of 120,959 passengers were expected to enter service between 2023 and 2028 (CLIA 2022, 2023). As for the size of the ships, whereas their capacity grew from an average of 500 passengers to 800 passengers in the 1970s, numerous cruise ships currently in service (running generally at 100 percent capacity or above) can accommodate between 3,000 and 6,680 passengers along with crews of 2,000 employees or more (Honey 2018a, 35). Albeit with different corporate structures, the same lines that launched the modern cruise business fifty years ago still dominate today's global cruise industry. With respect to the cruise market share distribution among major cruise players, Carnival, which acquired Princess Cruises in 2003, was (and still is) the largest cruise company in the world in 2021 with passenger and revenue market shares at 42 percent and 37.1 percent, respectively, followed by Royal Caribbean (23.6 percent and 21.2 percent), Norwegian (9.5 percent and 12.6 percent), and MSC (10.2 percent and 6.5 percent).[24] Of these four leading players, the Swiss-based MSC is the only non-US-based company. Although it accounts for a small portion of global tourism flows (around 2 percent), the cruise industry was one of the fastest-growing segments within the entire travel industry before the Covid-19 pandemic, moving nearly 30 million passengers in 2019 and generating an estimated $154.5 billion in total output contributions worldwide (CLIA 2021).

By the end of the 1970s, the Caribbean had already become an important cruise destination, and its popularity has remained very strong over the years notwithstanding tough competition from other regions of the world. Annual cruise passenger arrivals to the Caribbean increased from just 1.3 million in 1970 to 3.8 million in 1980, and from that moment they have witnessed a massive expansion. They doubled over the next decade to 7.7 million in 1990, nearly doubled again the following decade to 14.5 million in 2000, and reached 20.2 million in 2010 and 30.2 million in 2019 (CTO 2020, 2011).[25] In 2019, according to the Cruise Lines International Association, or CLIA, cruise ships transported an estimated 12 million passengers to the Caribbean. These travelers represented over 40 percent of global cruise passengers, a much higher share than travelers to other popular regions like the Mediterranean (14.9 percent), Asia and China (13.4 percent), Northern Europe (5.7 percent), Alaska (4.1 percent), Australia, New Zealand, and the South Pacific (4 percent), and the North American West Coast (3.9 percent). It is worth mentioning that because cruise ships make stopovers in multiple destinations in a single itinerary, passenger arrivals are always considerably higher than passenger volumes. It should also be underlined that the Caribbean accounted for 32 percent of total cruise ship capacity that was deployed globally in 2019, which suggests that Caribbean cruises had a higher ship capacity utilization than cruises in other parts of the world (CLIA 2023, 2019). All in all, the Caribbean is by far the most preferred cruise destination in the world, and it will likely remain so for quite some time. Although there are over eighty docking ports in the region,[26] the largest cruise ships dock in no more than thirty of them, to which it must be added a number of islands or enclaves that cruise companies have purchased to create their own tourist destinations for the exclusive use of cruise passengers. Exclusive private islands owned by cruise lines are located in the Bahamas (Great Stirrup Cay, CocoCay, Princess Cay, Castaway Cay, Ocean Cay Marine, Lighthouse Points, Beach Club at Bimini), Dominican Republic (Catalina Island, Cayo Levantado, Amber Cove), Belize (Half Moon Caye, Harvest Caye), and Haiti (Labadee).[27]

Caribbean cruises, most of them departing from Florida ports (Miami, Ft. Lauderdale, Port Canaveral, and Tampa), split their itineraries between Eastern and Western Caribbean destinations. In some cases, Southern Caribbean destinations are added to Eastern Caribbean itineraries. Usually shorter and less expensive, Western Caribbean voyages include destinations in the Greater Antilles like Jamaica and the Cayman Islands (and Cuba from 2016 until 2019) as well as Cozumel-Mexico and some Central American countries like Belize. Eastern Caribbean sailings feature Turks and Caicos, Puerto Rico, the Dominican Republic, and several locations in the Lesser Antilles. Only the Bahamas and Labadee-Haiti show up on itineraries for cruises in both sub-regions.[28] Concerning the flow of cruise tourists to the Caribbean, during the 1990s and early 2000s, there was a clear shift toward the Western Caribbean to the detriment of the Eastern and Southern Caribbean. The share of the Bahamas in the Caribbean cruise market also shrunk even though cruise arrivals to the country resumed strong growth in the 2000s. The Dominican Republic, Aruba and Bonaire (Southern Caribbean), Dominica, St. Kitts and Nevis, and St. Lucia recorded notable growth in cruise arrivals, while St. Vincent and the Grenadines, Martinique, and the US Virgin Islands witnessed declines (CEPAL 2005, 2003b). Since then, the market share of the Western Caribbean has had only minor fluctuations. As shown in Figure 4.1, the Caribbean welcomed nearly 24 million cruise tourists in 2014. The region's top destinations were the Bahamas (4.8 million arrivals), Cozumel (3.4 million), US Virgin Islands (2.1 million), St. Maarten (2 million), Cayman Islands (1.6 million), Jamaica (1.4 million), Puerto Rico (1.3 million), Turks and Caicos (972,000), and Belize (968,000) (CTO 2015). Cruise passenger arrivals in Cuba stood at just 8,100 in 2014 as only a handful of small- and medium-size ship cruise lines, mostly European-based, included stops on the island in their Caribbean itineraries (ONEI 2016a). But Cuba was about to join the cruise business.

Between 2015 and 2019, the share of the Western Caribbean sub-region in the Caribbean cruise market remained virtually steady at slightly over 30 percent as arrivals in Cozumel, Belize, and Cuba rose considerably while visits to the Cayman Islands and especially Jamaica experienced little or no growth (Table 4.4). Apart from Cuba that started from an extremely low base,

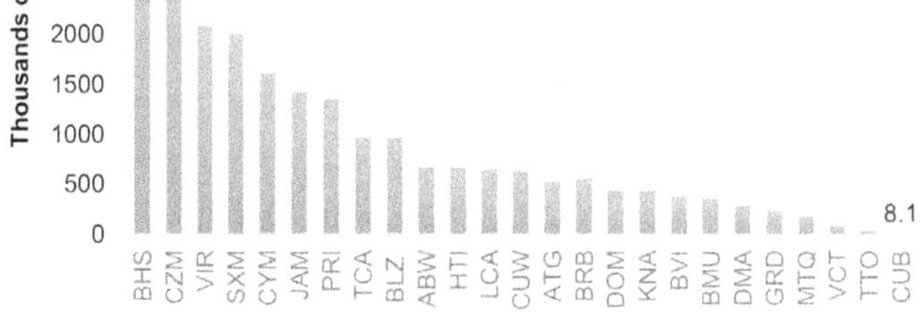

Figure 4.1 Caribbean cruise visitor arrivals by destination in 2014. Note: List of codes: BHS (Bahamas), CZM (Cozumel-Mexico), VIR (US Virgin Islands), SXM (St. Maarten), CYM (Cayman Islands), JAM (Jamaica), PRI (Puerto Rico), TCA (Turks & Caicos), BLZ (Belize), ABW (Aruba), HTI (Haiti), LCA (Saint Lucia), CUW (Curacao), ATG (Antigua & Barbuda), BRB (Barbados), DOM (Dominican Republic), KNA (St. Kitts & Nevis), BVI (British Virgin Islands), BMU (Bermuda), DMA (Dominica), GRD (Grenada), MTQ (Martinique), VCT (St. Vincent & the Grenadines), TTO (Trinidad & Tobago), CUB (Cuba). Generated by the authors, based on CTO 2015; ONEI 2016a.

Table 4.4 Cruise Visitor Arrivals to Principal Caribbean Destinations, 2015–19 (thousands)

	2015	2016	2017	2018	2019
Bahamas	4,513	4,690	4,626	4,878	5,430
Cozumel (Mexico)	3,396	3,637	4,104	4,265	4,569
Cayman Islands	1,717	1,712	1,728	1.921	1,831
Puerto Rico	1,457	1,402	1,229	1,659	1,785
St. Maarten	1,902	1,669	1,238	1,597	1,632
Jamaica	1,569	1,656	1,923	1,846	1,553
US Virgin Islands	1,879	1,777	1,317	1,431	1,433
Belize	958	1,005	1,014	1,208	1,171
Turks and Caicos	930	847	827	1,022	1,112
Dominican Republic	529	809	1,108	982	1,104
St. Kitts and Nevis	990	932	1,068	1,148	983
Aruba	607	656	792	815	832
Cuba	10	174	433	626	305

Sources: ONEI 2016a; MINTUR data obtained by the authors in May 2020; CTO 2020; 2018; 2016.

the Dominican Republic was the fastest-growing cruise destination in the Caribbean. The Bahamas, Puerto Rico, Turks and Caicos, and Aruba also saw a substantial increase in cruise arrivals. St. Maarten and the US Virgin Islands, instead, suffered the biggest drops. In 2019, the last "normal" year of cruise operations prior to Covid-19, the Bahamas remained the most visited cruise destination in the Caribbean with 5.4 million inbound visitors, followed by Cozumel (4.6 million), Cayman Islands (1.8 million), Puerto Rico (1.8 million), St. Maarten (1.6 million), Jamaica (1.5 million), US Virgin Islands (1.4 million), Belize (1.2 million), Turks and Caicos (1.1 million), the Dominican Republic (1.1 million), St. Kitts and Nevis (983,000), and Aruba (832,000). It must be noted that, based on the kinds of visitors they receive, seven of these twelve leading Caribbean cruise destinations depend more on cruise tourists than overnight tourists (González Núñez 2022, 130). In 2019, besides Cozumel, which is Mexico's largest Caribbean island with about 90 percent of its visitors arriving on cruise ships, the share of cruise tourist arrivals in the total number of inbound tourists exceeded 80 percent in St. Maarten and St. Kitts and Nevis, and was more than 70 percent in the Bahamas, Cayman Islands, US Virgin Islands, and Turks and Caicos.[29] With respect to Cuba, a key development was the Obama administration's decision in 2015 to allow cruise companies to bring US citizens to Cuba as long as they participated in cultural and educational activities onshore. As major cruise lines began to sell Cuba itineraries, cruise tourist arrivals on the communist island grew from a mere trickle in 2015 to a peak of about 626,000 in 2018, making Cuba a competitive cruise destination in the Caribbean. This short window of opportunity, however, closed abruptly in June 2019 when the Trump administration banned US cruise ships from sailing to Cuba.

Principal Source Markets

A critical feature of the Caribbean is the close proximity of most of its countries and territories to the United States, the world's largest economy, accounting for approximately a quarter of global GDP, and the world's second-largest source of international travelers and tourism expenditures after China (UNWTO 2023b).[30] The flying time from Miami, the main travel hub and gateway to the Caribbean, to any of the region's countries ranges from less than an hour to the Bahamas and Cuba to four and a half hours to Guyana. Almost every country is served by daily flights from the United States as major US airlines constantly launch new flights and routes to boost seat capacity and air connectivity to Caribbean destinations. The Caribbean is the most visited international region by US citizens after Mexico, Europe, and Canada, accounting for about 10 percent of total US outbound travel. Remarkably, the Caribbean receives annually more US visitors than South America and Central America combined, the Middle East, and the entire Asian region.[31] There are also over 150 cruise ships sailing the Caribbean every year, most of them departing from ports in South Florida.[32] It thus should come as no surprise that the United States is the leading source market for international tourists (overnight and cruise visitors) to the Caribbean. Although different islands attract different tourists, especially if we compare small island nations to larger destinations, additional key tourism source markets for the Caribbean region are Canada and European countries, in particular France and the UK and, to a lesser extent, Germany, Italy, Spain, and the Netherlands. Significant in some destinations but less numerous overall are arrivals from Latin American countries such as Argentina, Mexico, Brazil, and Chile, and from other Caribbean territories (WTTC 2022b).[33]

The United States is not only the principal source of tourists to the Caribbean as a whole but also to the vast majority of countries in different sub-regions, including the Hispanic Caribbean, the Commonwealth Caribbean, the French West Indies, the Dutch West Indies, and the US territories. For most of the

1980s, the US share of total annual international tourists to the Caribbean was close to 60 percent.[34] Nevertheless, as indicated in Table 4.5, during the dramatic expansion of inbound tourism in the region in the 1990s, the relative contribution of the United States plunged to approximately 50 percent as European tourist arrivals, especially because of their strong performance in fast-growing and rather large destinations like Cuba and the Dominican Republic, increased more rapidly than US arrivals to reach 25 percent of the total (González Núñez 2022, 109–10). Little more than seven million a decade earlier and three times the number of European tourists, by the early 2000s, over ten million American citizens traveled to the Caribbean every year as compared to around five million Europeans. During the 2000–2019 period, the US share of total overnight visitor arrivals to the region remained practically constant. In the meantime, Europe's share fell considerably while the share of Canada, whose favorite Caribbean destinations are Cuba, the Dominican Republic, Jamaica, the Bahamas, and Barbados, nearly doubled. In 2019, the United States sent 15.5 million tourists to the Caribbean, representing 49.2 percent of the total number of international stopover visitors to the region. Europe sent 5.9 million tourists or 18.5 percent of the total, followed by Canada with 3.4 million tourists or 10.7 percent of the total. Exhibiting virtually the same relative contribution for three decades, intra-Caribbean tourism accounted for 6.5 percent of all tourists to the Caribbean in 2019, which translated into a remarkable two million travelers. Even more consequential, tourist arrivals from regions other than North America and Europe have shown significant dynamism in recent years, a result that underscores a trend toward a much-needed diversification of the Caribbean's tourism source markets (CTO 2020).

Table 4.5 Overnight Visitor Arrivals to the Caribbean by Source Market, 1990–2019 (percentage distribution)

	1990	1995	2000	2004	2010	2014	2019
United States	57.7	55.8	51.6	52.5	49.6	50.3	49.2
Europe	17.2	22.3	25.5	24.3	21.3	18.3	18.5
Canada	6.6	6.0	6.0	8.0	11.7	12.3	10.7
Caribbean	6.9	6.5	6.3	6.6	6.5	6.0	6.5
Others	11.6	9.4	10.6	8.6	10.9	13.1	15.1

Sources: CTO 2020, 2015; Caribbean Tourism Organization, "International & Caribbean Tourist Arrivals 1970–2004," http://www.onecaribbean.org/content/files/intlcbbntouristarrivals1970to2004.pdf; Caribbean Tourism Organization, "Tourist Arrival in Caribbean by Main Market 1980–2004," http://www.onecaribbean.org/content/files/TAcbbnMainmarket1980to2004.pdf.

In effect, despite the aforementioned changes, Caribbean countries continue to be insufficiently diversified in their sources of inbound tourists. The inevitable consequence of this condition, which became evident in the aftermath of the global financial crisis of 2007–8, is that the performance of the Caribbean tourism industry is easily affected by the economic cycles in advanced economies. In 2009, as Canada, the largest European economies, and, most notably, the United States experienced sizable drops in GDP, rising unemployment, and plummeting household incomes, international tourist arrivals witnessed a plunge in almost every Caribbean country. Among the few exceptions, Cuba, the Dominican Republic, Guyana, and Jamaica managed to record a growth in visitors from abroad, but only thanks to price cuts that slashed profits and could not prevent a decrease in tourism revenues. The economic output of countries most reliant on tourism contracted sharply in 2009 as GDP fell nearly 12 percent in Aruba and in Antigua and Barbuda, 6.6 percent in Grenada, 5 percent in Barbados, and over 4 percent in the Bahamas and in Jamaica. Cuba and the Dominican Republic reported positive economic growth, albeit very small.[35] Overall, even if tourist arrivals rebounded quickly in several countries, tourism demand for the Caribbean remained weak until 2014, reducing hotel occupancy rates and thwarting new investments (Acevedo, Laframboise, and Wong 2017a, 43–4). Another key aspect is that the negative effects of economic crises in advanced economies are felt disproportionately by low- and middle-income households. This might be a problem for some Caribbean countries as low-cost tourism packages still make up a substantial portion of the Caribbean tourism market. While a focus on affordable all-inclusive vacation packages is likely to continue, the recent shift by leading hotel chains toward luxury all-inclusive offerings denote a strategy, in partnership with Caribbean authorities, to diversify not so much the tourism source markets but rather the type of customers.[36] The segment of the travel and hospitality industry that caters to wealthy people seeking exclusive and high-end travel experiences can be a profitable and reliable alternative to mass tourism.

To shed further light on the overreliance of the Caribbean tourism industry on the European and North American feeder markets, above all the United States, and to compare Cuba with other countries in the region, Figure 4.2 displays data on visitor arrivals to Caribbean countries by source market in 2014, the last full year before President Obama relaxed restrictions on US-based travel to Cuba. In 2014, the US market accounted for the lion's share of foreign international tourist arrivals to most destinations in the Caribbean, namely the US territories

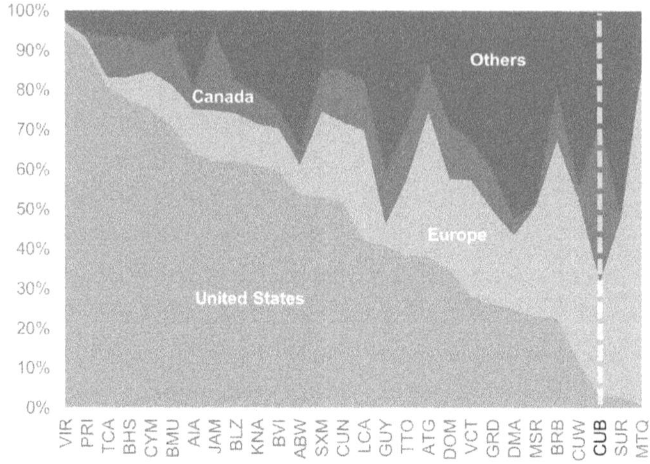

Figure 4.2 Visitor arrivals to Caribbean countries by source market in 2014. Note: List of codes: VIR (US Virgin Islands), PRI (Puerto Rico), TCA (Turks & Caicos), BHS (Bahamas), CYM (Cayman Islands), BMU (Bermuda), AIA (Anguilla), JAM (Jamaica), BLZ (Belize), KNA (St. Kitts & Nevis), BVI (British Virgin Islands), ABW (Aruba), SXM (St. Maarten), CUN (Cancún-Mexico), LCA (Saint Lucia), GUY (Guyana), TTO (Trinidad & Tobago), ATG (Antigua & Barbuda), DOM (Dominican Republic), VCT (St. Vincent & the Grenadines), GRD (Grenada), DMA (Dominica), MSR (Montserrat), BRB (Barbados), CUW (Curacao), CUB (Cuba), SUR (Suriname), MTQ (Martinique). Generated by the authors, based on CTO 2015; Werner and Alleyne 2015; ONEI 2015.

of Puerto Rico and US Virgin Islands (over 90 percent of inbound visitors were American citizens from the mainland), Turks and Caicos (about 80 percent), the Bahamas, Cayman Islands, and Bermuda (over 70 percent), Anguilla, Jamaica, Belize, and St. Kitts and Nevis (over 60 percent), and British Virgin Islands, Aruba, St. Maarten, and Cancún (over 50 percent). The remaining countries, among them the Dominican Republic and Barbados, had more diversified source market bases. European tourists had a major presence in Barbados and Curacao (over 40 percent of total arrivals), Antigua and Barbuda (over 30 percent), and in Cuba, Dominican Republic, Grenada, St. Lucia, St. Maarten, and Cancún (over 20 percent). Canadian tourists represented almost 40 percent of total visitors to Cuba, over 20 percent of total visitors to Jamaica, and over 10 percent of all visitors to Barbados, Dominican Republic, Trinidad and Tobago, Cancún, Turks and Caicos, and the Bahamas. As for Cuba, US citizens (excluding Cuban Americans) who traveled there in 2014 represented just 3 percent of total international arrivals, making Cuba one of the Caribbean countries least dependent on the United States as a tourism source market (CTO 2015; Werner and Alleyne 2015). Among the twenty-eight Caribbean destinations listed in Figure 4.2, only Suriname and Martinique had a lower reliance on US overnight visitors than Cuba. And even if Cuban Americans are included, Cuba received around 350,000 visitors from the United States in 2014, accounting for 11.6 percent of total international arrivals (ONEI 2016a). This share was still among the lowest in the Caribbean region.

When the Obama administration announced in early 2015 sweeping new rules to permit more travel by Americans to Cuba (yet trips for tourism purposes remained—and still are prohibited), some concerns emerged over the potential negative effects of these rules on the tourism sector of other Caribbean destinations.[37] In reality, despite subsequent US measures that reached their liberalizing peak in March 2016 when Americans were allowed to visit Cuba on their own for loosely defined educational activities,[38] and that ignited a remarkable surge in trips to the island by US citizens, the easing of restrictions on US-based-travel to Cuba by and large did not cut into the tourist traffic elsewhere in the Caribbean. It actually ended up encouraging US visits to the region as a whole. Between 2014 and 2017, as the annual volume of US visitors to Cuba quintupled (which translated into approximately 360,000 additional travelers per year), annual US overnight visitor arrivals in the Caribbean increased by more than 1.3 million. In 2019, the region welcomed 2.1 million more US tourists than five years earlier (CTO 2020, 2015).

As reported in Table 4.6, overnight US visitors (of non-Cuban descent) to Cuba increased from around 89,000 in 2014 to 448,000 in 2017, when they accounted for 11 percent of total international arrivals, before dropping to 298,000 in 2019 after the Trump administration rescinded several Obama-era policies toward the communist island (Pérez Villanueva and Perelló Cabrera 2020, 366).[39] During the 2014–17 period, competing Caribbean destinations like the Bahamas, Dominican Republic, Jamaica, and Puerto Rico all received growing numbers of US tourists. In the Bahamas and Jamaica, this trend continued until 2019. In the Dominican Republic, US tourist arrivals grew further in 2018 but fell noticeably the next year, largely because of safety concerns linked to a spate of US tourist fatalities in the country's resorts that grabbed a considerable amount of media attention.[40] Puerto Rico, for its part, witnessed a significant decrease in all international tourist arrivals in 2018 and struggled to recover in 2019 as it was reeling from two devastating hurricanes (UNWTO 2022b, 2020). Leaving aside Puerto Rico, the number of European tourists in the Bahamas, Dominican

Table 4.6 Overnight Visitors to Selected Caribbean Countries by Source Market, 2014–19 (thousands)

		2014	%	**2015**	%	**2016**	%	**2017**	%	**2018**	%	**2019**	%
Bahamas	USA	1,107	77.6	1,169	78.1	1,183	78.9	1,145	78.9	1,305	79.9	1,474	81.6
	Europe	83	5.8	102	6.8	110	7.4	116	8.0	122	7.5	119	6.6
	Canada	144	10.1	149	10.0	125	8.3	111	7.6	126	7.7	134	7.4
	Others	93	6.5	76	5.1	81	5.4	80	5.5	80	4.9	79	4.4
Cuba	USA[a]	89	3.0	160	4.6	189	5.0	448	11.0	297	7.7	298	7.8
	Europe	852	28.6	1,038	29.6	1,370	36.2	1,360	33.4	1,236	32.3	1,168	30.5
	Canada	1,175	39.4	1,300	37.0	1,205	31.8	1,132	27.8	1,095	28.6	1,112	29.0
	Others	867	29.0	1,013	28.8	1,025	27.0	1,136	27.8	1,206	31.4	1,250	32.7
Dominican Republic	USA	1,784	34.7	2,002	35.7	2,085	35.0	2,074	33.5	2,236	34.0	2,030	31.5
	Europe	1,184	23.0	1,134	20.3	1,262	21.2	1,431	23.1	1,377	21.0	1,327	20.6
	Canada	706	13.7	746	13.3	768	12.9	828	13.4	892	13.6	866	13.4
	Others	1,467	28.6	1,718	30.7	1,844	30.9	1,855	30.0	2,064	31.4	2,223	34.5
Jamaica	USA	1,296	62.3	1,344	63.3	1,406	64.4	1,510	64.2	1,628	65.8	1,839	68.6
	Europe	261	12.6	279	13.1	295	13.5	326	13.9	328	13.3	321	12.0
	Canada	420	20.2	391	18.5	372	17.1	405	17.2	400	16.2	396	14.8
	Others	103	4.9	109	5.1	109	5.0	112	4.7	117	4.7	125	4.6
Puerto Rico[b]	USA	1,485	90.8	1,588	91.0	1,599	89.1	1,642	90.4	1,030	85.3	1,362	91.0
	Europe	34	2.1	39	2.2	45	2.5	42	2.3	22	1.8	36	2.4
	Canada	23	1.4	22	1.3	20	1.1	20	1.1	13	1.1	18	1.2
	Others	93	5.7	96	5.5	131	7.3	112	6.2	143	11.8	81	5.4

Sources: UNWTO 2022b, 2020; Estimates of the authors from ONEI 2020a and MINTUR data obtained in May 2020; The Bahamas Ministry of Tourism, https://www.tourismtoday.com/statistics; Banco Central de la República Dominicana, https://www.bancentral.gov.do/a/d/2537-sector-turismo; Jamaica Tourist Board, https://www.jtbonline.org/report-and-statistics/.

Note:

[a] Exclude Cuban Americans.

[b] Non-resident tourists in hotels and similar establishments.

Republic, and Jamaica also increased between 2014 and 2019, while Canadian tourists grew in the Dominican Republic and remained virtually stable in the other two countries. Again, fears that the Obama administration's loosening of restrictions on Americans traveling to Cuba would cause disruption to the tourism industry of competing Caribbean countries turned out to be unfounded.

In all fairness, though, the volume of American tourists to Cuba unleashed by the removal of the tourism ban and of all additional restrictions on US-based travel to the island would certainly be greater than the spike in US visitors to Cuba triggered by Obama's measures. Thus, it could potentially entail a much bigger disruption to flows of US tourists to the Caribbean, with different effects across the region's destinations and over time. Some studies estimated that more than 1 million US citizens would visit Cuba in the short run if travel prohibitions were entirely abolished and between 2.7 million and 5.6 million would visit there annually once the market has fully adjusted (Acevedo and Wong 2017, Djiofack et al. 2015; González Núñez y Orro Fernández 2011; Romeu 2008; USITC 2007). No matter the actual timeline, the opening of Cuba to US tourism will be a major event since it will link for the first time in over six decades the region's largest potential recipient of tourists with the region's largest source market. It will bring enhanced competition and severe challenges to many tourism-dependent island countries of the Caribbean that for a long time have effectively enjoyed trade protection as a result of those restrictions. Because of their open borders and perfect trade integration with the mainland, the US Virgin Islands and Puerto Rico have benefited the most from this situation. But the very high costs of visiting Cuba caused by US sanctions also suggests that distant tourist destinations in the Caribbean may have received more American visitors than they would have without barriers to US-Cuba travel (Laframboise 2015; Romeu 2008). The expansion of the Cuban tourism industry with more visitors from the United States will threaten the US share of several Caribbean countries. Some will be more at risk than others even though there will be opportunities to gain from the redistribution of non-US tourists currently vacationing in Cuba, through cooperation, and especially from the continuous growth of overall tourism, including US tourism, to the Caribbean. After all, international tourist arrivals have increased almost everywhere in the region over the past three decades alongside the rapid growth of certain destinations like the Dominican Republic, Cuba, and Cancún.

Economists at the International Monetary Fund argued that the Caribbean countries facing the greatest challenges of open US-Cuba travel are those that rely primarily on the United States for their visitors, like the US Virgin Islands, Puerto Rico, Turks and Caicos, the Bahamas, Cayman Islands, Bermuda, Anguilla, and Jamaica, and those with an increased focus on the US market, such as the Dominican Republic and Antigua and Barbuda, among others.[41] Additional factors that could play a role in how successful certain countries will be in retaining US tourists include cultural and historical relations, as well as the physical distance, since Cuba's geographical proximity to the United States will be of great value, making it an easier, less time-consuming, and possibly a more affordable destination to visit for Americans.[42] Based on these parameters, the most vulnerable destinations would be Anguilla, Belize, St. Maarten, and the US Virgin Islands (Acevedo and Wong 2017, 71–2). Nevertheless, this does not mean that the other destinations are risk-free and that the most vulnerable are bound to suffer. Conditions in all of these countries will also depend on their ability to diversify their tourism source markets, reduce costs, improve the quality of services, provide alternatives to sun-and-beach tourism, promote multi-destination marketing and other collaborative approaches, and eventually absorb a portion of displaced Canadian and European tourists originally going to Cuba. Using a gravity model, the IMF economists found that a full removal of US-Cuba travel restrictions would significantly boost the flow of US tourists to the island (up to 5.6 million a year) and that the growth would be driven mostly by new travelers, which implies that aggregate tourism that flows to the Caribbean will increase. Taking into account that an estimated 1.5 million non-US tourists would leave Cuba for other places in the region, that an expansion of US-Cuba flights is not expected to undermine the availability of US flights to other countries, and that the whole process should be moving rather slowly, the IMF economists concluded that the elimination of all barriers to US-Cuba travel will not necessarily bring great disruption to the Caribbean tourism industry (Acevedo, Alleyne, and Romeu 2017b; Acevedo and Wong 2017, 74–5).

Needless to say, the United States is also the top source market for Caribbean cruise tourism, accounting for over 90 percent of total cruise passengers in the region (Honey 2018b, 48). With the lifting of the tourism ban, and after some ongoing lawsuits against US cruise lines under Title III of the Helms-Burton law are resolved,[43] Cuba will be included, again, in the cruise routes of US-based ships departing from the ports of Miami, Fort Lauderdale, and Tampa. Elaborated just a few years before the launch of the US-Cuba rapprochement under the Obama administration, the 2035 Master Plan of the port of Miami predicted that the opening of Cuba to cruise tourism would stimulate the development of shorter (3–5 days) pattern sailings from South Florida due to the proximity of key Cuban ports. Yet it specified that US cruise liners, unlike their European counterparts, would probably use Cuba as a port-of-call rather than a homeport operation.[44] Cuba might be easily added to schedules involving itineraries to nearby destinations (especially in the Western Caribbean) like the Bahamas, Cayman Islands, Jamaica, Dominican Republic, Puerto Rico, and the Yucatan Peninsula, thus providing a great opportunity for the region as a whole. There are some key challenges, though. Given that Cuba is a large island with multiple ports and attractions, the likelihood that cruise lines will drop some existing destinations to focus on all-Cuba itineraries is real. In fact, as Cuba started to become a popular cruise destination during the 2016–19 period, various cruise lines quickly modified their regional travel itineraries. For instance, in early 2017, Norwegian decided to drop its Grand Bahama stopover in favor of an overnight stop in Havana with its 2,000-passenger ship Norwegian Sky.[45] Moreover, the presence of Cuba in multi-destination cruises near the United States and attendant savings in fuel costs might negatively affect Southern Caribbean markets (Perelló Cabrera 2017; CHTA 2015). And most important of all, the emergence of Cuba as a truly competitive cruise destination in the Caribbean is contingent on the country's ability to carry out the major port infrastructure upgrades required by today's cruise ships that keep getting bigger and bigger (Fitzgerald 2017).[46]

Infrastructure and Revenues

Since the 1980s, significant inflows of foreign direct investment have helped tourism become a key contributor to economic growth and a crucial generator of foreign exchange and employment in the Caribbean. Along with sizable foreign investment in airports and other infrastructure for travelers, numerous international hotel chains have taken advantage of fiscal incentives to set up operations in the Caribbean, investing in new properties and remodeling and refurbishing existing ones. As a result, the hotel investment industry in the region witnessed remarkable growth and transformation. Today's hotel offer in the Caribbean is quite diverse and includes all-inclusive beach resorts, full-service hotels, integrated hotels and real estate complexes, condos and timeshares, vacation clubs, private villas, and more modest inns and bed and breakfasts. Over the past couple of decades, as various countries stepped up efforts to move away from mass tourism and promote more exclusive and costly tourism products, large hotel chains have invested heavily in major projects that combine luxury hotel infrastructure with real estate and golf courses, and, more recently, in luxury all-inclusive resorts offering a broad array of high-end leisure options (CEPAL 2019, 66, 2008a, 126). First introduced in the 1950s in Spain by the French company Club Méditerranée, or Club Med, the all-inclusive concept (pay one rate for lodging, meals, drinks, and select activities) appeared in the Caribbean in the second half of the 1970s, prospering initially in Jamaica thanks to Jamaican hotel firms like Sandals and Superclubs (Issa and Jayawardena 2003, 168; Ambrosie 2015, 19–20). Adopted by other major hotel chains and surviving various tourism trends, all-inclusive remains very popular today as it dominates many beach areas in the Caribbean. However, all-inclusive resorts in the Caribbean region have evolved considerably, targeting different types of holidaymakers such as couples, singles, families, and adults-only, offering a remarkable number of different vacation choices and experiences to travelers, and covering a wide range of budgets, from quite-affordable stays to extremely expensive luxury retreats.[47]

As substantial financial resources were devoted to boosting new accommodation capacity in hotels and similar facilities to meet rising tourism demand, the Caribbean's hotel room stock almost quadrupled during the 1980–2019 period. It grew at its fastest pace in the 1980s and 1990s, when all Caribbean countries, except Barbados and Haiti, and with the Dominican Republic and Cuba as the biggest contributors, significantly increased their supply of hotel rooms (González Núñez 2022, 110). In absolute terms, starting from a base of just 77,000 rooms in 1980, the region's hotel room stock incorporated about 47,000 new rooms in the 1980s, a record 84,000 in the 1990s, some 40,000 in the 2000s, and over 50,000 in the 2010s to reach a total of around 300,000 hotel rooms available for international tourists at the end of 2019. It should also be underscored that during this entire period, while there were notable differences at the country level and in different decades, the absolute percentage growth of the Caribbean's hotel room stock virtually mirrored that of total overnight visitors to the region.[48]

To underscore the investment efforts of individual countries to augment their lodging capacity, Table 4.7 provides data on the Caribbean's total hotel rooms by country between 1995 and 2019. During this period, the Dominican Republic and Cuba continued to be the largest suppliers of new hotel rooms in

Table 4.7 Hotel Rooms of Caribbean Countries, 1995–2019

	1995	2000	2005	2010	2014	2019
Anguilla	951	1,067	746	720	—	1,566
Antigua and Barbuda	3,317	3,185	3,276	5,657	5,815	6,841
Aruba	6,266	7,783	7,966	9,856	11,075	11,929
Bahamas	13,421	14,701	14,800	15,236	15,300	17,627
Barbados	5,084	6,456	6,353	6,659	6,324	6,573
Bermuda	4,141	3,339	3,067	2,691	2,415	2,434
British Virgin Islands	1,452	2,541	1,802	2,183	2,195	2,240
Cayman Islands	3,579	5,364	2,954	4,587	5,264	7,098
Cuba	27,256	37,178	45,999	56,432	62,090	74,211
Curacao	2,601	2,941	3,647	5,562	6,300	7,908
Dominica	588	890	787	1,036	1,141	645
Dominican Republic	32,846	51,916	60,088	66,968	69,607	85,694
Grenada	1,652	1,822	1,470	1,937	1,968	2,081
Guadeloupe	7,917	8,136	3,506	—	2,826	—
Haiti	1,758	1,758	—	—	—	—
Jamaica	17,649	19,908	22,528	24,667	24,140	29,634
Martinique	5,090	6,768	6,153	5,772	5,189	4,481
Montserrat	710	264	—	344	—	—
Puerto Rico	10,251	11,062	13,336	13,934	14,781	14,619
St. Kitts and Nevis	1,402	1,602	1,859	1,754	—	2,624
St. Lucia	3,974	4,428	4,511	4,735	4,847	5,057
St. Maarten	3,707	3,545	2,964	3,580	3,845	3,725
St. Vincent and Grenadines	1,176	1,747	1,692	2,113	2,133	2,181
Trinidad and Tobago	3,107	4,532	4,970	6,803	7,768	7,731
US Virgin Islands	5,154	5,008	4,761	5,008	4,975	2,526
Total	165,049	207,941	219,235	248,234	259,998	299,425

Sources: MINTUR data obtained by the authors in May 2022; MINTUR 2017, 2015a; UNWTO 2023c; United Nations World Tourism Organization, "145 Key Tourism Statistics," https://www.unwto.org/tourism-statistics/key-tourism-statistics.

the Caribbean. The Dominican Republic's hotel room stock soared dramatically from 32,846 rooms in 1995 to 85,694 rooms in 2019, an increase of nearly 53,000 units or an average of 2,200 new units per year. Similarly, Cuba's hotel room stock grew from 27,256 rooms to 74,211 rooms, an increase of some 47,000 units or an average of almost 2,000 new units per year. As a result, the combined share of hotel rooms in the Dominican Republic and in Cuba within the total room stock of the Caribbean region rose steadily from 36 percent in 1995 to 48.4 percent in 2005 and to 53 percent in 2019. Other countries invested heavily as well in the expansion of their hotel capacity. Among the largest suppliers of new hotel rooms, Jamaica added approximately 500 rooms per year or a total of 11,985 units to its stock between 1995 and 2019, followed by Aruba (5,663 new

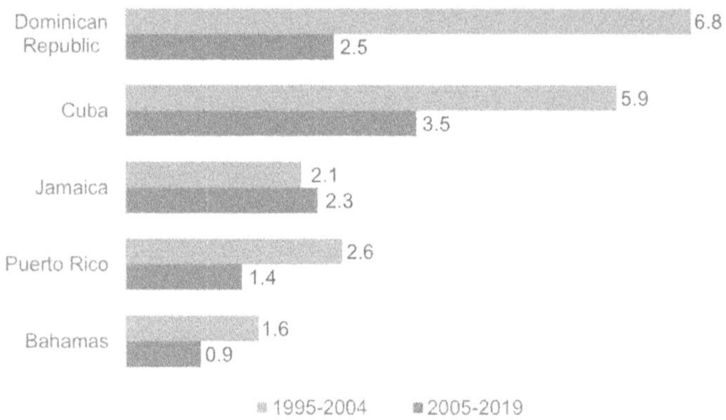

Figure 4.3 Average annual growth of hotel rooms in selected Caribbean countries, 1995–2019. MINTUR data obtained by the authors in May 2022; UNWTO 2023c; United Nations World Tourism Organization, "145 Key Tourism Statistics," https://www.unwto.org/tourism-statistics/key-tourism-statistics.

rooms), Curacao (5,307), Trinidad and Tobago (4,624), Puerto Rico (4,368), the Bahamas (4,206), Antigua and Barbuda (3,524), and the Cayman Islands (3,519). Once again, almost all Caribbean countries significantly increased their stock of hotel rooms, with only a few countries failing to do so because of hurricane damages that forced the closures of hotels and other lodging facilities. In any case, at the end of 2019, underlining the high concentration of hotel rooms, the five Caribbean countries with the largest tourism sectors in the region, namely the Dominican Republic, Cuba, Puerto Rico, Jamaica, and the Bahamas, boasted a combined 221,785 hotel rooms, representing close to three-fourths of the Caribbean's total rooms for international tourists (UNWTO 2023c).

With respect to the hotel infrastructure, Figure 4.3 compares the pace of expansion in room supply in the five leading Caribbean tourism destinations (excluding Cancún) between 1995 and 2019. It separates the 1995–2004 period from the 2005–19 period because it was around the mid-2000s when Cuba moved decisively to recentralize state control over its tourism sector. It was also the time when the Cuban military-run Gaviota group, clearly enjoying far more liquidity than the two major civilian-run corporations (Cubanacán and Gran Caribe) under the control of the Cuban Ministry of Tourism, began to assume a dominant role in financing hotel construction projects in Havana and other choice locations across Cuba (Figueras Pérez 2019; LeoGrande 2023, 177–8).[49] From 1995 to 2004, the average annual growth of hotel rooms in the Dominican Republic was 6.8 percent, the highest rate among the selected countries. In the midst of a dramatic surge in international visitor arrivals (an annual average growth of 13.2 percent in 1995–2004 as compared to 8.1 percent in the Dominican Republic), Cuba's hotel room stock also rose markedly at an annual average rate of 5.8 percent. The room supply in Puerto Rico (2.6 percent annual average expansion), Jamaica (2.1 percent), and the Bahamas (1.6 percent) grew at a much slower pace.

From 2005 to 2019, Cuba outpaced all its competitors in building new hotels for tourists. The annual average growth rate of hotel rooms in Cuba was 3.5 percent, as compared to 2.5 percent in the Dominican Republic, 2.3 percent in Jamaica, 1.4 percent in Puerto Rico, and 0.9 percent in the Bahamas. During this period, however, annual visitor arrivals in Cuba increased practically at the same average rate (4.4 percent versus 4.3 percent) as those in the Dominican Republic. Furthermore, international arrivals in Cuba received a critical boost in the Obama era from a short-lived spike in US visitors who were supposed to stay away from beach areas because of embargo rules and, for the most part, preferred private bed and breakfasts to hotels. In July 2018, amid falling arrivals of US overnight visitors (and European as well) due to stiffer travel restrictions imposed by the Trump administration, and with hotel occupancy rates on the island that barely reached 60 percent in good years since a very large number of rooms were either out of order or in very poor conditions, Gaviota announced plans to considerably expand its hotel capacity.[50] Raising serious doubts as to whether it was economically sound and financially sustainable to keep investing in new hotels given the sluggish performance of Cuba's tourism sector and the shortcomings of the existing infrastructure, Gaviota's hotel building spree, which continues to this day, would become even harder to comprehend in the Covid-19 era. As we will see in the next chapter, the pandemic crippled tourism and, in tandem with other factors, exacerbated Cuba's economic problems, wiping out the cash needed to import basic goods and pushing the island's economy into a deep crisis.

As a matter of fact, a critical contribution of tourism is its ability to generate foreign exchange earnings, which is particularly important in developing countries. As reported in Table 4.8, the Caribbean region's total gross revenues from tourism activities nearly doubled from about $8.8 billion in 1990 to $17.2 billion in 2000.[51] During this period, all Caribbean countries saw their income from tourism jump conspicuously except Antigua and Barbuda, Bermuda, Curacao, and St. Kitts and Nevis. Moreover, the Dominican Republic (tourism income of around $2.9 billion in 2000) emerged as the regional leader in terms

Table 4.8 Gross Tourism Revenues of Caribbean Countries, 1990–2014 ($US millions)

	1990	1995	2000	2005	2010	2014
Anguilla	35	48	55	86	99	136
Antigua and Barbuda	298	247	290	309	298	708
Aruba	353	521	807	1,095	1,242	1,736
Bahamas	1,333	1,346	1,734	2,069	2,147	2,316
Barbados	493	612	777	887	1,022	887
Bermuda	490	487	421	394	385	407
British Virgin Islands	133	195	367	413	389	459
Cayman Islands	236	489	559	371	404	565
Cuba	243	1,100	1,948	2,399	2,218	2,546
Curacao	238	175	220	244	385	820
Dominica	25	34	47	56	94	216
Dominican Republic	899	1,571	2,860	3,518	4,209	5,630
Grenada	37	58	178	94	97	388
Guadeloupe	197	380	454	306	510	—
Haiti	46	90	128	79	167	578
Jamaica	740	1,068	1,333	1,545	2,001	2,255
Martinique	240	384	302	282	389	483
Montserrat	7	17	9	9	6	9
Puerto Rico	1,377	1,828	2,388	3,239	3,211	3,439
St. Kitts and Nevis	58	65	58	121	89	110
St. Lucia	154	266	280	342	562	778
St. Maarten	316	349	483	613	681	922
St. Vincent and Grenadines	56	41	82	104	86	175
Trinidad and Tobago	95	72	213	453	449	447
US Virgin Islands	697	822	1,206	1,432	1,012	1,319
Total	8,796	12,484	17,199	20,883	23,274	27,329

Sources: ONEI 2015a, 2011a, 2006, 1998; UNWTO 2016, 2015, 2001, 1997; Caribbean Tourism Organization, "Historical Data (1970–2015)," https://www.onecaribbean.org/statistics/historical-data-1970-2015/; United Nations World Tourism Organization, "145 Key Tourism Statistics," https://www.unwto.org/tourism-statistics/key-tourism-statistics.

of tourism revenues, replacing Puerto Rico ($2.4 billion). Cuba ($1.9 billion) also improved its ranking, moving to third place ahead of the Bahamas ($1.7 billion) and Jamaica ($1.3 billion). The Caribbean's total tourism earnings grew more slowly in the 2000s but picked up pace the next decade as the tourism sector of a number of countries recovered strongly from the prolonged slump in the wake of the global financial crisis of 2007–8. In 2014, the region's annual gross tourism revenues were roughly $27.3 billion, more than three times higher than their level in 1990. The Dominican Republic confirmed itself as the top earner of tourism revenues in the Caribbean with approximately $5.6 billion in tourism receipts, followed by Puerto Rico ($3.4 billion), Cuba ($2.5 billion), the Bahamas ($2.3 billion), and Jamaica ($2.2 billion) (UNWTO 2016).

Table 4.9 Gross Tourism Revenues of Caribbean Countries, 2015–2019 ($US millions)

	2015	2016	2017	2018	2019
Anguilla	142	147	153	127	177
Antigua and Barbuda	802	844	820	855	994
Aruba	1,659	1,757	1,855	2,035	2,109
Bahamas	2,554	3,091	2,951	3,756	4,150
Barbados	974	1,071	1,081	1,118	1,250
Bermuda	370	440	513	583	601
British Virgin Islands	484	–	–	–	–
Cayman Islands	680	708	832	911	964
Cuba	2,819	3,069	3,302	2,782	2,645
Curacao	712	644	572	604	706
Dominica	202	197	161	88	135
Dominican Republic	6,116	6,720	7,184	7,548	7,472
Grenada	422	437	482	522	562
Guadeloupe	614	—	—	850	—
Haiti	609	549	588	620	450
Jamaica	2,401	2,539	2,809	3,099	3,599
Martinique	340	365	480	533	549
Montserrat	9	9	9	12	12
Puerto Rico	3,825	3,974	3,868	3,303	3,612
St. Kitts and Nevis	317	332	352	366	353
St. Lucia	810	776	875	963	1,064
St. Maarten	913	871	646	468	708
St. Vincent and Grenadines	207	216	210	239	242
Trinidad and Tobago	531	464	453	429	436
US Virgin Islands	1,324	1,343	1,202	993	1,031
Total	30,054	30,563	31,918	32,804	34,993

Sources: ONEI 2020a; UNWTO 2021a, 2020, 2019, 2018, 2017; United Nations World Tourism Organization, "145 Key Tourism Statistics," https://www.unwto.org/tourism-statistics/key-tourism-statistics.

In the second half of the 2010s, the Caribbean's total annual gross tourism revenues grew even faster than in the first half of that decade, reaching about $35 billion in 2019. Yet, against this general trend, St. Maarten recorded a significant drop in tourism earnings after it took a direct hit from hurricane Irma in early September 2017. The same happened to the US Virgin Islands, also hit by hurricane Maria just a couple of weeks later. Haiti's tourism sector performed poorly, especially in 2019, as the United States, Canada, and France issued travel advisories urging their citizens not to visit Haiti due to civil unrest, kidnappings, and crime.[52] Additionally, Cuba and Puerto Rico experienced a contraction in tourism receipts, Puerto Rico because of the widespread destruction and exacerbating challenges caused by the same hurricanes that struck St. Maarten and the US Virgin Islands, and Cuba because of the damaging effects of Trump's sanctions that came on top of widespread inefficiencies. Puerto Rico's gross tourism revenues plummeted 14.7 percent from some $3.9 billion in 2017 to $3.3 billion in 2018 before rebounding in 2019. Cuba's hard currency earnings from international visitors peaked at $3.3 billion in 2017 but fell 15.7 percent in 2018 to less than $2.8 billion and dropped an additional 5 percent in 2019 (ONEI 2020a). Most notably, Cuba's tourism earnings in 2019 stood close to their level in 2014, even if the island received about 845,000 more overnight visitors than in 2014. Conversely, among the best performances during the 2015–19 period, annual gross tourism revenues increased, on average, 21.5 percent in the Cayman Islands, 12.9 percent in the Bahamas, 9.8 percent in Jamaica, 6.6 percent in St. Lucia, 5.9 percent in the Dominican Republic despite a small drop in 2019, and 4.1 percent in Aruba. In 2019, the Dominican Republic remained, by far, the Caribbean's top tourism earner with around $7.5 billion worth of revenues, followed by the Bahamas with $4.1 billion, Puerto Rico and Jamaica with practically the same amount close to $3.6 billion, and Cuba with a little over $2.6 billion (UNWTO 2023a, 2016).

Given their importance in the Caribbean tourism market, and to highlight some key differences, the five Caribbean countries with the largest tourism sectors should be discussed further. As a group, they have been gaining market share since 1990 in terms of tourists received and income from tourism activities, but with one critical distinction. Their market share of total inbound overnight visitors to the Caribbean grew faster than their share of the region's total tourism revenues. More specifically, these five countries' ratio of total international visitors to the Caribbean increased from 58.6 percent in 1990 to 62.7 percent in 2000, 66.7 percent in 2010, and 68.1 percent in 2019. During the same period, their ratio of the region's total tourism receipts rose from 52.2 percent in 1990 to 59.7 percent in 2000, and afterward, it showed little variation as it stood at 59.2 percent in 2010 and 60.8 percent in 2019.[53] All of this suggests that the most visited countries with the largest tourism sectors in the Caribbean, while offering a range of accommodations and products for mass and elite tourism, continue to offer less expensive stays and vacation packages and target foreign travelers with lower disposable income than other (smaller) Caribbean destinations. As noted before, the luxury travel and tourism market, with its unique, hyper-personalized, and exclusive experiences for high-income holidaymakers, is a growing segment of Caribbean tourism. Over the past couple of decades, the Caribbean market has also expanded from traditional sun-and-beach vacation tourism to include niche tourism sectors like ecotourism, all-inclusive wellness resorts, adventure tourism, agritourism, and medical tourism, among others (CEPAL 2008a, 123). When it comes to luxury and niche tourism, small Caribbean islands still have an edge on the region's largest destinations, but the latter have been making inroads into the high-end tourism segment in recent years.[54]

In effect, a closer look at the Caribbean's five largest tourism markets during the 1990–2019 period reveals quite different individual performances with some notable achievements. On the one hand, the Dominican Republic was the only country that steadily improved its relative position within the Caribbean with respect to both market indicators previously described. Between 1990 and 2019, the Dominican Republic increased its share of total international visitor arrivals in the Caribbean from 13.1 percent to 24.5 percent, and its share of the region's total tourism receipts from 10.2 percent to 21.1 percent. The fact that the latter indicator grew more in relative terms than the former implies that over the analyzed period, tourism receipts per capita in the Dominican Republic increased more than the Caribbean average. The same can be said for the Bahamas (where tourism expenditures per capita, nonetheless, have traditionally been above the Caribbean average) and Jamaica, but for more recent periods. Between 2000 and 2019, the Bahamas' share of the Caribbean's total tourism receipts rose from 10.1 percent to 11.7 percent while its share of the region's total visitors declined from 8.8 percent to 6.9 percent. Between 2010 and 2019, Jamaica's share of the Caribbean's total tourism revenues jumped from 8.6 percent to 10.2 percent while its share of the region's total inbound visitors remained virtually constant at around 10 percent. On the other hand, Puerto Rico experienced a deterioration of both market indicators, but its share of the Caribbean's total tourism revenues decreased from 15.6 percent in 1990 to 10.3 percent in 2019, while its share of the region's total visitor arrivals plunged more markedly from 22 percent to 12.1 percent, which means that tourist spending per capita still grew during that period.[55] As for Cuba, the fall of its share of the Caribbean's total tourism receipts during the 2000–2019 period from 11.3 percent to 7.5 percent was quite consequential given that it coincided with an increase in its share of total arrivals to the region from 9.9 percent to 14.5 percent.[56] In simple words, when it comes to tapping tourism segments that generally produce higher revenues, the Dominican Republic, the Bahamas, and Jamaica have made significant progress, Puerto Rico has performed below the Caribbean average, and Cuba is lagging far behind all of its principal competitors.

Finally, albeit not as lucrative for the country of destination as overnight tourism, cruise tourism can provide substantial economic benefits to a local economy. In all fairness, cruise tourists generate as little as one-tenth or less the revenues generated by overnight visitors, as they leave, on average, between 20 and 30 dollars in destinations with poor supporting facilities and as much as 150–200 dollars in destinations with more adequate infrastructure, organization, and tourism services. Cruise line private islands, in particular, contribute very little to the local economy (Figueras Pérez and Pajón Espina 2021, 121; Honey 2018c, 133). Additionally, the vast majority of income generated by cruise activities goes to the cruise companies in the form of ticket sales and onboard purchases by passengers as major cruise lines have been adding more opportunities for onboard discretionary spending.[57] Still, even though some Caribbean countries (along with concerns about the cruise industry's negative environmental and health impacts) have been complaining about the minimal onshore spending by today's cruise ship passengers,[58] cruise tourism contributes to destination economies through onshore passenger expenditures, mainly in excursions and retail purchases, onshore expenditures by crew, and expenditures by the cruise lines for supplies, port services, and port fees and taxes. Furthermore, cruise activities positively affect tourism destinations by requiring a modern port infrastructure, upgraded services and facilities to provide comfortable shopping and entertainment amenities to passengers, and by creating substantial employment (Perelló Cabrera 2017). Carried out by the Business Research and Economic Advisors (BREA) on behalf of the Florida-Caribbean Cruise Association (FCCA), a 2018 study assessed the impact of cruise tourism on the economies of thirty-six destinations located in the Caribbean (the vast majority), Mexico, and Central and South America.[59] The BREA study estimated that for the 2017–18 season, cruise tourism generated close to $3.4 billion in direct expenditures, 79,000 jobs, and $903 million in employee wages. Passenger expenditures accounted for over three-quarters of total direct expenditures (BREA 2018).[60]

To give a sense of the economic impact of cruise tourism on the economies of the most popular Caribbean destinations, Table 4.10 collects data from the aforementioned BREA study on cruise tourism spending in each of the region's top twelve destinations during the 2017–18 cruise season. Cozumel (Mexico) was the leader with $474.1 million in total direct expenditures, of which $378 million were passenger expenditures, $45.5 million were crew expenditures, and $50.6 million were cruise line expenditures. The Bahamas ranked second

Table 4.10 Cruise Tourism Expenditures in Principal Caribbean Destinations, 2018[a] ($US millions)

	Passenger Expenditures	Average Spending per Passenger (dollars)	Crew Expenditures	Cruise Line Expenditures	Total Expenditures
Cozumel (Mexico)	378.0	108.4	45.5	50.6	474.1
Bahamas[b]	322.6	131.9	28.9	54.2	405.7
Jamaica[c]	205.6	116.0	12.8	26.1	244.5
Cayman Islands	172.1	105.2	17.5	34.9	224.5
US Virgin Islands	156.1	165.4	10.5	18.1	184.7
Puerto Rico (San Juan)[c]	100.3	90.8	22.6	28.3	151.2
St. Kitts and Nevis	133.0	135.4	8.7	7.6	149.3
St. Maarten	111.5	142.2	17.8	13.9	143.2
Dominican Republic[c]	77.8	88.3	4.0	52.9	134.7
Aruba	82.5	121.9	9.1	11.1	102.7
Turks and Caicos	66.8	95.5	7.4	12.3	86.5
Belize	68.3	77.9	5.1	12.7	86.1

Source: BREA 2018.

Notes:

[a] May 2017 through April 2018.

[b] Only includes visits to Nassau and Freeport.

[c] Includes passengers and crew aboard both port-of-call and homeport cruise ships.

with $405.7 million in total spending, followed by Jamaica ($244.5 million), Cayman Islands ($224.5 million), US Virgin Islands ($184.7 million), and San Juan, Puerto Rico ($151.2 million). Most telling, given the key role of onshore expenditures by passengers, there were significant differences in terms of spending per passenger. The US Virgin Islands led all destinations with an average spending per passenger of $165.4, followed by St. Maarten ($142.2), St. Kitts and Nevis ($135.4), the Bahamas ($131.9), Aruba ($121.9), Jamaica ($116), Cozumel ($108.4), and Cayman Islands ($105.2). In the Dominican Republic, Puerto Rico, and the remaining destinations, cruise passengers spent onshore an average of less than $100 (BREA 2018). It should be noted that total cruise expenditures in the Bahamas are underreported since the BREA study includes only visits to Nassau and Freeport, and that the average spending per passenger in the Bahamas is likely to be lower than reported since the study did not collect data on private islands owned by cruise lines where onshore spending is minimal. Regarding Cuba, which is not on the list of destinations surveyed by BREA, the US-Cuba Trade and Economic Council estimated in early 2018 that cruise lines would deliver over 624,000 cruise passengers to Cuba from 2017 until 2020, generating $89 million in revenues for Cuba from passenger onshore spending (an average of about $140 per person) and an additional $23 million from port fees. The cruise lines' gross revenues instead were projected to exceed $834 million (USCTEC 2018). In short, there is little doubt that cruise tourism is a particularly lucrative business for cruise lines, but regardless, its economic impact on certain Caribbean destinations is quite significant.

Key Performance Indicators

Contribution to GDP, total exports of goods and services, and employment, trends in overnight and cruise visitor arrivals, and expenditures by international inbound visitors (as well as international transportation receipts) are important indicators of the strength of a country's tourism sector. There are several additional indicators that are used to evaluate the performance of the tourism industry and its competitiveness, but accurate and meaningful comparisons between Caribbean countries are problematic given the difficulty of finding reliable and consistent data. At any rate, this section collects available data from the United Nations World Tourism Organization, the Caribbean Tourism Organization, and official sources from selected countries to shed further light on the Caribbean tourism performance, with a focus on the region's largest tourism markets. It provides data on and analyzes four critical indicators, namely the length of stay of international visitors, hotel occupancy rates, gross revenues per room, and gross revenues per tourist.

The length of stay is a key parameter that determines the number of guest nights (days) of inbound stopover visitors in accommodation establishments. It is one of the most useful indicators utilized to assess tourism demand along with visitor arrivals, visitor expenditures, and visitor profile statistics. Longer visitor stays in a particular tourist destination are generally correlated with higher hotel occupancy rates and increased revenues. While the pace of growth of hotel rooms is also a factor, the relationship between the duration of tourist trips and hotel occupancy levels is straightforward, as more nights spent by tourists in hotels allow the latter to maximize their occupancy. Moreover, an increased length of stay has a substantial impact on the economic contribution made by tourists since it allows them to undertake a higher number of experiences and activities that should stimulate their overall spending. Longer stays may also result in enhanced guest loyalty that can foster repeat bookings and travel, may allow hotels to forecast their earnings and occupancy rates more accurately, and may lead to cost savings for hotels in various areas. Put simply, longer visitor stays offer a number of advantages that can boost considerably hotel earnings and profitability (Davies and Mangan 1992; Gomes de Menezes, Moniz, and Cabral Vieira 2008).

A historical time series data on the average length of stay (ALOS) of international visitors to the Caribbean (regional average) is not available. Yet, information from the Caribbean Tourism Organization suggests that the ALOS in the region has remained close to 7 days for the past decade or so (CTO 2021, 2015).[61] Based on calculations of the authors from UNWTO statistics and official data from individual countries, Figure 4.4 compares the average length of stay of tourists in hotels in the Caribbean's leading tourism destinations between 1995 and 2019. Before analyzing the information, though, a few important things must be specified about the Cuban data included in this figure. The values of ALOS of international visitors to Cuba are significantly lower than those reported in other parts of this book (annual values for the 2014–22 period) because, as a matter of consistency, they are calculated using only the total number of nights spent by guests in Cuban hotels and similar establishments. In 2015, Cuba's Office of National Statistics and Information started to add overnight stays by foreign tourists in private accommodations to its counts of total guest nights of

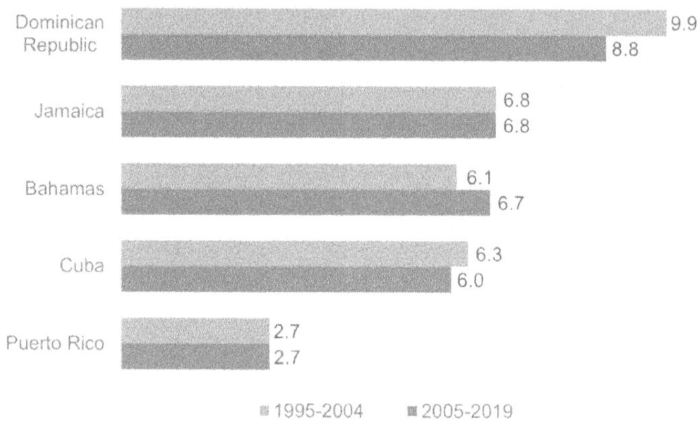

Figure 4.4 Average length of stay of tourists in selected Caribbean countries, 1995–2019. Calculations of the authors from ONEI 2020a, 2015a, 2011a, 2006, 1998; Banco Central de la Republica Dominicana, "Estadísticas: Sector Turismo," https://www.bancentral.gov.do/a/d/2537-sector-turismo; The Bahamas Ministry of Tourism, https://www.tourismtoday.com/statistics; United Nations World Tourism Organization, "145 Key Tourism Statistics," https://www.unwto.org/tourism-statistics/key-tourism-statistics. Note: Hotels only.

international visitors (ONEI 2015a). Since ONEI revised its annual data back to 2012 with the added counts of overnight stays in *casas particulares* but has continued to publish time series with data back to 1985 and a sharp increase in total guest nights from 2012 onwards,[62] the only way to meaningfully compare the ALOS of visitors to Cuba before and after that year is to focus only on overnight stays in hotels.

Noticeably, with respect to the average length of stay of international tourists in hotels between 1995 and 2019, the Dominican Republic was well ahead of its principal competitors. The ALOS of inbound visitors to the Dominican Republic during the 1995–2004 period was 9.9 days, considerably longer than the ALOS of visitors to Jamaica (6.8 days), the Bahamas (6.7 days), Cuba (6.3 days), and Puerto Rico (2.7 days). The performance of the Dominican Republic was particularly impressive given that international overnight visitor arrivals virtually doubled during this period (UNWTO 2006, 1997). In 2005–19, the Bahamas was the only major Caribbean destination to record a longer ALOS (6.7 days) of its tourists than in the previous period, but it still trailed the ALOS of tourists in Jamaica (6.8 days) and the Dominican Republic (8.8 days). The ALOS of tourists in Cuba shrank slightly to 6 days while in Puerto Rico it remained stable at 2.7 days. It is worth noting that the decreased ALOS of tourists in the Dominican Republic in 2005–19, although still longer than in competing destinations, most likely was the result of growing arrivals of US stopover visitors who generally tend to have shorter stays than Canadian and European vacationers.[63] By the same token, it is telling that the ALOS of tourists in Cuba's hotels was shorter than the ALOS of tourists in hotels in the Dominican Republic, Jamaica, and the Bahamas, even if Canadians and Europeans represented a much larger share of total inbound tourists in Cuba than they did in the other countries. Only tourists in Puerto Rico had a shorter average length of stay than those in Cuba, but that was because many hotels in Puerto Rico tend to cater to short-stay American visitors who can travel to the US territory without a passport.

Another important tourism indicator is the hotel occupancy rate, which indicates the percentage of total hotel rooms that are occupied at any given time. This directly affects earnings and overall profitability. In fact, hotel occupancy offers valuable insights into the efficiency of hotel operations, revealing how effectively the hotel is utilizing its capacity and maximizing revenues. Tracking occupancy rates over time helps to identify changes in tourism demand and implement strategic decisions to increase profitability as well as to forecast future occupancy patterns. Mainly due to the region's dependence on leisure travel, international tourism in the Caribbean exhibits a highly seasonal pattern reflected in the volatility of the average hotel occupancy rates throughout the year. Typically, hotel occupancy in the Caribbean reaches its highest levels during the first four months of the year that coincide with the cold winter months in the United States, Canada, and Europe. Occupancy then declines through late spring and early summer (with some upward fluctuations in July-August) to hit its lowest levels in September and October when the risk of hurricanes increases considerably and grows again during the last two months of the year.[64] Recently, though, when Covid-19 complicated summer trips from the United States to Europe, the Caribbean reportedly started to be reconsidered as a year-round tourism destination whose attractiveness was also helped by lower prices and special travel packages during the summer season along with a growing demand for luxury islands.[65]

Hotel room occupancy rates in the Caribbean, on average, have exhibited relatively little variation over the past three decades, mostly hovering around 60 percent to 65 percent except for a few years (primarily from 2012 to 2017) when they stood slightly above 65 percent (WTTC 2004; CEPAL 2008b; CTO 2013, 2015, 2016, 2018).[66] As displayed in Figure 4.5, the Dominican Republic

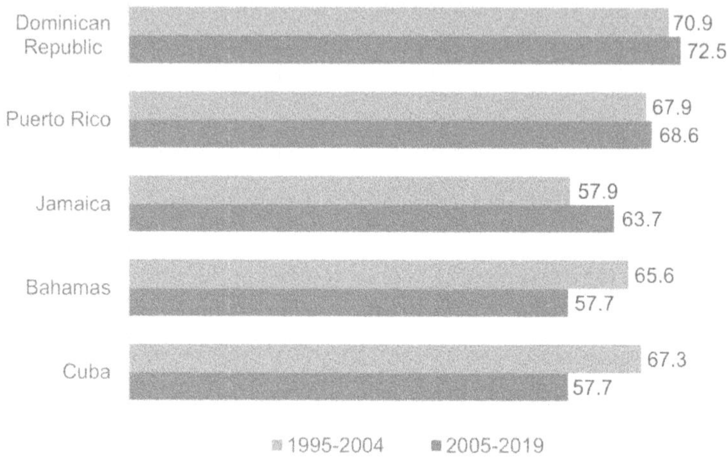

Figure 4.5 Average hotel occupancy rates in selected Caribbean countries, 1995–2019. Generated by the authors, based on ONEI 2020a, 2015a, 2011a, 2006, 1998; United Nations World Tourism Organization, "145 Key Tourism Statistics," https://www.unwto.org/tourism-statistics/key-tourism-statistics.

outperformed its principal competitors in terms of utilization of accommodation capacity, doing so while building new hotels at a very fast pace. Mainly the result of mass-market tourism strategies of affordable prices at large all-inclusive beach resorts (Morales, Barinas, Ramírez, and Martinez 2021), the average hotel occupancy rate in the Dominican Republic was 70.9 percent during the 1995–2004 period, substantially higher than the rate in Puerto Rico (67.9 percent), Cuba (67.3 percent), the Bahamas (65.6 percent), and Jamaica (57.9 percent). Between 2005 and 2019, the Dominican Republic achieved even better results as it recorded an average hotel occupancy of 72.9 percent. Average hotel occupancy rates also improved in Puerto Rico (68.6 percent) and especially in Jamaica (63.7 percent) while falling in the Bahamas (57.7 percent) and most conspicuously in Cuba (57.7 percent). The performance of Jamaica's hotel industry in 2005–19 was noteworthy since the country was the only one among the leading tourism destinations in the Caribbean that expanded hotel room capacity at a faster pace than in the previous period.[67] Cuba, on the other hand, continued its massive hotel construction program but performed notably worse than competitors like the Dominican Republic and Jamaica whose tourism models, like that of Cuba, focus to a significant degree on mass tourism and (particularly the Dominican Republic) reasonably priced hotel deals.[68] Furthermore, casting additional doubts on the effectiveness of Cuba's tourism marketing strategies, hotel occupancy in the Dominican Republic and Jamaica tends to rebound distinctly in July and August while in Cuba normally occupancy falls steadily during the summer with little or no increase until November (ONEI 2019a, 2016b, 2013; Clarke, Mayo, and Cornwall 2015).[69]

When it comes to assessing the performance of a country's accommodation sector, in particular the effectiveness of pricing strategies and hotel room sales efforts, two of the most commonly used indicators are the average daily rate (ADR), which measures the average rental revenue per occupied room per day, and the revenue per available room (RevPAR), which refers to the average revenue earned per available room, whether booked or not. Unfortunately, to complicate cross-country comparisons, these indicators are not readily available for Cuba, and even for other Caribbean countries, the coverage is often incomplete and not always accurate. Nevertheless, some limited yet meaningful comparisons from sparsely available information are still possible.

Based upon independent surveys and research carried out by the hospitality consulting firm STR (formerly known as Smith Travel Research), the Caribbean Tourism Organization reported that the average RevPAR in the Caribbean region was $137.85 in 2017 (CTO 2018). Calculations by the authors from the same unpublished database of the Cuban Ministry of Tourism that was featured in Chapter 3 and that ranked the financial performance of 144 Cuban hotels in 2017 yielded a RevPAR of $80.48 for the Cuban market that year. Even if it was measured from the financial results of the best-performing hotels in the country (during the best year of inbound overnight tourism) representing less than half of the total number of hotels, Cuba's RevPAR was still considerably lower than the regional average. A look at the financial results in 2017 of Spain-based Meliá Hotels International, which manages more hotels in Cuba than any other foreign chain, also revealed that Meliá's Cuban hotels recorded an average occupancy rate of 59.5 percent, an average room rate or ARR (essentially ADR over a longer period of time) of 97.6 euros, and RevPAR of 58 euros. In contrast, the same indicators for Meliá's hotels in the Dominican Republic were, respectively, 74 percent, 114.6 euros, and 84.8 euros.[70] Additionally, providing some rare evidence of Cuba's key hotel metrics during the low (summer) season and how they compare to those of other major Caribbean destinations, STR reported that hotel occupancy and RevPAR in Cuba in August 2019 were, respectively, 50 percent and $31.77. On the other hand, the same indicators for the Dominican Republic were 65.5 percent and $69.82 despite major drops following a spate of tourist deaths in the country earlier that year, and for Puerto Rico were 68.1 percent and $123.12.[71]

At any rate, two valuable tourism indicators that can easily be produced from available data are gross revenues per room and gross revenues per tourist. They both include all expenditures by international inbound visitors. Ideally, the first indicator should only include all revenues generated from hotel operations, such as room rentals, food and beverages, and other sources (i.e., stores, spa treatments, entertainment, and other paid services), to complement RevPAR and measure the overall performance of a country's hotel sector in terms of revenues. But what is known as TRevPAR (total revenue per available room) is even less available than RevPAR. Therefore, albeit admittedly with limited accuracy, dividing international tourism receipts by the total number of hotel rooms, and repeating this measurement over time and for different countries, can be a useful way to show how effective a country's hotel development has been in boosting tourism receipts and how that compares to the experience of other countries. As disclosed in Table 4.11, among the five largest tourism markets in the Caribbean, Cuba not only has consistently had the lowest amount of gross revenues per room, but also is the country where this indicator has exhibited the worst results over time. From 1995 until 2000, gross tourism revenues per room grew in Cuba as they did in Puerto Rico, the Bahamas, Jamaica, and the Dominican Republic, but since then they have dropped steadily to a level in 2019 that was lower than their level in 1995. Conversely, this indicator has continued to improve in the other countries. During the 1995–2019 period, gross revenues per room more than doubled in the Bahamas and Jamaica, increased over 80 percent in the Dominican Republic, and rose nearly 40 percent in Puerto Rico.

As far as expenditures are concerned, gross tourism revenues per tourist is a crucial indicator that captures the average spending per visitor. In essence, it reveals how efficient a country's tourism industry is in extracting income from inbound visitors (Quintana et al. 2004, 171), and sheds some light on the kind of tourists, in terms of income level, that a particular destination tends to attract.[72] Obviously, the length of stay also affects the average expenditure per visitor and should be taken into account when analyzing tourism performance and international competitiveness. Table 4.12 presents data on gross revenues per tourist (overnight visitors) for practically every Caribbean country between 1995 and 2019. Again, Cuba stood out as the only country in the region where this indicator suffered a decline. Whereas the average spending per visitor in the Caribbean increased from $884 in 1995 to $1,380 in 2019, the same figure for Cuba plummeted from $1,482 to just $691 even if the average length of stay of international visitors to the island experienced little variation during that period.

Table 4.11 Gross Revenues per Room[a] of Selected Caribbean Countries, 1995–2019 ($US dollars)

	1995	2000	2005	2010	2014	2019
Puerto Rico	178,324	215,874	242,876	230,443	232,663	247,076
Bahamas	100,290	117,951	139,797	140,916	151,372	235,434
Jamaica	60,513	66,958	68,581	81,120	93,413	121,448
Dominican Republic	47,829	55,089	58,547	62,851	80,883	87,194
Cuba	40,358	52,397	52,153	39,304	41,006	35,643

Sources: Calculations of the authors from MINTUR data obtained in May 2022 and from ONEI 2015a, 2011a, 2006, 1998; Caribbean Tourism Organization, "Historical Data (1970–2015)," https://www.onecaribbean.org/statistics/historical-data-1970-2015/; United Nations World Tourism Organization, "145 Key Tourism Statistics," https://www.unwto.org/tourism-statistics/key-tourism-statistics.

Note:

[a] Hotel rooms only.

Table 4.12 Gross Revenues per Tourist[a] of Caribbean Countries, 1995–2019 ($US dollars)

	1995	2000	2005	2010	2014	2019
Anguilla	1,263	1,250	1,387	1,597	1,915	1,863
Antigua and Barbuda	1,123	1,224	1,261	993	2,843	3,302
Aruba	842	1,119	1,496	1,507	1,619	1,885
Bahamas	842	1,123	1,287	1,567	1,629	2,297
Barbados	1,385	1,426	1,621	1,921	1,706	1,753
Bermuda	1,258	1,268	1,459	1,659	1,817	2,234
British Virgin Islands	890	1,306	1,225	1,179	1,189	—
Cayman Islands	1,354	1,579	2,208	1,403	1,475	1,916
Cuba	1,482	1,119	1,061	885	853	691
Curacao	781	1,152	1,099	1,126	1,806	1,521
Dominica	567	671	709	1,237	2,667	1,500
Dominican Republic	884	962	953	1,021	1,095	1,159
Grenada	537	1,380	959	882	2,917	2,989
Guadeloupe	594	562	354	1,301	-	-
Haiti	621	914	705	655	1,243	1,573
Jamaica	931	1,007	1,045	1,041	1,084	1,342
Martinique	840	574	583	814	986	987
Montserrat	944	900	900	1,000	900	1,200
Puerto Rico	582	689	875	1,035	1,059	1,136
St. Kitts and Nevis	823	794	953	908	1,048	2,942
St. Lucia	1,146	1,037	1,075	1,837	2,302	2,509
St. Maarten	755	1,118	1,310	1,537	1,844	2,212
St. Vincent and Grenadines	683	1,123	1,095	1,194	2,465	2,881
Trinidad and Tobago	277	534	978	1,157	1,085	1,121
US Virgin Islands	1,810	2,209	2,411	1,715	2,191	2,006
Caribbean	884	998	1,075	1,157	1,246	1,380

Sources: ONEI 2015a, 2011a, 2006, 1998; Caribbean Tourism Organization, "Historical Data (1970–2015)," https://www.onecaribbean.org/statistics/historical-data-1970-2015/; United Nations World Tourism Organization, "145 Key Tourism Statistics," https://www.unwto.org/tourism-statistics/key-tourism-statistics.

Note:

[a] Overnight visitors.

Cuba's poor performance was mainly triggered by a combination of factors such as widespread inefficiencies in the tourism sector, an insufficient diversification of complementary services, and the socioeconomic profile of inbound tourists who tend to be medium-to-low income travelers (Ayala Castro 2020). During the 1995–2019 period, apart from some major fluctuations in the Cayman Islands, Dominica, Grenada, Haiti, and the US Virgin Islands, gross revenues per tourist grew steadily almost everywhere in the Caribbean, including in the leading destinations. In 2019, this indicator was $1,342 in Jamaica, $1,159 in the Dominican Republic, and $1,136 in Puerto Rico, on the rise in all three countries but lower than the Caribbean average. On the other hand, suggesting a greater inflow of higher-income tourists, the average spending per visitor was over $2,000 in the Bahamas, Bermuda, St. Maarten, and the US Virgin Islands, close to $3,000 in St. Vincent and Grenadines, St. Kitts and Nevis, and Grenada, and approximately $3,300 in Antigua and Barbuda.

Conclusion

Albeit with notable differences between destinations, tourism is an important economic activity in most Caribbean countries, making a significant contribution to GDP, creating employment for numerous people, generating much-needed foreign exchange, stimulating foreign investment, and enhancing backward productive linkages with other economic sectors, among various benefits. Notwithstanding the United States' long-standing ban on US tourism in Cuba and additional travel restrictions that severely limited (and still do) the flow of US visitors to the island, Cuba has become one of the largest tourism markets in the Caribbean. In 2019, Cuba ranked second in the region after the Dominican Republic in terms of the number of inbound overnight visitors (despite a decline since 2017), had the second-largest stock of hotel rooms, most of them in four- and five-star establishments, and ranked fifth in terms of hard currency revenues earned from tourism activities after the Dominican Republic, the Bahamas, Puerto Rico, and Jamaica. Following the launch in late 2014 of a historic rapprochement with the United States during the Obama administration, Cuba also emerged as a competitive player in the cruise tourism business. While in terms of passenger arrivals, it remained behind leading destinations like the Bahamas, Cozumel (Mexico), the Cayman Islands, Puerto Rico, St. Maarten, and a few others, Cuba grew considerably as a recipient of cruise tourists until the Trump administration banned US cruises to Cuba in early June 2019.

Quite concerning, though, this chapter's brief review of some key determinants of tourism efficiency in the Caribbean, especially in the region's largest destinations, from the mid-1990s until 2019, revealed clearly that the Cuban tourism industry was outperformed by its main competitors. Furthermore, it does not seem to be a coincidence that the deterioration of tourism efficiency in Cuba intensified after Fidel Castro launched a process of recentralization of the island's tourism sector and of the overall economy in the mid-2000s. Indeed, since then, whether it was the length of stay of international inbound visitors, the hotel occupancy rates and

RevPAR, gross tourism revenues per room, or the average expenditure per visitor, Cuba has fallen behind the Dominican Republic, Jamaica, Puerto Rico (except for ALOS), and the Bahamas. The Dominican Republic, most notably, exhibited the best performance as it recorded a longer ALOS by inbound tourists and higher hotel occupancy rates than its competitors. It also saw its hard currency revenues from tourism activities expand at a faster pace than international visitor arrivals, which translated into a higher amount of visitor spending per capita. Unlike Cuba, where gross revenues per tourist fell steadily, virtually all Caribbean countries recorded an increase in the average spending per visitor during the 1995–2019 period. In short, by the time the novel coronavirus (Covid-19) outbreak was declared a global pandemic in early March 2020, Cuba's tourism sector was already mired in serious structural problems that not even the relaxation of US economic sanctions against the island during the Obama presidency had been able to mitigate. As we will see in the next chapter, Covid-19 made the situation much worse.

Notes

1 In 2019, the top ten leading destinations for international tourists in Latin America were Mexico (around 45 million tourists), Argentina (7.4 million), Dominican Republic (6.4 million), Brazil (6.3 million), Chile (4.5 million), Peru (4.4 million), Cuba (4.3 million), Colombia (4.2 million), Puerto Rico (3.2 million), and Costa Rica (3.1 million) (UNWTO 2021a).

2 For statistics on international travel and tourism by Canadian residents, see https://www.statcan.gc.ca/en/subjects-start/travel_and_tourism.

3 During the 2015–19 period, tourism accounted for an average of about 15 percent of GDP and 13 percent of employment in Mexico. It also represented approximately 10 percent of GDP in Chile, 9 percent in Peru and Argentina, and close to 8 percent in Brazil, with a substantial contribution to employment in all of these countries. In Peru, tourism revenues accounted for an average of nearly 10 percent of total exports of goods and services (Gómez García et al. 2021, 22).

4 The International Monetary Fund identifies the following as Caribbean Small Developing States (SDS): Antigua and Barbuda, the Bahamas, Barbados, Belize, Dominica, Grenada, Guyana, St. Kitts and Nevis, St. Lucia, St. Vincent and the Grenadines, Suriname, and Trinidad and Tobago.

5 Based on the TDI for 2015–19, Trinidad and Tobago (no. 107 ranking), Guyana (141), and Suriname (158) were all outside of the top 100 list of the most tourism-dependent countries in the world (Gómez García et al. 2021, 21).

6 For the most recent World Bank classification of countries by income level, see: https://blogs.worldbank.org/opendata/new-world-bank-country-classifications-income-level-2022-2023.

7 Latin America and the Caribbean is the world's second-most disaster-prone region after Asia and the Pacific. For a thorough review of the disasters (over 1,500 events—tropical storms and hurricanes, earthquakes, volcanic eruptions, droughts, floods, and others—affecting more than 190 million people since 2000) in Latin America and the Caribbean during the 2000–2022 period, see United Nations Office for the Coordination of Humanitarian Affairs (OCHA) and United Nations Office for Disaster Risk Reduction (UNDRR), "Overview of Disasters in Latin America and the Caribbean 2000–2022," June 2023, https://www.undrr.org/publication/overview-disasters-latin-america-and-caribbean-2000-2022.

8 In 2015–19, average annual revenues generated by the tourism sector in the Dominican Republic amounted to approximately $7 billion. During the same period, the country received about $5.9 billion in remittances from abroad and exported $5.7 billion worth of products from its free trade zones. For additional information, see the economic statistics of the Central Bank of the Dominican Republic at: https://www.bancentral.gov.do/.

9 Unlike the Dominican Republic, Cuba does not report official data on remittances from abroad. The figures provided by Morales (2023) are his estimates.

10 The Economist Intelligence Unit (EIU) ranked Cuba's business environment as one of the worst in the world (seventy-ninth out of eighty-two countries) for the 2018–22 period. In Latin America, out of twelve countries, only Venezuela was ranked lower than Cuba (EIU 2023).

11 For a detailed analysis of the pros and cons of investment incentives, see Lise Johnson and Perrine Toledano, "Investment Incentives: A Survey of Policies and Approaches for Sustainable Investment." Columbia Center on Sustainable Investment, October 2022, https://academiccommons.columbia.edu/doi/10.7916/qzyx-1c57.

12 See the UNWTO Glossary of Tourism Terms at https://www.unwto.org/glossary-tourism-terms. Also see United Nations 2010b, 10.

13 Historical data (1970–2015) from the Caribbean Tourism Organization (CTO) on overnight visitor arrivals and cruise passenger arrivals to the Caribbean are available at: https://www.onecaribbean.org/statistics/historical-data-1970-2015/. The CTO annual numbers of total stay-over tourist arrivals to the Caribbean are higher than those reported by the World Tourism Organization because CTO figures include arrivals to Cancún and Cozumel, two popular tourist destinations in the Mexican Caribbean.

14 Unlike overnight visitors, annual cruise visitors to the Caribbean increased in 2001–2.

15 The fifteen member states of the Caribbean Community or CARICOM are the following: Antigua and Barbuda, Bahamas, Barbados, Belize, Dominica, Grenada, Guyana, Haiti, Jamaica, Montserrat, St. Lucia, St. Kitts and Nevis, St. Vincent and Grenadines, Suriname, and Trinidad and Tobago. Haiti, which secured full membership in 2002, is the only French-speaking country and the only one with a relatively large population (over 11 million people).

16 See CTO statistics at http://www.onecaribbean.org/content/files/DistributionTAC bbnSub-Region2004to2004.pdf.

17 Calculations of the authors from UNWTO 2021b and Secretaría de Turismo de Quintana Roo (Mexico), http://sedeturqroo.gob.mx/ARCHIVOS/indicadores/Indicador-Tur-Dic-2019_actualizado.pdf.

18 Simon Romero and Nicholas Bogel-Burroughs, "Crisis Hits Dominican Republic Over Deaths of U.S. Tourists," *New York Times*, June 23, 2019.

19 Gina Martinez, "Dominican Republic Announces New Safety Measures After Deaths of Eleven American Tourists," *TIME*, September 13, 2019.

20 All figures for the Dominican Republic in this paragraph come from tourism statistics of the Central Bank of the Dominican Republic, available at: https://www.bancentral.gov.do/a/d/2537-sector-turismo.

21 Doyle Rice, "Dorian's Legacy: The Slowest, Strongest Hurricane to Ever Hit the Bahamas," *USA TODAY*, September 6, 2019.

22 The famous TV series "The Love Boat," which ran from 1977 to 1986 and popularized the concept of contemporary cruise vacations, was filmed aboard various Princess ships. For more information, see https://www.princess.com/news/the-love-boat/.

23 Cruise Market Watch, "2021 Worldwide Cruise Line Passenger Capacity," https://cruisemarketwatch.com/capacity/.

24 Cruise Market Watch, "2021 Worldwide Cruise Line Market Share," https://cruisemarketwatch.com/market-share/.

25 Caribbean Tourism Organization, "Cruise Passenger Arrivals in the Caribbean, 1980-2004" http://www.onecaribbean.org/content/files/cruisearrivals1980to2004.pdf.

26 A list of Caribbean cruise ports can be found at: https://www.vacationstogo.com/cruise_port/Caribbean.cfm.

27 Cruise lines that own private islands in the Caribbean are Carnival, Royal Caribbean, Norwegian, Disney Cruise Line, Holland America Line, Costa Cruises, Mediterranean Cruise Line, and Virgin Voyages.

28 Ashley Kosciolek, "Eastern Caribbean vs. Western Caribbean Cruises," Cruise Critic, October 9, 2019.

29 Calculations of the authors from the United Nations World Tourism Organization, "145 Key Tourism Statistics," https://www.unwto.org/tourism-statistics/key-tourism-statistics.

30 For additional information on global outbound tourism, see UNWTO Tourism Dashboard at: https://www.unwto.org/tourism-data/global-and-regional-tourism-performance. For the global GDP by country in 2023, see Pallavi Rao, "Visualizing the $105 Trillion World Economy in One Chart," Visual Capitalist, August 9, 2023, https://www.visualcapitalist.com/visualizing-the-105-trillion-world-economy-in-one-chart/.

31 "U.S. Outbound Travel to World Regions," International Trade Administration (ITA), US Department of Commerce, https://www.trade.gov/sites/default/files/2023-10/US-Outbound-to-World-Regions.xlsx.

32 Andrea Zelinski, "Cruise Demand Grows for the Caribbean, and Lines Answer the Call," Travel Weekly, August 28, 2023.

33 "Tourism in the Caribbean Grew in 2019," ForwardKeys, January 23, 2020.

34 Calculations of the authors from sources of Table 4.5.

35 World Bank's World Integrate Trade Solution, "GDP Growth By Country and Region 2009," https://wits.worldbank.org/CountryProfile/en/Country/ALL/StartYear/2009/EndYear/2009/Indicator/NY-GDP-MKTP-KD-ZG.

36 "How Marriott Is Redefining Luxury Travel in the Caribbean and Latin America," Skift, September 7, 2022; "How Marriott Is Transforming the All-Inclusive Segment Through Loyalty," Skift, July 20, 2022; Laura Dowrich-Philips, "Five Travel Trends That Will Impact the Caribbean," Loop News, July 27, 2021.

37 Harriet Baskas, "U.S.-Cuba Travel Shouldn't Faze Other Caribbean Destinations," NBC News, January 24, 2015; Andrew O'Reilly, "Cuba's Cold War Charm Has D.R. Wary About Losing U.S. Tourists to Next-Door Neighbor," Fox News Latino, March 15, 2015.

38 Josh Lederman and Michael Weissenstein, "Obama Abolishes Last Major Restrictions on US Travel to Cuba," Associated Press, March 15, 2016.

39 Felicia Schwartz, "Trump Announces Rollback of Obama's Cuba Policy," *Wall Street Journal*, June 17, 2017; Karen DeYoung, "Trump Administration Ends Group Travel to Cuba by Americans," *Washington Post*, June 4, 2019.

40 Gina Martinez and Josiah Bates, "11 U.S. Tourists Have Died in the Dominican Republic in 2019. Should You Cancel Your Trip?" *TIME*, June 24, 2019.

41 Among the countries that depends heavily on the US market for inbound tourism, the Bahamas and Bermuda reportedly have been diversifying away from the US market since the early 2000s (Acevedo and Wong 2017, 67).

42 As the Caribbean Hotel & Tourism Association (CHTA 2015) noted, Cuba's proximity might cut into the market for spontaneous and impulse vacations, a growing area facilitated by online offers and instant booking services. This could negatively affect in particular the Bahamas (which relies on Florida for a significant portion of its international arrivals) as well as other countries like Puerto Rico, the

US Virgin Islands, Belize, the Cayman Islands, Cancún, and Jamaica that have been beneficiaries of last-minute and impulse vacation decisions.

43 Mengqi Sun, "Judge Orders Cruise Companies to Pay More Than $436 Million for Docking at Seized Terminal in Cuba," *Wall Street Journal*, January 6, 2023.

44 PortMiami, "*2035 Master Plan*," November 2011, www.miamidade.gov/portmiami/library/2035-master-plan/complete-master-plan.pdf.

45 "Cruise Liner Pulls Out of Bahamas for Cuba," *Jamaica Observer*, February 10, 2017.

46 Tom Stieghorst, "Focus on Cruise: ¡Cuba abierta!," Travel Weekly, April 12, 2017.

47 Andrea Sachs, "All-Inclusive Resorts Want You To Forget Their Cheesy Reputation," *Washington Post*, January 26, 2023.

48 Calculations of the authors from UNWTO 2023, 1997. The Caribbean's total hotel room stock grew slower than total stopover visitors to the region in the 1980s, faster in the 1990s and 2000s, and again slower in the 2010s.

49 "La Planta Hotelera Cubana Experimenta un Importante Crecimiento," Hosteltur, May 17, 2005.

50 "Gaviota Abrirá 2.750 Habitaciones en 2018," EFE, July 13, 2018.

51 The Caribbean region's gross tourism earnings were around $3.2 billion in 1980 (González Núñez 2022, 113).

52 "St. Maarten Had 8 Per Cent Economic Decline in 2018," *The Daily Herald*, July 30, 2019; Alma Beauvais, "How a Lavish Pop-Up Picnic Is Bringing Hundreds of Tourists to Haiti This Weekend," *Miami Herald*, August 12, 2022.

53 Calculations of the authors from sources of Table 4.8 and Table 4.9.

54 "Las Reservas en Viajes de Lujos Crecerán 47% en el 2023," Forbes, November 18, 2022; "Expedia: Luxury Travel Demand Is Rising in the Caribbean," April 11, 2017.

55 It should be pointed out that the recent performance of Puerto Rico's tourism sector has been negatively affected by natural disasters. In 2016, the year before hurricanes Irma and Maria struck the island, Puerto Rico's share of the Caribbean's total tourism receipts was 13 percent while its share of total inbound visitors to the region was 14.8 percent.

56 All figures in this section are calculations of the authors, *supra note* 53.

57 Zachary Crockett, "The Economics of Cruise Ships," The Hustle, March 15, 2020; Andrea Zelinski, "Onboard Spending Keeps Flowing," Travel Weekly, May 29, 2023.

58 Ron, Hurtibise, "Bahamas Unhappy with Cruise Passenger Spending; No Longer Paying Incentives to Cruise Lines," *South Florida Sun-Sentinel*, October 22, 2018.

59 The 2018 BREA study was based on surveys (involving more than 40,000 passengers and 18,000 crew members) conducted onboard ships of the FCCA

member lines, namely Carnival, Royal Caribbean, Norwegian, and eight additional cruise lines.

60 Worth nearly $2.6 billion, passenger expenditures represented 76.2 percent of total direct expenditures. Cruise line expenditures ($534 million) accounted for 15.9 percent of the total, and crew expenditures ($266 million) for 7.9 percent.

61 According to the Caribbean Tourism Organization, the average length of stay of international visitors to the Caribbean was 7.2 days in 2013, 7.3 days in 2014, and about 7 days in 2019–20 (CTO 2021, 2015).

62 See Oficina Nacional de Estadísticas e Información, "Series Estadísticas 1985–2022: Turismo," https://www.onei.gob.cu/turismo-0.

63 Horwath HTL, "Market Report: The Caribbean," September 2019, https://horwathhtl.com/publication/market-report-caribbean/.

64 September is generally the least popular month for leisure travel to the Caribbean.

65 Lily Girma, "There's No Such Thing as Low Season in the Caribbean Anymore," Bloomberg, April 13, 2023.

66 Among the best years for the Caribbean hotel industry, the average hotel occupancy rate in the region was 67.7 percent in 2000, 66.3 percent in 2012, 67.1 percent in 2013, 68.0 percent in 2014, 68.8 percent in 2015, and 66.6 percent in 2017 (WTTC 2004; CTO 2013, 2015, 2016, 2018).

67 Jena Tesse Fox, "Through Foreign Investment, Jamaica Ramps Up Hotel Development," Hotel Management, June 15, 2017.

68 To shed some light about different hotel prices in the Caribbean, the average daily rate (ADR) of hotels in the Dominican Republic during the first five months of 2019 was $154. The same indicator in Jamaica was $265. As for other countries, the ADR was $240 in Puerto Rico, $303 in the Bahamas, $333 in Bermuda, $339 in Barbados, $397 in Aruba, $438 in St. Lucia, and $552 in the Cayman Islands (STR 2019).

69 For comprehensive statistics on hotel occupancy in the Dominican Republic, see "Banco Central de la Republica Dominicana, "Sector Turismo: Tasa de Ocupación," https://bancentral.gov.do/a/d/2537-sector-turismo.

70 Meliá Hotels International, "2017 Integrated Annual Report," https://www.meliahotelsinternational.com/en/shareholdersAndInvestors/AnnualReportDocs/2017/2017%20Integrated%20Annual%20Report_0.pdf.

71 Christina Jelski, "Caribbean Resorts Now Seeing Big Drops in RevPAR, Occupancy," Travel Weekly, October 6, 2019.

72 While there are significant measurement challenges, the average spending per visitor by country of origin is an even more valuable indicator that allows policymakers to identify the most productive international tourism source markets. Although many countries are unable to offer comprehensive statistics

to produce this indicator, gross tourism revenues per visitor by source market can be measured from information collected through visitors' surveys, tourism expenditures surveys, and balance of payments, along with data provided by national statistical agencies, central banks, and international organizations (Dupeyras and MacCallum 2013, 20–1).

5

International Tourism in Cuba and the Caribbean in the Post-Covid Era

On March 11, 2020, only a few months after a new contagious pneumonia-like viral illness was first identified in Wuhan, China, the World Health Organization (WHO) declared the novel coronavirus (Covid-19) outbreak a global pandemic. As the transmission of the virus by respiratory and extra-respiratory routes fueled the rapid spread of the disease worldwide, one of the most challenging and traumatic periods in recent memory began. Amid startling daily counts of new coronavirus cases that overwhelmed hospitals, and of deaths caused by the disease, many countries widely recommended and mandated mask-wearing in public. They also implemented a variety of mandatory and non-mandatory social distancing measures, in particular the closing of schools and workplaces, the cancelation of sports and entertainment events, and limitations on internal movement or shelter-in-place orders.

Not surprisingly, the Covid-19 pandemic hit international travel and tourism extremely hard as virtually all destinations around the world were forced to put in place some kind of travel restrictions, including border closures, quarantines, curfews, and mandatory testing. Serious gaps in multilateral cooperation and coordination with respect to travel policies, health protocols, and research in testing, treatments, and vaccines added to the uncertainty and weak tourism demand. Moreover, tourism-dependent countries were poised to experience the negative impact of the pandemic for much longer than other nations due to the dependency of the tourism sector on contact-intensive services that could not thrive until large numbers of people brushed off their fears of contagion and felt safe to travel again. Before Covid-19, travel and tourism contributed directly and indirectly to approximately 10 percent of global GDP and accounted for more than 320 million jobs worldwide (Behsudi 2020). In 2019, an astonishing 1.5 billion people took foreign trips, generating $1.7 trillion in export revenues (including some $200 billion in receipts for passenger transportation) from tourism. But an unprecedented disruption to tourism caused by plummeting demand and widespread travel restrictions was about to occur. According to the World Tourism Organization, international tourist arrivals (overnight visitors) plunged over 70 percent to just 407 million in 2020, which went down as the "worst year on record for global tourism." Export revenues from tourism also plummeted 63 percent to $0.6 trillion (UNWTO 2023b). Obviously, the collapse of international visitor spending in destinations and of earnings from international transportation services led to massive losses in hard currency revenues for tourism-dependent economies.

Notwithstanding a race to vaccinate after several Covid vaccines, among them those produced by Pfizer-BioNTech and Moderna, received emergency use authorization,[1] international tourist arrivals increased by just 13 percent in 2021 to 458 million as different sorts of travel restrictions remained in place for long periods in various countries to cope with spikes in new cases, hospitalizations, and deaths triggered by the circulation of major variants of the virus. It was only in late spring 2022, mainly because of the removal of most travel restrictions amid declining numbers of severe Covid outcomes, but also because of the fact that many people simply quit following pandemic rules and regulations as they had had enough of them, that tourism started to bounce back strongly on the back of pent-up demand.[2] At 28 percent and 31 percent, respectively, in 2020 and 2021, international tourism ended 2022 at 66 percent of its pre-pandemic levels with 969 million international arrivals. It also generated $1.3 trillion in export revenues and contributed to 7.6 percent of global GDP, hitting about 76 percent of pre-pandemic levels in both indicators. But the negative economic impact of Covid-19 had been brutal. Between 2020 and 2022, the pandemic caused aggregate losses of 2.6 billion international arrivals, $2.5 trillion in export revenues from tourism, and $4.2 trillion in the direct contribution of tourism to global GDP. Most importantly, international tourism continued its recovery in 2023 as it recorded 1.286 billion overnight visitor arrivals, generated $1.6 trillion in export revenues, and directly contributed $3.3 trillion to global GDP. The last two indicators were similar to their levels in 2019. In terms of arrivals, international tourism closed 2023 at 88 percent of pre-pandemic levels, with the Caribbean being among four sub-regions (including Southern Mediterranean Europe, Central America, and North Africa) that exceeded their 2019 levels, and was expected to make a full recovery by the end of 2024 (UNWTO 2024, 2023b; WTTC 2023a).

This chapter provides an analysis of international tourism in Cuba and the Caribbean in a post-pandemic world. It begins with a review of GDP trends and the measures enacted by Caribbean governments to combat Covid-19 and mitigate its adverse economic and social impacts. Special emphasis is given to Cuba and its main competitors in tourism, like the Dominican Republic, Puerto Rico, Jamaica, and the Bahamas. Trends in international visitor arrivals and tourism revenues are also presented to compare the recovery of tourism across the Caribbean region. Moreover, after examining the current situation of US economic sanctions against Cuba, this chapter sheds further light on the recent performance of Cuba's tourism sector and identifies critical problems that negatively affect the Cuban tourism industry.

Covid Shock and Policy Responses in the Caribbean

Besides being among the world's most impacted regions in terms of Covid-19 cases and related deaths,[3] Latin America and the Caribbean was the region in the developing world hardest hit by the pandemic in terms of economic contraction, with the Caribbean's tourism-dependent economies suffering the most as they saw their economic lifeblood nearly disappear overnight. Already on a low-growth trajectory in the decade prior to the Covid outbreak,[4] the Caribbean faced unprecedented supply and demand shocks and saw its GDP shrink by 9.2 percent in 2020 as opposed to an average GDP reduction of 6.8 percent for Latin America as a whole. With the need to adopt lockdown and physical distancing policies and to halt production activities to fight a frightening pandemic with its many complications and variants, the Caribbean's GDP grew 5.9 percent in 2021 while Latin America's GDP expanded, on average, 6.8 percent to fully erase the decline of the previous year. It was only in 2022, after many islands had reopened to international tourism and dropped pre-arrival Covid testing and vaccination requirements for all travelers, that economic growth in the Caribbean (12.9 percent) started to outpace overall growth in Latin America (3.8 percent). This trend reportedly continued in 2023 with the Caribbean's GDP projected to increase 9.4 percent as opposed to an estimated growth of 2.2 percent for Latin America. Most importantly, the economic expansion of the Caribbean outpaced that of the entire Latin American region even if we exclude the performance of Guyana, whose GDP rose dramatically after a massive oil production boom started in late 2019 following a slew of quality oil discoveries made by a consortium led by the US company Exxon Mobil.[5] Excluding Guyana, the Caribbean's GDP increased 4.5 percent in 2022 and was projected to grow 3.4 percent in 2023 (CEPAL 2023a, 2023b, 2022).

All Caribbean economies suffered from the implementation of various Covid-19 mitigation measures involving, among other things, border closures, curfews, increases in health spending, the expansion of safety net programs, restrictions on business operations, subsidies, and tax relief. Yet the deleterious effects of the pandemic were not felt the same way across the region. Table 5.1 shows the annual GDP growth rates of fifteen Caribbean countries during the 2020–23 period. The selected group includes several small islands highly dependent on tourism that unsurprisingly experienced the sharpest contractions of economic activity (and spikes in unemployment) because of the Covid-19 outbreak, and bigger countries with more diversified economies that, overall, were less impacted by the pandemic. Aruba, the world's most tourism-dependent economy,[6] witnessed the greatest decrease of its gross domestic product (−24.0 percent) in 2020, followed by St. Lucia (−23.6 percent), the Bahamas (−23.5 percent), Antigua and Barbuda (−17.5 percent), Dominica (−16.6 percent), St. Kitts and Nevis (−14.6 percent), Grenada (−13.8 percent), and Barbados (−12.7 percent). Another tiny island, St. Vincent and the Grenadines, underwent a rather moderate economic

Table 5.1 GDP Growth Rates of Selected Caribbean Countries, 2020–23

	2020	**2021**	**2022**	**2023**	**Average 2020–23**
Aruba	−24.0	27.6	10.5	5.3	4.8
Dominican Republic	−6.7	12.3	4.9	2.4	3.2
Bahamas	−23.5	17.0	14.4	4.3	3.0
St. Vincent and Grenadines	−3.7	0.8	5.5	6.2	2.2
St. Lucia	−23.6	11.3	15.7	3.0	1.6
Barbados	−12.7	−1.3	13.8	4.4	1.0
Antigua and Barbuda	−17.5	6.6	8.5	5.9	0.9
Grenada	−13.8	4.7	7.3	4.8	0.7
Jamaica	−9.9	4.6	5.2	2.2	0.5
Dominica	−16.6	6.9	5.6	4.7	0.1
Puerto Rico	−4.2	0.4	3.2	−0.7	−0.3
St. Kitts and Nevis	−14.6	−0.9	8.8	3.4	−0.8
Trinidad and Tobago	−9.1	−1.0	1.5	2.1	−1.6
Haiti	−3.3	−1.8	−1.7	−1.9	−2.2
Cuba	−10.9	1.3	1.8	−1.9	−2.4

Sources: ONEI 2024a, 2023a; IMF 2024; CEPAL 2023b.

downturn at −3.7 percent despite a considerable decline in inbound tourist arrivals as tourism's relatively low weight in the country's economic output limited its contributing impact to GDP reduction.[7] Aside from the Bahamas, among the Caribbean countries with the largest tourism sectors, Cuba recorded the steepest drop in GDP in 2020 (−10.9 percent), followed by Jamaica (−9.9 percent), the Dominican Republic (−6.7 percent), and Puerto Rico (−4.2 percent). As for the two remaining countries, both with a fairly low dependence on tourism, Trinidad and Tobago's GDP plunged 9.1 percent due to a great extent to plummeting global demand and prices in the energy sector. Haiti, the poorest country in the Western Hemisphere, did not suffer a major setback from Covid as its GDP fell 3.3 percent in 2020, but nonetheless this was the country's worst economic performance in years (IMF 2024; ONEI 2024a; CEPAL 2023b).

As 2020 came to a close and some of the most stringent anti-Covid measures (government lockdowns and total entry bans, in particular) were lifted in a number of countries, most Caribbean economies resumed growth, albeit with significant differences. In effect, with remarkable GDP growth rates of 27.6 percent and 12.3 percent, respectively, Aruba and the Dominican Republic, among the first destinations in the region to officially reopen to international tourists, were the only Caribbean countries whose economic output had surpassed its pre-pandemic levels by the end of 2021.[8] That year, amid the resurgence of new Covid-19 cases, slow vaccine rollouts, and persistent fears of contagion, the GDP of countries like Barbados, St. Kitts and Nevis, Trinidad and Tobago, and Haiti actually continued to decline. The Caribbean's post-pandemic economic recovery, though, strengthened during 2022 as Covid protocols were eased further and key economic sectors, above all tourism, bounced back. Economies largely driven by tourism indeed recovered more rapidly than expected from the Covid-induced shock of 2020 (IDB 2023). Except for Haiti, all Caribbean economies grew in 2022, with the GDP of tourism-oriented countries like St. Lucia (15.7 percent), the Bahamas (14.4 percent), Barbados (13.8 percent), and Aruba (10.5 percent) recording double-digit growth. A look at average GDP growth rates in the Caribbean from 2020 until 2023 also reveals that the best-performing economies during this period, led by Aruba (4.8 percent) and the Bahamas (3.0 percent), and with the only exception of the Dominican Republic (3.2 percent), were those of countries with a relatively high level of dependence on tourism. Among this latter group, which includes St. Vincent and Grenadines, Antigua and Barbuda, St. Lucia, Grenada, Barbados, Jamaica, and Dominica, only St. Kitts and Nevis had yet to reach its pre-pandemic level of economic output by the end of 2023. The same can be said for Puerto Rico and Cuba. During the 2020–23 period, Puerto Rico's GDP averaged −0.3 percent while Cuba's GDP averaged −2.4 percent (ONEI 2024a). Facing a serious economic crisis caused by a combination of internal (systemic problems) and external (Covid-19 and other shocks) factors (Mesa-Lago 2023),[9] Cuba had the worst economic performance of all Caribbean countries.

Since GDP and tourism trends in the Caribbean in the post-pandemic era to a substantial degree reflected the impact of policy measures to curb the propagation of Covid-19, it is useful to review such measures, especially those implemented by Cuba and by its principal tourism competitors in the region. Before doing that, though, a brief analysis of the evolution of the Covid-19 pandemic in the Caribbean, with statistics on total case and death counts and some country data, is necessary to fully understand the struggle of Caribbean governments to safeguard the health of their citizens while trying to minimize the negative effects of containment and mitigation measures on economic activities.

The first Covid-19 case in the Caribbean was recorded in the Dominican Republic on March 1, 2020,[10] followed by a rapid spread across the region with multiple contagion waves over the next several months amid the emergence of new variants of the virus and slow vaccination campaigns.[11] Already more than 50,000 in late June 2020, the total number of confirmed Covid cases in the Caribbean had reached some 370,000 by the end of 2020, with nearly 6,000 related deaths. The Dominican Republic accounted for the biggest share (47.1 percent) of all cases in the region, followed by Puerto Rico (23.1 percent), Jamaica (3.6 percent), Cuba (3.5 percent), Belize (2.9 percent), Haiti (2.7 percent), the Bahamas (2.1 percent), and Trinidad and Tobago (1.9 percent). The remaining Caribbean countries combined had only 13.1 percent of total cases, but they too experienced Covid outbreaks, hospitalizations, and deaths. Because of greater international air connectivity, the largest tourist destination markets in the Caribbean had a higher percentage of imported cases, particularly if compared with small islands in the Southern Caribbean where positive cases arrived primarily from cruise ships and were followed by limited cases of local transmission (Umakanthan et al. 2021, 669). The total number of confirmed Covid cases in the Caribbean was 2.3 million at the end of 2021, 4.4 million at the end of 2022, and about 4.5 million (with over 37,000 related deaths) in early May 2023 when the World Health Organization declared that Covid-19 was no longer a global health emergency.[12] By then, Puerto Rico had 27.4 percent of all cases, followed by Cuba (24.8 percent), the Dominican Republic (14.7 percent), Trinidad and Tobago (4.3 percent), Jamaica (3.4 percent), Belize (1.6 percent), the Bahamas (0.8 percent),[13] and Haiti (0.7 percent). Compared with the early stages of the pandemic, Trinidad and Tobago, Puerto Rico, and in particular Cuba, had increased their share of total confirmed cases in the Caribbean, Jamaica's share remained virtually unchanged, and the shares of Belize, Haiti, the Bahamas, and especially the Dominican Republic had fallen. The rest of the Caribbean now accounted for 22.3 percent of total cases. It is also worth noting that Cuba, Puerto Rico, and above all the Dominican Republic managed to keep the case fatality rate lower than other countries (Schwalb et al. 2022). Needless to say, the long economic hangover of the Covid-19 pandemic is still going on.

Although the Covid-19 pandemic heightened human suffering in the Caribbean and undermined the region's economies, Caribbean governments, by and large, acted with speed and determination to prevent their citizens from contracting the coronavirus disease and to protect people and firms from the harmful economic impact of the global health crisis. Containment measures followed the same pattern almost everywhere in the world, but the timing of their implementation in the Caribbean was similar to that of countries with the fastest responses, such as New Zealand and Iceland. On average, Caribbean governments began controlling movements into their countries twenty-seven days before the first confirmed Covid-19 case, started to enact measures to control mass gatherings one day before the first case, and began controlling movements within their countries nine days after the first case. Moreover, compared with other international locations, they relied generally on stricter anti-Covid measures, particularly with respect to travel requirements and entry restrictions.[14] During the initial outbreak period between March and May 2020, most Caribbean countries (among them the Dominican Republic, Jamaica, and the Bahamas, respectively on March 19, 21, and 24) implemented full border closures, except a few small islands like Antigua and Barbuda, Barbados, Grenada, Montserrat, and St. Vincent and Grenadines, which opted for partial closures. Technically, Puerto Rico, Cuba, Trinidad and Tobago, Aruba, and St. Maarten also left their borders partially open, but only because they allowed their own residents to return. At any rate, partial closures included flight suspensions from certain destinations, bans on cruise ship arrivals, visa restrictions, and health checks at borders. Along with social distancing, quarantines, sanitary measures, and the mandatory use of face masks, a combination of some forms of lockdown and stay-at-home orders, curfews (evening and overnight), and various mobility restrictions accompanied strict border controls and limitations on gatherings (Murphy et al. 2020, 4–7).

After an early success in slowing down the rates of Covid-19 infections and deaths, the Caribbean region struggled to contain the spread of the virus, notably due to extensive violations of social distancing requirements fueled by high levels of informality, and the limitations of health infrastructures and social protection systems (OECD 2020b). And yet, facing the dilemma of preventing disease transmission while preventing the collapse of their economies as their primary economic engine, the tourism sector, was at a standstill, many Caribbean countries moved quickly to reopen their borders (albeit in stages in some cases and with bans on arrivals from certain countries) and relax domestic restrictions on people and businesses. Antigua and Barbuda, Aruba, Jamaica, St. Lucia, and the US Virgin Islands were the first countries to reopen for international tourism in June 2020. In a critical move for the region's tourism sector, a leading destination market like Jamaica began to welcome back foreign visitors less than two months after it closed all air (and sea) ports to incoming traffic. It did so with updated protocols that included health screening and risk assessment on arrivals and the requirement for passengers from high-risk areas to upload a valid molecular (PCR) Covid-19 test prior to departure.[15] The Bahamas, Barbados, Bermuda, Curacao, the Dominican Republic, Haiti, Puerto Rico, St. Maarten, St. Vincent and Grenadines, and Turks and Caicos reopened to international tourists in July 2020, followed by Anguilla, Grenada, and Dominica in August, Belize in October, St. Kitts and Nevis in November, and the British Virgin Islands in December. As for Cuba, its government began a gradual reopening to international tourism in the summer and fall of 2020, first in remote all-inclusive resorts in the northern keys in July, then in Varadero and other areas of the country in October, and then by reopening Havana's airport in mid-November after a nearly eight-month shutdown.[16] With a unique community-based medical system, strict internal containment measures, and tough Covid protocols for international travelers including a PCR test upon arrival, two weeks in quarantine at state isolation centers, and possibly a second PCR test after five days, Cuba reportedly managed the pandemic better than most Caribbean countries during 2020 (Wylie 2021, 357).

The reopening of borders to international travelers to relaunch tourism and economic activity in the Caribbean, however, was not without challenges since it increased the risk of introducing new chains of infections and new variants of Covid-19. In the second half of 2020, confronted with sudden surges of

Covid cases and deaths, many Caribbean countries were forced to implement stricter travel protocols and escalate measures to control the movement of people and public gatherings. When it came to welcoming back tourists, some countries partially reversed or paused their opening, others took a piecemeal approach, and a few decided to keep their borders closed with no tourism. Most remarkably, in late July 2020, only three weeks after reopening to foreign travelers, the Bahamas first banned commercial flights from the United States while still accepting visitors from Canada and Europe, and then, a few days later, changed course by announcing a fourteen-day quarantine for all arrivals into the country.[17] Around the same time, Puerto Rico's government announced major rollbacks of its Covid-related domestic restrictions just one day after allowing tourists back on the island and unsuccessfully asked the US Federal Aviation Administration to temporarily suspend flights from Florida and Texas where Covid cases had snowballed.[18] In August, the US Virgin Islands declared a one-month pause on leisure travel to respond more effectively to a Covid outbreak. Jamaica and St. Lucia kept foreign visitors confined to certain areas to better track outbreaks, Aruba and Grenada took a targeted approach by establishing tighter testing requirements for people traveling from high-risk countries, and Trinidad and Tobago and the Cayman Islands essentially continued to have their borders closed to international tourists.[19] The Caribbean country hardest hit by Covid-19 in the initial stage of the pandemic, the Dominican Republic, instead implemented a flexible strategy of open borders, relaxed entry regulations, and free health and travel insurance provided to tourists (shielding them from additional lodging expenses if they fell ill) while maintaining tight controls at hotels and enacting an array of policy measures that severely limited human mobility within the country (Maloney et al. 2023, 2).[20]

In dealing with the Covid-19 emergency, if 2020 was a difficult year for Caribbean countries, 2021 was by no means less difficult as a number of variants of the virus (Alpha and other variants in the first half of the year, then Delta and Omicron) wreaked havoc across the region against a backdrop of slow vaccine rollouts due to vaccine skepticism, logistics delays, and insufficient supply. Cuba started the year with a record number of Covid cases that forced Havana's government to strengthen internal containment measures, dramatically cut flights from the United States and other key source markets, and add the requirement for inbound travelers (besides the PCR test upon arrival) to demonstrate proof of a negative PCR test within three days before arrival. During summer 2021, Cuba also confronted the strongest wave of infection from the Delta variant of the coronavirus disease, which was brought under control only toward the end of the year by mass application of the homegrown vaccines Soberana 02, Soberana Plus, and Abdala. Given the strict Covid protocols for overseas visitors after it reopened its tourist hubs in October 2020, and unlike its principal competitors in the Caribbean, Cuba practically kept its borders closed to international tourists from March 2020 until mid-November 2021, when it finally welcomed back tourists with eased and competitive entry regulations that removed the testing requirement or quarantine period for vaccinated travelers.[21] Meanwhile, other Caribbean countries, too, experienced spikes in Covid cases and modified entry rules. The Dominican Republic, in particular, had no pre-arrival testing requirement until summer 2021, when it started again to mandate a PCR test amid soaring cases. By late 2021, the main difference among Caribbean countries in terms of Covid-related entry requirements for international passengers was the reliance on PCR or Antigen tests with vaccination as an alternative (Dominican Republic, Jamaica, Bahamas, Puerto Rico, US Virgin Islands, Aruba, Barbados, and Cuba, among others) as opposed to mandatory proof of vaccination (Cayman Islands, British Virgin Islands, Grenada, St. Kitts and Nevis, and Trinidad and Tobago). For a while, this reshaped travel trends in the Caribbean by channeling many unvaccinated travelers to the islands that let them in and vaccinated ones to destinations that banned the unimmunized.[22]

At any rate, notwithstanding the arrival of the Omicron variant that caused several coronavirus outbreaks in the Caribbean at the beginning of 2022, it did not take long before all Covid-19 restrictions for foreign travelers were lifted across the region. Cuba did so in April 2022, and by the early fall of that year, most Caribbean countries had done the same amid significant progress (albeit uneven) with vaccination plans.[23] The Caribbean's reopening to international tourism was also facilitated by critical Covid-related policy changes enacted by the region's leading source markets for tourist arrivals. Most importantly, the United States dropped pre-arrival Covid testing for returning citizens and all other incoming travelers on June 12, 2022. Countries of the European Union and Canada followed suit between June and October.[24] Above all, this meant that American, European, and Canadian vacationers in the Caribbean, unless they got really sick and could not travel back home, no longer had to worry about being forced to prolong their stay in the destination country, with considerable additional expenses in case that they had no travel insurance if they tested positive for Covid before their return trip. Since then, the entire Caribbean has remained open for business, and since the pandemic apparently entered a new phase with the public health emergency ending in early May 2023, many countries of the region have stopped reporting new Covid-19 cases.

Impact of Covid-19 on Caribbean Tourism

Aside from the serious risks of death, hospitalization, and long-term health effects (lasting months or even years after the illness) that remain largely unexplained and understudied, Covid-19 has exposed Caribbean people and their economies in a way that they had not experienced in quite some time, causing a severe and unprecedented economic crisis in the region. The economic fallout from the pandemic affected a number of sectors and activities. It also lowered incomes for Caribbean households and worsened living standards through a combination of job losses, business restrictions and closures, and rent loss. Some Caribbean countries suffered severely from the lockdown measures and travel restrictions around the world that triggered an unprecedented decline in global energy demand and oil prices. But the most consequential development was the halt in foreign tourist arrivals to the Caribbean. The collapse in international travel pulled the rug out from under the tourism industry, the mainstay of most Caribbean economies, magnifying the economic shock of Covid-19 and the urgent challenge before Caribbean policymakers to balance the need to keep their citizens safe and enact policies to bring back tourists and replace lost revenues (Arteaga Garavito et al. 2020).

To highlight the negative impact of Covid-19 on Caribbean tourism and the progress made in spearheading a recovery, Table 5.2 presents data on annual overnight visitor arrivals to the Caribbean by destination country for the 2019–23 period. In 2020, when most governments closed their country's borders and instituted emergency public health measures during the peak of the pandemic, and then cautiously reopened with strict entry protocols, the Caribbean region witnessed a precipitous fall in inbound tourism from 26.3 million international stay-over visitors the previous year to 10.3 million, a decrease of around 16 million visitors. And this decline does not even capture the full extent of the shock to Caribbean tourism induced by Covid-19 as the first two months of 2020 were normal high season months that preceded the coronavirus outbreak. Moreover, the collapse in tourist numbers was not of the same magnitude everywhere. As they benefited from US territory status in terms of airlift connectivity when they

Table 5.2 Overnight Visitor Arrivals to Caribbean Countries, 2019–2023 (thousands)

	2019	2020	2021	2022	2023[a]	2023/2019 (%)
Anguilla	95	25	28	74	96	101.0
Antigua and Barbuda	301	125	170	265	282	93.7
Aruba	1,119	368	807	1,101	1,260	112.6
Bahamas	1,807	453	892	1,452	1,720	95.2
Barbados	713	195	145	540	637	89.2
Bermuda	269	42	72	146	185	68.8
British Virgin Islands	302	83	56	173	261	86.4
Cayman Islands	503	122	17	284	429	85.3
Cuba	3,828	1,055	345	1,586	2,402	62.7
Curacao	464	175	265	490	582	125.4
Dominica	90	22	15	61	67	75.5
Dominican Republic	6,446	2,405	4,994	7,163	8,059	125.0
Grenada	188	54	47	149	178	94.7
Jamaica	2,681	880	1,464	2,478	2,876	107.3
Puerto Rico	3,180	2,617	2,755	3,273	3,466	109.0
St. Kitts and Nevis	120	30	20	80	104	86.7
St. Lucia	424	131	199	356	381	89.8
St. Maarten	320	106	249	373	395	123.4
St. Vincent and Grenadines	86	27	24	59	78	90.7
Trinidad and Tobago	389	95	41	226	309	79.4
US Virgin Islands	514	303	685	663	632	123.0
Total	26,341	10,312	14,506	22,870	26,500	100.6

Sources: ONEI 2023a; MINTUR data obtained by the authors in January 2024; UNWTO 2024, 2023a; United Nations World Tourism Organization, "145 Key Tourism Statistics," https://www.unwto.org/tourism-statistics/key-tourism-statistics; Tourism Analytics, https://tourismanalytics.com/caribbean-statistics.html; ForwardKeys 2023a.

Note:
[a] Preliminary estimates

stepped up efforts to get back to business, but also lacked the authority to close their airports and ports as such decision falls under the jurisdiction of the US federal government,[25] Puerto Rico (less than 20 percent) and the US Virgin Islands (around 40 percent) were the Caribbean countries with the smallest percentage drops in visitor arrivals. All other countries recorded percentage drops well over 50 percent to as much as 70–80 percent. In absolute terms, the Dominican Republic was the country that lost the highest number of overnight visitors (some 4 million) in 2020 compared to 2019, more than Cuba (2.8 million), Jamaica (1.8 million), the Bahamas (1.35 million), and Puerto Rico (563,000) (UNWTO 2022c). The situation was so dire that the average hotel occupancy rate across the region hit its lowest point in June 2020 at just 13.5 percent and stood at 30.2 percent for the full year (neither the Dominican Republic nor its main competitors reached 40 percent occupancy rates),[26] which resulted in huge revenue losses for the Caribbean tourism industry, temporary and permanent closures of many tourism-related businesses, and large drops in GDP (documented above) and employment levels (CHTA 2021a, 2021b). According to the World Travel and Tourism Council, as a result of the pandemic, the Caribbean's travel and tourism sector lost $32.7 billion (direct, indirect, and induced impacts) in revenues in 2020 and its contribution to GDP plunged by 53.2 percent that year, worse than the global average decrease of 50.4 percent. Over 700,000 tourism-related jobs were lost, representing close to one-fourth of all jobs in the sector (WTTC 2022b, 2022c).

With many borders reopened in the midst of tremendous difficulties, Caribbean tourism experienced a mild upturn in 2021, with total overnight visitor arrivals in the region reaching $14.5 million, an increase of approximately 4.2 million inbound travelers from 2020. Except for Cuba, where a long wait to end tough entry restrictions linked to the coronavirus caused tourist visits to nosedive, the Caribbean's largest destination markets regained a significant portion of the tourists they had lost the previous year. International tourist arrivals in the Dominican Republic rose from 2.4 million in 2020 to around 5 million in 2021. Similarly, overseas visitors to Jamaica and the Bahamas jumped considerably in 2021, and even Puerto Rico managed to achieve a growth in inbound tourism. Tourism inflows also rebounded in a few other countries, among them Aruba, St. Maarten, St. Lucia, and, above all, the US Virgin Islands. In 2021, with flight capacity ahead of pre-pandemic levels as a result of a major increase in airlift from the US mainland,[27] the US Virgin Islands was the only Caribbean country where visitor arrivals had already surpassed their 2019 levels. Reflecting the moderate resurgence in the Caribbean tourism industry, the region's average hotel occupancy soared to 44.4 percent in 2021, with the highest rates recorded in the US Virgin Islands (71 percent), Puerto Rico (63.9 percent), Aruba (56.5 percent), Curacao (51.6 percent), and the Dominican Republic (46.6 percent).[28] As another encouraging signal, the Caribbean's travel and tourism sector added 311,000 new jobs that year and saw its total contribution to GDP growing by $10.5 billion or 36.6 percent (WTTC 2022c). Nevertheless, tourist arrivals continued to decline in several Caribbean countries in 2021, and this happened not only in countries that had delayed the reopening of their borders to tourists like the Cayman Islands, Trinidad and Tobago, and Cuba, but also in those like Grenada, St. Kitts and Nevis, and the British Virgin Islands that had been welcoming tourists (though with travel restrictions in place) the entire year (UNWTO 2023a). Albeit short-lived, a discernible pattern emerged.

For the most part, international visitor arrivals fell in countries that required mandatory proof of vaccination for incoming travelers, yet tourists bounced back in countries (with Barbados as an exception) where vaccination was just an option in alternative to testing.

In effect, in 2022, especially in the second half of the year, Caribbean tourism showed strong signs of recovery from the pandemic and a return to normalcy. With the gradual removal of testing and vaccination requirements for entry into practically all Caribbean countries, and into the United States, Canada, the European Union, and other key source markets (facilitating return-home travel for international tourists), total stopover visitor arrivals in the Caribbean reached 22.9 million, an increase of nearly 8.4 million travelers from the previous year. The US market, which accounted for more than 50 percent of all overseas visitors to the region, drove the recovery of Caribbean tourism by sending 3.2 million more tourists than in 2021. Arrivals from the European market also rose conspicuously (2.4 million), whereas those from the Canadian market grew more slowly due to restrictions on Canadian outbound travel that remained in effect until October (CTO 2023).[29] Curacao, the Dominican Republic, Puerto Rico, and St. Maarten joined the US Virgin Islands in the group of countries that ended 2022 with international arrivals exceeding their 2019 levels. Aruba, Jamaica, and some other countries were getting close to their pre-Covid numbers while overnight visitors to Cuba were not even half their level in 2019 (UNWTO 2023a). Overall, the average hotel occupancy in the Caribbean jumped to 60.7 percent in 2022, fueled by the positive rates in Aruba (75 percent), Curacao (70.7 percent), Turks and Caicos (70.3 percent), Puerto Rico (69 percent), and, to a lesser extent, the US Virgin Islands (64.7 percent), the Dominican Republic (64.6 percent), and Jamaica (63.1 percent).[30] Still below pre-pandemic levels in both indicators, the Caribbean's travel and tourism sector generated about 303,000 new jobs in 2022 and its total contribution to GDP increased by $20.3 billion or 33.2 percent (WTCC 2023b).

Another critical aspect of the recovery of Caribbean tourism was the steady improvement in air connectivity from the sharp decline in 2020 following the enactment of Covid-19 travel restrictions (IATA 2023). Air connectivity is positively correlated with both tourism development and economic growth, as is tourism activity with economic output. By the end of 2022, major airlines had reestablished most of their Caribbean flight schedules, but they had yet to resume around 25 percent of Caribbean-serving routes.[31] Furthermore, already in decline prior to the pandemic (Ram, Reeves, and James 2018), connectivity between Caribbean countries was (and still is) hampered by insufficient intra-regional airlift as regional airlines did not have enough planes or flight frequency to meet the travel needs of tourists and other types of travelers. Regional airlines can play a particularly important role in boosting multi-destination travel by seeking codeshare and interline agreements to offer more routes and frequencies, establish more complex itineraries, tap new markets, and reduce operational costs resulting in lower passenger fares.[32]

Underscoring the challenges of expanding air connectivity and the unequal progress with it across the Caribbean, the Spanish travel analytics company ForwardKeys informed that seat capacity in international flights to Punta Cana and Santo Domingo, Dominican Republic, in 2022 was, respectively, 24 percent and 10 percent above the 2019 level, while the same indicator was −1 percent for Montego Bay, Jamaica, −4 percent for Aruba, and −14 percent for Nassau, Bahamas. Given that these destinations (less true for the Bahamas) were among the best performers in 2021–22 in terms of tourist arrivals, it is safe to assume that the Caribbean as a whole continued to struggle to restore extra-regional and intra-regional air linkages. In any case, along with an expansion of intra-Caribbean air traffic fueled by new routes launched by regional carriers, several countries such as the Dominican Republic, Jamaica, Puerto Rico, Barbados, the US Virgin Islands, Turks and Caicos, Curacao, and Grenada, reportedly significantly widened their flight networks and international connections in 2023, announcing new flights from major carriers (ForwardKeys 2023b, 2022).[33] In its tourism forecast analysis for the high season from November 2023 to March 2024, another Spanish travel intelligence firm, Mabrian, reported that seat capacity in direct flights to the Caribbean from Europe and North America was expected to increase 15 percent over the same period a year before, and exceed the November 2019–March 2020 level. Regarding specific countries, the growth of seat capacity was 18 percent for the Dominican Republic, 10 percent for Jamaica, and only 1 percent for Cuba. As for the actual values, the estimated seat capacity in direct flights to Cuba (1.49 million) was two and a half times lower than the capacity in flights to the Dominican Republic (3.89 million), and even lower than the capacity in flights to Jamaica (1.71 million) (Mabrian 2023).

As a result of enhanced air connectivity, the elimination of Covid-19 travel barriers, and increased global air travel demand, the Caribbean ended 2023 with international overnight visitor arrivals reaching approximately 26.5 million and finally surpassing their level in 2019 (UNWTO 2024). Such a robust tourism revival was driven by buoyant US wanderlust and by a surge in arrivals from Canada, with the latter, interestingly, being the top growing origin market for premium cabin travelers embarking on long stays and creating opportunities for the promotion of multi-destination trips. Canadian affluent travelers apparently visited mostly the Dominican Republic, Cuba, and Jamaica.[34] The European market remained key despite losing ground, while Latin American source markets were on the rise (ForwardKeys 2023c). However, the pace of tourism growth differed vastly across countries. In 2023, overseas visitors had returned to pre-pandemic levels in eight Caribbean countries, namely Anguilla, Aruba, Curacao, the Dominican Republic, Jamaica, Puerto Rico, St. Maarten, and the US Virgin Islands. With its excellent network of intra and extra-regional air connections, the Dominican Republic had arguably the most noteworthy tourism performance in the region since it received 8.06 million stopover visitors in 2023 and shattered by almost one million the previous record of 7.16 million set in 2022. Stopover tourists in the Bahamas, Grenada, Antigua and Barbuda, St. Vincent and Grenadines, St. Lucia, and Barbados also reached about 90–95 percent of their pre-Covid levels. On the other hand, international tourism in Trinidad and Tobago, Dominica, Bermuda, and above all Cuba, was still far from a complete recovery. Overnight visitors to Cuba stood at a little over 60 percent of their 2019 level in 2023. Fully open to foreign tourists since November 2021, Cuba had the worst recovery performance in the Caribbean region, possibly with the only exception of crisis-torn Haiti.[35]

The pandemic had a devastating effect not only on air travel but also on the cruise business, the fastest-growing sector of the travel industry until things began to unravel in 2020. As a vital region claiming over 40 percent of the cruise market globally (CLIA 2023), the Caribbean took a major economic hit from Covid-19. Cruise lines operating in the Caribbean, above all the US giants Carnival and Royal Caribbean, and to a lesser extent Norwegian, that hold a dominant position in the region, took an even bigger hit. As the first images of cruise ships with stranded passengers and crew, many sick with Covid-19, made headlines around the globe in the early months of 2020, several Caribbean countries prevented cruise ships from docking in their ports and creating contagion hotspots. Consequently, cruise activity in the Caribbean was paralyzed from March 2020 until June 2021, and the resumption of operations was not without setbacks. In late December 2021, most notably, some Caribbean ports started once again to reject cruise ships amid fears of Omicron-related coronavirus infections.[36] It was only in the second half of 2022, and definitely in 2023, that the cruise business resumed strong growth. A remarkable 30.4 million in 2019, the total number of cruise visitors in the Caribbean plunged to 8.5 million in 2020 and to about 3.8 million in 2021, before bouncing back to 19.2 million in 2022 and to an estimated record of 31.1 million in 2023 (CTO 2024, 2023).[37] Based on the level in 2019, Carnival and Royal Caribbean lost a combined $52 billion in revenues in 2020–21, but as the Covid-19 emergency gradually faded, both firms' revenues jumped back to record levels in 2023.[38] There was also a sizable decline in the economic benefits generated by cruises to destination countries across the Caribbean, which amounted to some $2 billion a year prior to the pandemic (González Moncada 2020, 24). A quick look at two popular cruise destinations, the Bahamas and the Dominican Republic, for which data are available, sheds light on the direct negative impact the cessation of cruise operations had on Caribbean economies. The Bahamas' annual revenues from cruise tourism plummeted from $392.8 million in 2019 to $78.3 million in 2020 and $53.8 million in 2021, while the revenues of the Dominican Republic shrunk from $100.8 million in 2019 to $35 million in 2020 and $30.8 million in 2021. Yet both countries' revenues grew considerably in 2022, respectively to $380.7 million and $154.1 million, and they surely increased spectacularly in 2023 given the record numbers of cruise arrivals.[39]

Presented in Table 5.3, data on annual cruise visitor arrivals to the twelve most visited Caribbean destinations (and Cuba) in 2019–23 reveal, as a matter of fact, that the Bahamas and the Dominican Republic outperformed their main competitors in the post-Covid era, experiencing a less steep reduction in passenger visits during the worst phase of the pandemic, and a much faster recovery afterward. During 2020–21, all leading destinations witnessed a sharp decrease in cruise arrivals, with drops ranging from 100 percent in the Cayman Islands (fully closed to cruise tourism from April 2020 until March 2022),[40] to over 90 percent in Turks and Caicos, Jamaica, and Puerto Rico, over 80 percent in St. Kitts and Nevis, Cozumel, St. Maarten, Aruba, Belize, and the US Virgin Islands, nearly 80 percent in the Bahamas, and slightly less than 70 percent in the Dominican Republic (CTO 2022, 2020). All principal Caribbean destinations for cruise tourism have staged a vigorous recovery since 2021, but only three of them ended 2023 with inbound cruise tourists above their pre-pandemic levels. The Caribbean's top cruise destination, the Bahamas, received about 7.8 million cruise passengers in 2023, some 2.3 million (or 43.1 percent) more than in 2019. Now the region's third largest recipient of cruise tourists after Cozumel, Mexico, the Dominican Republic welcomed over 2.2 million cruise passengers in 2023, more than double the level

Table 5.3 Cruise Visitor Arrivals to Principal Caribbean Destinations, 2019–23 (thousands)

	2019	2020	2021	2022	2023	2023/2019 (%)
Bahamas	5,430	1,327	1,115	5,390	7,773	143.1
Cozumel (Mexico)	4,569	1,132	650	2,938	4,077	89.2
Dominican Republic	1,104	343	333	1,325	2,259	204.6
US Virgin Islands	1,433	440	246	1,021	1,613	112.6
St. Maarten	1,632	435	232	844	1,318	80.8
Jamaica	1,553	449	71	852	1,266	81.5
Cayman Islands[a]	1,831	538	0	743	1,271	69.4
Puerto Rico	1,785	465	160	543[b]	1,270	71.1
Turks and Caicos	1,112	206	27	734	939	84.4
Belize	1,171	343	210	618	904	77.2
Aruba	832	259	136	610	818	98.3
St. Kitts and Nevis	983	268	102	481	774	78.7
Cuba	305	14	0	11	13	4.3

Sources: MINTUR data obtained by the authors in January 2024; CTO 2022, 2021, 2020; Tourism Analytics, https://tourismanalytics.com/cruisevisitors-statistics.html.

Notes:

[a] Covid-19-related cruise ship ban from March 2020 until March 2022.

[b] January-November.

in 2019. The US Virgin Islands also surpassed the pre-pandemic levels of cruise tourism, and Aruba was extremely close to doing the same. Meanwhile, Cuba's cruise business did not suffer much from Covid-19 because the administration of Donald Trump had practically killed that nascent business in June 2019 by prohibiting, effective immediately, cruise ships and recreational vessels from departing the United States on temporary sojourn to Cuba. From a record number of more than 600,000 passengers in 2018, annual cruise visitors to Cuba dropped to a trickle during the pandemic. In late 2022, the British ship Marella Discovery 2, an almost thirty-year-old vessel with a capacity of approximately 2,000 guests, was the first ship to resume regular cruises to Havana from Montego Bay, Jamaica (Marella Cruises had already sailed to the Cuban capital in March and April with a smaller vessel),[41] but it appears that only a few small cruise lines have started operations in Cuba since then. Of the roughly 13,000 cruise travelers who visited Cuba in 2023, about 12,500 were British tourists.[42]

To conclude on the adverse impact of the Covid shock on international tourism in the Caribbean, Table 5.4 presents data on the annual gross tourism revenues (overnight and cruise tourism) of Caribbean countries for the 2019–22 period. As a whole, the Caribbean region's total international tourism receipts fell from around $35 billion in 2019 to $14.3 billion in 2020, a nearly 60 percent

Table 5.4 Gross Tourism Revenues of Caribbean Countries, 2019–22 ($US millions)

	2019	2020	2021	2022	2022/2019 (%)
Anguilla	177	47	59	142	80.2
Antigua and Barbuda	994	350	488	695	69.9
Aruba	2,109	946	1,650	2,311	109.6
Bahamas	4,150	967	2,322	4,222	101.7
Barbados	1,250	574	614	918	73.4
Bermuda	601	94	143	342	56.9
Cayman Islands	964	251	39	583	60.5
Cuba	2,645	1,152	417	1,049	39.6
Curacao	706	281	550	988	139.9
Dominica	135	21	16	60	44.4
Dominican Republic	7,472	2,675	5,697	8,406	112.5
Grenada	562	58	59	322	57.3
Jamaica	3,599	1,256	2,095	3,621	100.6
Puerto Rico	3,612	2,921	2,787	3,330	92.2
St. Kitts and Nevis	353	107	153	302	85.5
St. Lucia	1,064	318	564	1,093	102.7
St. Maarten	708	234	492	956	135.0
St. Vincent and Grenadines	242	83	48	149	61.6
Trinidad and Tobago	436	144	53	324	74.3
US Virgin Islands	1,031	686	1,004	1,023	99.2
Total	34,993	14,287	20,870	32,954	94.2

Sources: ONEI 2023a; UNWTO 2023a, 2021a, 2020; United Nations World Tourism Organization, "145 Key Tourism Statistics," https://www.unwto.org/tourism-statistics/key-tourism-statistics; WTTC 2023b.

drop worth over $20 billion. Several small islands saw their tourism revenues shrink by more than 70 percent, and some islands like Bermuda, Dominica, and Grenada recorded an even bigger plunge. Among the principal tourism destinations, the Bahamas suffered the biggest decrease in revenues in 2020 (almost 80 percent or $3.2 billion), followed by Jamaica (65 percent, $2.3 billion), the Dominican Republic (64 percent, $4.8 billion), Cuba (56 percent, $1.5 billion), and Puerto Rico (19 percent, $560 million). Reflecting growing overseas visitor arrivals, most Caribbean countries obtained higher revenues from tourism activities in 2021 compared to 2020, and all countries of the region recorded an increase in tourism revenues the next year. In 2022, along with Aruba, Curacao, St. Lucia, and St. Maarten, the Caribbean's top tourism markets, with the exception of Cuba, had already surpassed their pre-pandemic levels in terms of tourism earnings or, as in the case of Puerto Rico, were approaching those levels. More specifically, in 2022, gross tourism revenues of the Bahamas and Jamaica were virtually at the same level they were in 2019, those of the

Dominican Republic were over 10 percent above, and those of Puerto Rico were around 8 percent below. On the other hand, although the revenues of some small islands were also far below their pre-pandemic levels, Cuba was the Caribbean country most behind in the recovery in 2022, with its tourism revenues not even reaching 40 percent of their level in 2019 (UNWTO 2024; ONEI 2023a). Given the large increases in visitor arrivals virtually everywhere in the Caribbean, it is safe to assume that most countries of the region enjoyed better results in terms of tourism receipts in 2023, yet Cuba most likely was still nowhere near its pre-Covid levels.

The Extra Burden of US Sanctions against Cuba

The remaining part of this chapter digs deeper into the principal trends and results of international tourism in Cuba in the post-Covid era and examines key limitations of the Cuban tourism industry that undermine its performance and competitiveness. A comprehensive analysis of the Cuban tourism sector, though, cannot overlook the adverse effects of toughened US economic sanctions against Cuba under the administration of Donald Trump, first enacted in late 2017 and then intensified further during the last two years of his presidency, that added to the Covid shock by creating more and more problems for prospective visitors to the island, by making it riskier to do business with Cuba, and by curtailing Cuba's hard currency revenues and access to external financing. Among other things, Trump placed sweeping restrictions on US-based travel to Cuba and strict caps on family remittances, prohibited Americans and US companies from doing business with the Cuban military, threatened and waged fines on US and foreign firms for illegal commercial transactions with the Cuban government, allowed US lawsuits over seized property in Cuba, and placed Cuba back on the US list of state sponsors of terrorism, which activated a little-known provision making it more troublesome for international travelers to visit the United States if they had previously visited Cuba. Trump's sanctions against Cuba were eased under the administration of Joe Biden, but the latter's only moves were a minor relaxation of travel restrictions and a few other modest changes that kept Trump's hardline stance largely unaltered. Along with the coronavirus pandemic, Cuban authorities have blamed what they call "the economic, financial and commercial blockade" imposed by the United States, commonly referred to as the US embargo against Cuba, for crippling tourism and the country's domestic industry and for depleting the foreign exchange the island needs to import basic products.[43] While neither Covid nor the embargo are ultimately responsible for critical failings of the Cuban tourism industry that stifle its growth and development, it is undeniable that Cuba's tourism sector has suffered, and continues to suffer, because of its lack of access to the US market, the leading tourism source market for the Caribbean, and because of Washington's extraterritorial policies targeting dealings with Cuba by non-US individuals and companies.

Of all Trump's measures with respect to Cuba, travel restrictions have been the most consequential. As shown in Chapter 3, the ban on individual educational trips (people-to-people travel) to Cuba by US citizens, enacted in November 2017, and the requirement for those trips to be made in groups overseen by a licensed US-based tour guide, cut US overnight visitors to the island by around one-third in 2018. Furthermore, over one hundred hotels, as well as marinas, stores, and rum factories owned by the Cuban military, were now off-limits to Americans since they were included in a US State Department's list of Cuban entities with which financial transactions were prohibited. That list was updated eight times during the Trump presidency to include entities in virtually every sector of the Cuban economy.[44] Trump's measures in early June 2019, which eliminated the people-to-people travel category altogether and banned cruises to Cuba, had a significant impact as well. Although Covid-19 would become an issue in a matter of months and force Cuba to close its doors to foreign tourism, US stopover visitor arrivals to Cuba fell more than 20 percent in the second half of 2019 (after growing in the first half) compared to the same period a year earlier, and saw a similar drop in the first two months of 2020. Cuba also received some 320,000 cruise visitors less in 2019 (after a record number of cruise arrivals in the first half of the year) than in 2018, of which almost half were US visitors.[45] And to increase economic pressure on Cuba, between October 2019 and August 2020, Trump first banned regularly scheduled US flights to all Cuban cities except Havana and then barred public and private charter flights to all Cuban destinations other than Havana.[46] In late September 2020, besides prohibiting US travelers from importing Cuban alcohol and tobacco products and attending conferences and other events in Cuba, the Trump administration announced the creation of a specific list of more than four hundred "Cuba prohibited accommodations" (hotels and homes marketed as private businesses but controlled by the Cuban government or certain well-connected insiders) that Americans could not patronize.[47] All of these measures were still in effect when Cuba reopened its borders to tourists in November 2021, complicating

Cuba's plans to revive its tourism industry and stimulate economic growth in the difficult post-Covid era.

Compared with the Obama years, when US and foreign firms were eagerly exploring commercial opportunities in the Cuban market, doing business with Cuba became more difficult with Trump in office as his administration stepped up civil penalties and enforcement actions for violations of US sanctions against Cuba. Between 2017 and 2020, among several actions, the US Treasury Department fined the US-based travel service providers Expedia Group, Hotelbeds USA, and Cubasphere, the travel insurance firms Allianz Global Risks US Insurance Company, Ace Limited (Switzerland), and Generali Global Assistance, and financial entities such as Toronto Dominion Bank (Canada), Société Générale (France), JPMorgan Chase Bank, Unicredit Bank AG (Germany), and Standard Chartered Bank (UK). Companies headquartered outside the United States were fined either because they were subsidiaries of US firms or, in the case of some banks, for processing transactions to or through the United States. In certain cases, companies were fined for violating multiple sanctions programs.[48] Moreover, Havana's officials claimed that in October 2019, the US government forced third-country companies to terminate their lease contracts with Cuba's flag carrier Cubana de Aviación, which caused the cancelation of various flights to the Dominican Republic, Haiti, Mexico, Venezuela, Martinique, and Guadeloupe.[49] In early June 2020, the Trump administration ordered Marriott International (the owner of Starwood Hotels) to end its hotel operations in Cuba, removing another Obama-era symbol of engagement with the island.[50] Ironically, US media reported that Trump himself violated the US embargo against Cuba in 1998 when executives of his company secretly conducted business in Cuba and funneled the cash to cover expenses for their Cuba trip (over $68,000) through the US consulting firm Seven Arrows Investment and Development Corp., making the expenses appear legal by linking them after the fact to a charitable operation. As recently as 2016, as revealed by Miguel Fluxà, the executive chairman of the Spanish hotel group Iberostar, Trump (already involved in the presidential campaign) apparently was looking at buying hotels in Cuba. Indeed, there has been speculation that Trump's executives violated the embargo through frequent trips to Cuba from 2010 until 2016 to seek golf-related business opportunities, an allegation corroborated by the fact that Trump registered his Trump trademark in Cuba in 2010 to pursue investments in hotels and golf courses on the island.[51]

From a legal standpoint, avoiding being fined by the US Treasury Department for violating US sanctions rules is not the only concern for US and foreign companies doing business with Cuba. They have also been facing lawsuits for running afoul of the Helms-Burton law since the Trump administration activated Title III of that legislation in May 2019, allowing US nationals to sue in US courts any persons or entities that traffic in expropriated properties in Cuba. According to a database compiled by the US-Cuba Trade and Economic Council, plaintiffs so far have filed forty-five lawsuits under Title III against eighty-three defendants, twenty-seven from the United States, fifteen from Cuba, and thirty-one from a number of countries including Canada, Chile, China, Denmark, France, Germany, Netherlands, Panama, Singapore, Spain, Switzerland, Thailand, and the UK. Around half (41) of the defendants were sued for their Cuba-related activities in the areas of travel and tourism, among them cruise lines (Carnival, Royal Caribbean, Norwegian, MSC Cruises), airlines (Air Europa, American Airlines, Iberia, LATAM), hotel chains (Accor, Barceló, Blue Diamond, Iberostar, Jolly, Meliá, NH Hotels, and the Cuban groups Cubanacán, Gaviota, and Gran Caribe), online booking companies (Airbnb, Booking.com, Cheaptickets, Expedia, Hotels.com, Kayak, Orbitz, Travelocity, Tripadvisor, Trivago) and tour operators (Havanatur, Sunwing Travel Group). In the list of defendants, there were also credit card companies (Mastercard and Visa), banks (BNP Paribas, BBVA, National Bank of Canada, Société Générale) and a mix of firms in industries such as mining, sugar, tobacco, and others.[52] Of the lawsuits filed, more than one-third were dismissed in whole, a few were dismissed in courts of appeal, and less than one-fourth entered the discovery phase of litigation. Until now, there has been only one settlement, that of the Swiss manufacturer of building materials LafargeHolcim with an American family that once owned a sugar plantation in Cuba,[53] and one stunning court judgment in favor of a plaintiff. In late December 2022, a US District Judge in Miami ordered the cruise lines Carnival, Royal Caribbean, Norwegian, and MSC to pay a total of $440 million to Delaware-registered Havana Docks Corporation for docking at a seized terminal in Cuba, a decision that has been appealed.[54] All in all, with one notable exception, plaintiffs have not had success thus far in bringing lawsuits under Title III of Helms-Burton, but some cases are still being litigated and the ruling against the cruise lines could fuel more lawsuits by Cuban exiles and other US claimants, with foreign firms in Cuba's tourism sector as likely targets (Fox and Delgado 2023; Freyre, Domb, and Maxwell 2022; Bellinger, III et al. 2021).

As if all of this were not enough, right before leaving office, on January 12, 2021, Trump rolled back further Obama's orchestrated détente with Cuba by returning the island to the US list of state sponsors of terrorism,[55] a designation that would require lengthy deliberations in the future to be rescinded. The designation itself had limited practical effects since it carried prohibitions on US economic aid and exports to Cuba that, to a large degree, were already in place under embargo rules. It did, however, produce three impactful changes. First, it made it more difficult for Cuba to collect and make international payments and obtain credit from banking institutions and prohibited clients of US banks from using credit cards in Cuba. Second, because of Cuba's inclusion on the US terrorism list, the maximum amount of US-origin content in foreign-made items destined for Cuba decreased from 25 percent to 10 percent, which complicated Havana's efforts to import certain products. Third, and most importantly for the Cuban tourism sector, the terrorist tag added Cuba to the list of countries covered under the Visa Waiver Program Improvement and Terrorist Travel Prevention Act of 2015. As a result of this, foreign travelers who have visited Cuba on or after January 12, 2021, are no longer eligible to participate in the Visa Waiver Program (VWP) and use the swift online application process known as ESTA (Electronic System for Travel Authorization) to visit the United States. In its place, they must apply for a visa, which makes the process more cumbersome and more expensive.[56] Instead of paying $21 to obtain travel authorization through ESTA, travelers must apply in person at a designated location and pay $185, and may experience a significant delay between paying for the visa fee and scheduling an interview appointment. Citizens of European countries like France, Germany, Italy, Spain, and the UK that are among the largest source markets of tourists to Cuba are eligible to travel to the United States under the VWP. Since Cuba is still designated as a state sponsor of terrorism, prospective overseas visitors to Cuba in the post-Covid era must assess whether a trip to the island is worth facing tightened US entry requirements in the future, and once they do so, many of them may just decide to go somewhere else. Of course, they may have no interest in traveling to the United States or may be willing to take a chance and hope that US immigration officials will never know they went to Cuba, especially because Cuban authorities reportedly have recently stopped stamping the passports of inbound international travelers, but it is hard to believe that this kind of uncertainty has no impact on potential vacationers in Cuba.[57]

In effect, despite his early promises, President Biden has done little to roll back Trump's sanctions against Cuba. During his 2020 campaign, Biden tried to avoid as much as he could to comment on Cuba on the assumption that he could end up losing Cuban American votes in South Florida while gaining nothing elsewhere. But eventually, he did make his views known. Responding to a question about his vision for US-Cuba relations included in a March 2020 survey of the *Americas Quarterly* magazine, Biden unequivocally wrote, "Americans—and especially Cuban Americans—can be our best ambassadors for freedom in Cuba. Therefore, as president, I will promptly reverse the failed Trump policies that have inflicted harm on the Cuban people and done nothing to advance democracy and human rights." As election day drew nearer, Biden made campaign stops in Miami where he vowed to relax restrictions on travel and remittances to Cuba that Trump had tightened and pursue greater engagement with Havana.[58] Yet, asides from saying that the Cuba policy was under review, he did nothing during his first six months in office as Cuba ranked low on the list of priorities of his administration while hypersensitivity about the political cost of any softening of sanctions against the island ran high among Democrats after Trump's success with Cuban American voters in the 2020 elections (Leogrande 2021). Cuba did not help either, moving up on Biden's agenda for the wrong reasons after Cuban authorities' harsh response to nationwide anti-government protests on July 11, 2021, which led the US president to impose targeted financial sanctions on certain Cuban officials for their role in the repression and promise to find ways to get remittances directly into the hands of Cubans and help them circumvent internet censorship.[59] For the next several months, Biden's Cuba policy remained on hold. Cuba was in the news for its punitive mass trials against the July 11 protesters, its deteriorating economic conditions, and a huge migrant exodus to the United States that continues to this day. As far as travel to the island was concerned, 2022 began with the news of the settlement (involving a payment of around $91,000) Airbnb had reached with the US Treasury Department for processing payments related to US guests who had traveled to Cuba unauthorized, and for failing to keep required records linked to Cuba-related transactions.[60]

After nearly a year and a half since he had entered the White House as chief executive, Biden finally announced various changes to US policy toward Cuba in May 2022. These changes, which took effect a month later, facilitated family reunification and enhanced the capacity of US consular services to process immigrant visas for Cubans, removed some limitations (particularly the cap of

$1,000 per quarter) on family remittances, increased support for independent Cuban entrepreneurs by authorizing greater access to US internet services, microfinance, and training, and reauthorized scheduled and charter flights to locations beyond Havana as well as group (not individual) people-to-people educational travel and attendance at professional meetings and conferences in Cuba with a general license (Sullivan 2022).[61] Regarding restrictions on US-based travel to Cuba, Biden's measures essentially reestablished the rules that were in place after Trump completed his first major tightening of Cuba sanctions in November 2017. Thus, Cuba's tourism sector has taken a beating not only from the Covid-19 pandemic but also from tough US sanctions during both the Trump and Biden administrations. And more generally, considering that Cuba is still on the US list of state sponsors of terrorism, that US-based remittances are not allowed to go to non-family members in Cuba, which would expand their use for entrepreneurs and humanitarian assistance, and that there is no real intention in the White House to pursue a robust policy of engagement with Havana, it is fair to argue that Biden has done more to stick with Trump's Cuba policy than to reverse it. Even the long-awaited US measures to support Cuban small private businesses and internet freedom on the island, which were enacted in late May 2024,[62] remained a far cry from the Obama-era policy changes that sought to chart a new course on Cuba.

Cuba's Tourism Performance

As noted before, Cuba has struggled to reawaken its tourism sector after keeping the country's borders practically closed to international tourists for more than a year and a half (beginning in March 2020) due to Covid-19. By the end of 2023, unlike its main competitors like the Dominican Republic, Jamaica, and Puerto Rico that had surpassed their pre-pandemic levels in terms of overnight visitor arrivals, and unlike the Bahamas and several other Caribbean countries that were approaching those levels, Cuba had only recovered a little over 60 percent of the overnight visitors it received annually before Covid-19 wreaked havoc on the global travel and tourism industry. The recent poor performance of Cuba's tourism sector is threatening to derail the government's plans to haul the Cuban economy out of a deepening crisis that started over a decade ago when economic support from key ally Venezuela began to fade. Following the death of Venezuelan president Hugo Chávez in early 2013, a major economic crisis in Venezuela significantly weakened the economic ties between Caracas and Havana, basically focused on Venezuelan oil shipments to the island and Cuban exports of medical and other professional services to Venezuela, besides lesser elements such as direct investments by Venezuelan state companies in Cuba, as well as military and security cooperation (Fonseca and Polga-Hecimovich 2020). After growing, on average, 5.3 percent per year in 2004–13, Cuba's annual GDP averaged just 1.6 percent in 2014–19 despite the US-Cuba rapprochement launched in December 2014 during the Obama administration. Since then, along with the Cuban government's failed policies springing from miscalculation and unwillingness to change the country's inefficient centrally planned economic system, the coronavirus pandemic, crippling US sanctions, and the severely pinched Venezuelan lifeline have exacerbated Cuba's economic woes (Spadoni 2023, 638). Russia's war in Ukraine created additional problems in the short term by disrupting the rising flow of Russian tourists to Cuba, but difficult times eventually brought Moscow and Havana closer together both economically and politically.[63] Cuba's GDP shrank by an average of around 2.3 percent in 2020–23 (ONEI 2023a; MEP 2023).

Cuba responded to the global coronavirus outbreak by deploying medical brigades to a number of virus-ravaged nations to treat Covid-19 patients and earn hard currency to make up for plummeting tourism revenues and help paying for much-needed imports. By September 2020, Cuba had sent medical personnel to fight Covid-19 to almost forty countries across five continents, among them Italy, Andorra, China, Kenya, Togo, Jamaica, Peru, Saudi Arabia, South Africa, and Turkey. Some low-income nations received the medics for free, but many others agreed to pay cash for the doctors' services, with the Cuban government retaining a sizable share of those payments.[64] Even more important, since the start of the pandemic, and especially since the launch of Russia's military offensive in Ukraine, Cuba has significantly strengthened economic (and political) ties with Russia in an attempt to revitalize its lagging economy with the infusion of Russian goods, investment, and other financial resources. Albeit nowhere near the level of cooperation that existed in the Soviet era, top Russian and Cuban officials met numerous times and, beginning in 2023, signed a series of agreements under which Russian firms, among other things, would ensure a steady supply of critically needed oil products as well as wheat and fertilizers to Cuba, carry out investments (with special concessions) to increase Cuban sugar and rum output, create a Cuba-based marketplace for Russian food and home goods, and overhaul parts of the island's crumbling tourism infrastructure while promoting Russian tourism in Cuba. In essence, Cuba would become once again a subsidiary market of Russia, yet through public-private business relationships.[65] Regarding tourism, it is worth underscoring that the annual number of inbound Russian visitors to Cuba had increased steadily in the years prior to the pandemic from only 44,000 in 2015 to a peak of about 178,000 in 2019, making Russia the third largest source market for international visitors to Cuba after Canada and the United States, and ahead of traditional Western European markets. In the first two months of 2020 before Covid-19 became an issue, compared to the same period a year earlier, overseas visitor arrivals to Cuba from all main source markets declined except from Canada (a small 5 percent increase) and, most notably from Russia whose tourists to Cuba grew more than 50 percent (ONEI 2021a, 2020a, 2020d).

According to official Cuban data shown in Table 5.5, international visitor arrivals (overnight and cruise visitors) to Cuba plummeted dramatically during the worst stage of the pandemic from approximately 4.3 million in 2019 to less than 1.1 million in 2020 and just about 356,000 in 2021. During the first ten months of 2021, when its borders were essentially closed to foreign tourists, Cuba received some 204,000 international visitors. Around 122,000 or 60 percent of those visitors were Russian tourists who traveled to Cuba, mainly on direct flights to Varadero, under strict Covid-related entry requirements and running the risk of being quarantined over coronavirus infections, which for several of them became a reality (ONEI 2021b).[66] In late January 2022, reflecting the rising importance of Russia as a tourism feeder market, Cuba relaxed Covid rules for Russian tourists arriving in groups, who no longer had to present their Covid vaccination certificates or negative results of their PCR tests.[67] But only a month later, Russia's invasion of Ukraine unleashed fresh problems for crisis-stricken Cuba, in particular higher commodity and shipping costs, and an abrupt fall in Russian tourist arrivals as the closing of western airspace to punish Moscow for its war of aggression made Cuba-bound flights much more expensive and caused the loss of many tourist bookings for the island. Nevertheless, Cuba's strengthening of relations with Russia, including its rejection of Ukraine's joining of the North Atlantic Treaty Organization (NATO) and the apparent involvement of Cuban fighters alongside Russian forces in the fight against Ukraine, did bring some tangible benefits to Havana. Besides various business deals and sizable supplies of Russian oil to Cuba providing relief to debilitating power outages, inbound Russian tourists started to grow strongly again in the second half of 2023 after Russia's flag carrier Aeroflot resumed scheduled

Table 5.5 International Visitors to Cuba by Origin, 2019′–23

	2019	2020	2021	2022	2023	2023/2019 (%)
Canada	1,120,077	413,409	68,944	532,487	936,436	83.6
Cuban residents abroad	623,972	150,388	52,804	333,191	358,480	57.4
Cuban Americans	552,895	129,865	29,451	286,443	320,193	57.9
Russia	177,977	74,019	146,151	54,383	184,819	103.8
United States	498,538	58,147	7,039	100,494	159,032	31.9
Spain	146,339	24,667	12,520	83,025	89,285	61.0
Germany	174,956	45,074	9,159	58,715	69,475	39.7
Mexico	166,995	27,886	2,924	31,123	62,359	37.3
France	169,394	46,537	5,434	44,412	54,493	32.2
Italy	133,436	30,269	5,462	42,854	52,865	39.6
UK	123,745	21,091	3,417	51,275	49,622	40.1
Argentina	95,578	15,955	866	26,086	46,750	48.9
Colombia	43,048	7,315	2,154	22,476	41,103	95.5
Others	786,583	169,971	39,596	233,566	332,260	42.2
Total	4,275,588	1,084,728	356,470	1,614,087	2,436,979	55.9

Sources: ONEI 2024b, 2023a, 2023b, 2022b, 2021a, 2020a; MINTUR data obtained by the authors in January 2024.

flights to Cuba, skirting the airspace of hostile countries.[68] After being the main provider of tourists to Cuba during the critical phase of the Covid-19 pandemic, Russia retained a key role in boosting the island's ailing tourism sector in the post-pandemic era as well.

With its borders reopened to tourists, Cuba welcomed some 1.6 million international visitors in 2022 and little over 2.4 million visitors in 2023, representing approximately 56 percent of their level in 2019. Still, by far, Cuba's largest tourism source market, Canada sent about 936,000 tourists to the island in 2023. Canadian tourists remain the central focus of Cuba's recent tourism marketing campaign, which highlights not only beach vacations but also cultural tours, different kinds of adventure experiences, and wellness travel.[69] Roughly the same number in 2017–19, Cuba's inbound visitor arrivals from Canada were twice the number of arrivals from the United States (Cuban Americans and US citizens of non-Cuban descent combined) in 2023. That year, compared with pre-pandemic levels, Cuba received less visitors from almost all countries, among them Canada (83.6 percent of the 2019 level), the United States (Cuban Americans stood at 57.9 percent and US citizens at just 31.9 percent), Spain (61 percent), the UK (40.1 percent), Germany (39.7 percent), Italy (39.6 percent), Mexico (37.3 percent), and France (32.2 percent). Among Cuba's top twenty tourism source markets, Russia was the only one that had fully recovered its pre-Covid levels, sending a record number of around 185,000 tourists to Cuba in 2023 (ONEI 2024a, 2023a). In the early months of 2024, while the US Department of State reissued an increased caution warning for travelers to Cuba due to petty crime threatening tourists and growing violent crime,[70] and while international tourism in Cuba continued to recover very slowly, winter-weary Russians kept flying to the island in growing numbers. Their trips had been made easier by Russia's launch in December 2023 of its Mir credit card for use in Cuba by Russian citizens.[71] A new round of agreements between Moscow and Havana signed in March 2024 called for an expansion of direct flights from Russia to various Cuban airports and for additional tourism projects. That same month, speaking to Russian reporters at the international tourism fair in Moscow, Cuba's tourism minister Juan Carlos García Granda said that Cuba aimed to receive more than 200,000 Russian tourists in 2024 and hoped their number would reach half a million in the near future.[72] Put simply, the Cuban government currently is pinning its hopes on Russian and Canadian tourists to provide a desperately needed shot in the arm to the Cuban tourism industry that has struggled to revive following the Covid-19 pandemic.

Whereas effective promotion strategies are certainly needed to jumpstart Cuba's tourism sector and allow it to keep up with its competitors in the Caribbean region, more difficult to understand, from a rational standpoint, is Cuba's aggressive hotel-building program that has continued in the post-Covid era despite shortages of international visitors and low hotel occupancy rates. As reported in Table 5.6, Cuba incorporated over 8,000 new rooms (some 2,100 per year) to its hotel room stock during the 2020–23 period as total hotel rooms on the island increased from 74,211 to 82,592. During the 2020–22 period, to keep things in perspective, the hotel room stock of the Dominican Republic grew by little more than 1,000 units from 86,279 to 87,317, and this occurred while international tourist arrivals were rebounding strongly to record levels (more than three times higher than Cuba's levels) and occupancy rates surpassed 70 percent. As the Dominican Republic hit another tourism record in 2023 with nearly 8.1 million overnight visitor arrivals (and 2.2 million cruise passenger arrivals that brought the total to over 10 million visitors), the total number of hotel rooms in the country actually decreased to less than 84,000 as several hotels were either being remodeled or in the process of changing owners.[73]

In Cuba, on the other hand, the military-run Gaviota hotel group alone added 7,511 rooms to its portfolio in 2020–23. Gaviota was responsible for around 90 percent of total room growth on the island during that period, with Cubanacán, Gran Caribe, Islazul, and other Cuban entities making very small contributions. Only 10 percent some two decades earlier, Gaviota's share of Cuba's total hotel rooms had surpassed 50 percent by the end of 2023. At the annual tourism fair in Varadero in May 2022, Cuban prime minister, and former tourism minister and president of Gaviota, Manuel Marrero Cruz, defended Cuba's hotel construction pipeline by saying that "each hotel that is opened is

Table 5.6 Cuba's Hotel Capacity by Entity in 2023 (year-end, existing rooms)

Entity	Rooms	Percentage of Total (%)	2019–23 Growth
Gaviota	41,596	50.4	7,511
Cubanacán	16,576	20.1	639
Gran Caribe	13,926	16.9	67
Islazul	9,457	11.4	141
Other entities	1,037	1.2	23
Total rooms	82,592	100.0	8,381

Source: MINTUR data obtained by the authors in August 2024

a factory that produces an income for the country."[74] Although this is true, the key issue is that a large share of Cuba's hotel rooms are simply not needed given the number of foreign tourists who travel to the island each year. Comparing total international visitor arrivals per existing room in 2022 (based on Cuba's official goal of receiving 2.5 million tourists that year, a goal that was not even met) and in 2018, Cuban economist Pedro Monreal calculated that over half of the country's hotel rooms were unnecessary.[75] And based on the average cost per room (about $200,000) of hotel building projects listed in the portfolio of opportunities for foreign investors unveiled by the Cuban Ministry of Foreign Trade and Investment in late 2021 (MINCEX 2021), the total cost of the roughly 8,400 hotel rooms that were added to Cuba's stock in 2020–23, and primarily featured in Gaviota's properties built with Cuban capital, would be close to $1.7 billion. At a time when Cuba is going through its worst economic crisis in thirty years characterized by plummeting real wages, constant blackouts, shortages of food, medicines, and fuel, and a severe liquidity crunch epitomized by Havana's unprecedented move in late 2023 to seek help from the United Nations' World Food Programme to guarantee the supply of subsidized powdered milk for Cuban children (Carranza 2024; Morris 2024; Mesa-Lago 2023),[76] it would be wise for the Cuban government to hit the pause button on hotel investments and allocate those financial resources to other strategic priorities.

It is worth noting that nearly 63,000 hotel rooms, or around three-fourths (75.8 percent) of Cuba's total room stock at the end of 2023, were in four- and five-star hotels. With 23,445 rooms (28.4 percent of the country's total) and 14,301 rooms (17.3 percent), respectively, Varadero and Havana remained the tourist areas with the largest shares of hotel rooms in Cuba. In fact, some new hotels, including a few ultra-luxury establishments, were inaugurated in these two areas during the 2020–23 period. In Havana, a number of old properties were also refurbished. Since the Covid-19 outbreak, however, more than two-thirds of new hotel rooms have been built in three other tourism poles,[77] namely the northern keys of the province of Villa Clara, where Gaviota has made its biggest investments, the keys of the archipelago of Jardines del Rey in the northern part of the province of Ciego de Ávila, and the northern beaches of the province of Holguín. Additionally, a brand new resort opened its doors near the colonial town of Trinidad, and various properties were renovated in Cayo Largo del Sur, a small island off Cuba's southern coast. As noted in Chapter 3, unveiled in 2016 when tourism prospects looked far brighter, Cuba's 2030 Vision Plan that contemplated the construction of more than 100,000 hotel rooms by 2030 to reach a total of around 166,000 units and envisaged boosting the number of international overnight visitor arrivals to more than 10 million by that year, was replaced in the post-Covid era with a more modest yet still challenging plan to expand Cuba's hotel capacity to 95,000 rooms and attract over six million visitors annually by 2030.[78] Even so, Cuba's strategy to foster tourism development has not changed much over the years. Except for recent efforts to tap more upscale tourism markets and the launch of some advertising campaigns to promote different kinds of tourism (Reimondo Barrios and Reis 2019), Cuban authorities have continued to emphasize sun-and-beach tourism, to rely on the internal cash flow of state-owned corporations like Gaviota, which belongs to the Cuban armed forces' holding company GAESA, to build the vast majority of the projected hotel rooms, and to favor hotel management contracts with foreign chains over mixed enterprises, not to mention their opposition to 100 percent foreign ownership in Cuban hotels.

At the end of 2022, as reported in Table 5.7, seventeen foreign chains administered 108 hotels in Cuba with a combined 43,065 rooms, representing 53.6 percent of the island's total hotel room stock. There were also twenty-two hotels featuring a combined 8,166 rooms with pending management contracts, and 312 hotels managed by Cuban companies, among them the hotels Comodoro, Deuville, Neptuno, Saratoga (heavily damaged by a massive explosion in May 2022), Triton, and the famous Nacional in Havana, several additional historic properties in Old Havana, and a number of beach resorts, most of which are run by Cubanacán. Compared with the situation at the end of 2019, foreign chains managed significantly more hotels and more rooms in Cuba in 2022 (but their share of total rooms was still lower than in 2017 and 2018), and there were twenty-two newly built hotels.[79] New foreign chains with hotel management contracts in Cuba included Indonesia's Archipelago International, Singapore's Banyan Tree, Italy's Elite Club Vacanze, and Spain's Sirenis, NH Hotel Group, and Axel Corporation, while Spain's Sercotel Hoteles and the US company Marriott had left.

Still the foreign chain with the largest hotel portfolio in Cuba, Spain's Meliá Hotels International administered thirty-four hotels on the island with 14,559 rooms at the end of 2022. Meliá had the same number of hotels and roughly the same number of rooms as in 2019, but it had reshuffled its portfolio and renovated various properties. In 2023, it opened the Meliá Trinidad Península hotel, the first five-star hotel in Trinidad, and the Sol Turquesa Beach on Yaraguanal beach in the Holguín tourism pole, took over the management of

Table 5.7 Foreign Hotel Operators in Cuba in 2022 (year-end)

	Company	Country of Origin	Number of Hotels	Number of Rooms	% of Total Rooms
1	Meliá Hotels International	Spain	34	14,559	33.82
2	Blue Diamond Hotels & Resorts	Canada	20	9,396	21.81
3	Iberostar Hotels & Resorts	Spain	15	6,288	14.60
4	Valentín Hotels	Spain	3	1,986	4.61
5	Archipelago	Indonesia	3	1,669	3.87
6	MGM Muthu Hotels	India	4	1,633	3.79
7	Be Live Hotels—Globalia	Spain	7	1,502	3.49
8	Barceló Hotels & Resorts	Spain	3	1,349	3.13
9	Banyan Tree	Singapore	2	768	1.78
10	Roc Hotels	Spain	3	761	1.77
11	Kempinski Hotels S.A.	Switzerland	3	653	1.52
12	Sirenis Hotels & Resorts	Spain	1	633	1.47
13	Accor Hotels	France	1	566	1.31
14	Blau Hotels & Resorts	Spain	2	544	1.26
15	Elite Club Vacanze	Italy	4	444	1.03
16	NH Hotel Group	Spain	2	251	0.58
17	Axel Corporation	Spain	1	63	0.15
	Total		108	43,065	100.00

Source: MINTUR data obtained by the authors in May 2023.

the hotels Sevilla and Habana Catedral in Havana, and disaffiliated from two properties (Gran Hotel and Hotel Camagüey Colon) in the city of Camagüey.[80] Among the leading foreign hotel operators in Cuba, Canada's Blue Diamond has been the most active in recent years. In 2022, with the main purpose of offering Canadian tourists intimate travel experiences in Cuba with vacation packages booked through its partner Sunwing Travel Group, Blue Diamond launched a project to essentially take over Cayo Largo del Sur by renovating and managing four all-inclusive resorts and a group of seven individual villas featuring a total of 1,348 rooms.[81] Since then, Blue Diamond has absorbed quite a few tourist facilities in Cuba, including the hotels Inglaterra, Regis, and Paseo del Prado (an ultra-luxury property renamed Royalton Habana and previously administered by France's Accor Hotels) in Havana, the all-inclusive adults-only Sanctuary White Sands resort in Cayo Cruz in Jardines del Rey, and the hotels Saratoga, Caballeriza, and Esmeralda, under the brand Mystique, in the city of Holguín. Interestingly, in early 2021, Sunwing signed a long-term agreement with Marriott to combine their brand concepts and market at least twenty Blue Diamond's hotels in Mexico, Dominican Republic, Jamaica, and other Caribbean locations under Marriott's Autograph Collection. Once the US embargo against Cuba is lifted, the same marketing arrangement could be extended to Blue Diamond's properties in Cuba, including the Hotel Inglaterra that Marriott reportedly was planning to manage before it was ordered by the Trump administration to close down its business operations on the island.[82] Rivaling that of Meliá, Blue Diamond's current Cuba portfolio consists of thirty-four hotels and resorts, and eight villas.[83]

As for the other foreign hotel chains operating in Cuba, most of them have signed new management contracts since 2019, especially Spain's Iberostar, whose current portfolio includes 18 four- and five-star hotels in locations such as Havana, Trinidad, Varadero, and other beach areas. Two additional chains, Turkey's ATG (owned by the tour operator Anex Tour) and Portugal's Vila Gale, entered the Cuban market in 2023. It is also worth mentioning that the construction of ultra-luxury hotels in Havana continued with the opening of the Gran Hotel Bristol, managed by Switzerland's Kempinski Hotels, in February 2020, and of Archipelago's Grand Aston La Habana in March 2022. Another high-end property, the majestic hotel Tower K in Havana's Vedado district that upon completion will be the tallest building in Cuba, is set to be administered by Iberostar.[84] Ultra-luxury beach resorts, like the Cayo Guillermo Kempinski Resort on the shores of Playa Pilar in Jardines del Rey, are being built as well. All in all, with more hotels in the pipeline, foreign chains still seem quite keen on Cuba and unfazed by the slow recovery of the island's tourism sector from the Covid-19 shock.

Indeed, some key indicators for the 2019–22 period presented in Table 5.8 exemplify Cuba's poor tourism performance in the post-Covid era, a performance that, for the most part, showed only small signs of improvement in 2023. Leaving aside their massive decline in 2020–21 when Cuba's borders were virtually closed, international overnight visitor arrivals to Cuba had recovered to just about 40 percent of their pre-pandemic levels in 2022 and little over 60 percent in 2023.[85] The average length of stay (ALOS) of stopover visitors to Cuba increased considerably from 7.1 days in 2019 to 11.5 days in 2021, when very few people (primarily Russian tourists) visited the island, but by 2022 it was down to 7.9 days, lower than the ALOS in 2014–17. The average year-round hotel occupancy in Cuba plummeted to a meager 11.4 percent in 2021, rebounded weakly to 27.7 percent in 2022, and grew modestly the next year. Occupancy rates in four- and five-star hotels, where international tourists by and large spend their nights while vacationing in Cuba, exhibited

Table 5.8 Selected Indicators of International Tourism in Cuba, 2019–22

	2019	2020	2021	2022
International overnight visitors	3,827,935	1,055,492	344,877	1,585.712
Average length of stay (days)	7.1	8.2	11.5	7.9
Hotel rooms	74,211	75,342	77,809	80,279
Average hotel occupancy rates (%)	49.8	24.0	11.4	27.7
Five-star hotels	57.4	27.5	6.5	29.2
Four-star hotels	53.2	25.4	12.0	24.8
Gross revenues (US millions)	2,645.1	1,152.4	416.9	1,049.5
State sector	2,223.1	1,031.5	365.9	800.0
Private sector	373.1	105.6	38.2	237.0
International transportation	48.9	15.3	12.9	12.5
Gross revenues per tourist (US dollars)	619	1,062	1,169	650

Sources: ONEI 2023a; MINTUR data obtained by the authors in May 2023.

similar disappointing results (ONEI 2024b, 2023a). As for earnings, following a dramatic plunge in 2020–21, gross tourism revenues in both the state and private sectors bounced back in 2022, yet that year they still amounted to 36 percent and 63 percent, respectively, of their level in 2019. Gross revenues per tourist, on the other hand, rose markedly in the aftermath of Covid-19 and then started to drop amid rising inbound visitors. Cuban officials revealed that earnings from tourism activities grew by some $400 million in 2023 (MEP 2023), which would put Cuba's total gross tourism revenues ($1.45 billion) at a little over 50 percent of their level in 2019, and gross revenues per tourist (around $600) back to the low levels in 2018–19. To make things worse, the main indicators of national tourism were significantly worse in 2023 compared to the previous year (ONEI 2024b). Given the eroded purchasing power of Cuban salaries amid a prolonged and near-unprecedented economic crisis, national tourism simply cannot provide a viable alternative to international tourism under current conditions in Cuba. In sum, it is hard to find encouraging aspects of Cuba's tourism recovery in the post-pandemic era.

Principal Problems of Cuba's Tourism Sector

It cannot be denied that the coronavirus pandemic, US economic sanctions, and Russia's war on Ukraine have negatively affected the flow of overseas visitors to Cuba and the overall performance of Cuba's tourism sector. While practically all Covid-related restrictions that hampered international travel have been dropped, and Russian tourist arrivals on the island have resumed strongly after coming to a virtual halt in the immediate aftermath of Russia's invasion of Ukraine, Cuba is still dealing with the adverse impact of stiffer sanctions imposed by the Trump administration and largely kept in place by the Biden administration, and the long-standing US ban on tourism in Cuba. Furthermore, as a result of a prolonged economic and political crisis that began toward the end of the presidency of Hugo Chávez in the early 2010s and has worsened during the presidency of his successor, Nicolás Maduro, Venezuela has drastically reduced its purchases of Cuban professional services, its oil shipments to Havana (also due to US sanctions on vessels transporting Venezuelan oil to Cuba), and its investments in Cuba. Reduced economic ties with Venezuela meant unreliable oil supplies at higher prices and less hard currency revenues for Cuba, which impacted transportation and other costs of the Cuban tourism industry, and Cuba's ability to import goods for the tourism sector and invest in domestic production to meet the sector's demand for goods and services. However, even if the aforementioned developments are important, it would be misleading to say that Cuba's overall tourism performance depends primarily on external factors over which Cuban authorities have little or no control. In reality, the main problems of Cuba's tourism sector have to do more with misguided policies and decisions, the island's tourism model, and critical shortcomings of the Cuban economy. The remaining part of this chapter identifies and analyzes these problems.

First, Cuba does not belong to some key regional organizations that, among other things, foster cooperation among members on tourism matters and could help the island's tourism sector become more fully integrated within the Caribbean region. Besides being a member of the United Nations World Tourism Organization (UNWTO), whose members include the Caribbean countries with the largest tourism sectors (Puerto Rico is an associate member), Cuba belongs to the Association of Caribbean States (ACS) and has a trade and economic agreement with the Caribbean Community (CARICOM). Yet, it is not a member of the Caribbean Tourism Organization (CTO), the Caribbean Hotel & Tourism Association (CHTA), and the Secretariat for Central American Tourism Integration (Secretaría de Integración Turística Centroamericana, or SITCA). Valuable tourism services provided by the CTO and the CHTA, and their affiliations to regional and international entities, may provide certain benefits to Cuba, but membership fees for these organizations are expensive, and Cuba, like the Dominican Republic, unsubscribed from the CTO a few years ago after being a member in the past.[86] Membership of SITCA, instead, should definitely be pursued since it is through that organization that multi-destination tourism programs could be effectively implemented. Whereas the CTO and the CHTA have advocated for some time for special arrangements to position the Caribbean as a multi-destination experience that goes beyond sun-and-beach tourism, SITCA offers a unique opportunity for Cuba since its principal competitor, the Dominican Republic, is a member of that organization along with various Central American countries including the highly connected Panama. In March 2023, at an official event in Havana, the Dominican Ambassador in Cuba, Flavio Amaury Rondón, said that Cuba and the Dominican Republic should market multi-destination vacation packages allowing tourists going to Varadero to also visit Dominican locations like Jarabacoa or Pedernales, and allowing those who visit the Dominican Republic to visit Cuba's northern keys or Santiago de Cuba as well. Dominican media reported that Cuba and Dominican tour operators and travel agents met in Havana in mid-2023 to explore possibilities for multi-destination tourism, and that multi-destination packages with charter flights to Cuba, the Dominican Republic, and Cancún were being promoted.[87] Furthermore, Quintana Roo's authorities have stepped up efforts to attract Russian tourists going to Cuba with multi-destination packages using the daily air connection between Havana and Cancún, which could also open up opportunities for Cuba to receive foreign tourists visiting the Mexican Caribbean.[88]

Second, unlike its Caribbean competitors, Cuba does not have a tourism law, which is essential to establish the fundamental principles of tourism policy, the legal framework that regulates the promotion, development, and control of tourism activities, and the powers of the state together with the rights and obligations of tourism providers and users. Basically, this kind of legislation oversees the various aspects and functions of the tourism industry, and protects the rights and interests of both tourists and businesses. Although it has faced criticism for not keeping up with the times and for limiting the development of emerging tourist destinations in the country, the Dominican Republic's tourism law (*Ley* Orgánica de *Turismo de la República Dominicana, No. 541*) was enacted on December 31, 1969.[89] Subsequently, the Dominican Republic introduced a number of regulations in the 1980s for the classification and operations of hotel establishments (Law 351 of 1964 already allowed casino gambling in hotels), restaurants, gift shops, travel agencies and tour operators, and vehicles used for tourism transportation, and passed Law 158 in 2001 granting special tax incentives to individuals and companies that promote or invest capital in tourism activities in the Dominican market. As for other countries, Mexico enacted its tourism law (*Ley General de Turismo DOF 17-06–2009*) in 2009 and revised it the last time in 2017, Puerto Rico passed similar legislation (*Ley de Desarrollo Turístico de Puerto Rico de 2010, No.74*) on July 10, 2010, and Jamaica has a number of laws regulating various aspects of the tourism industry.[90] Cuba, on the other hand, started working on a preliminary draft of its tourism law in 2012 under Raúl Castro, but the draft never became law.[91] In any case, Cuba's tourism law should be accompanied, if not preceded, by a law on enterprises (*ley de empresas*) that overcomes the dichotomy between the state and non-state sectors of the Cuban economy, and ensures that all kinds of companies (state-owned, private, or mixed) are given the necessary business autonomy, legal security, and incentives to meet their goals. Cuban authorities, reportedly, have been debating and finalizing the future law on enterprises for quite some time, but the legislation has yet to be approved[92]

Third, in stark contrast with other Caribbean countries' tourism officials who share a market-oriented mindset exemplified by their view that foreign capital is critical to financing hotel investments, Cuban officials who are in charge of devising strategies and policies to promote international tourism and maximize its contribution to Cuba's economic development are, by and large, ideologues with a rather uncompromising and dogmatic hostility toward market capitalism, entrenched prejudices about foreign investment, and an aversion to the creation of private wealth. The shortcomings of the Cuban tourism industry, which contributed to its slow post-Covid recovery, are to a significant degree a result of Cuban authorities' persistent commitment to highly centralized decision-making and to a dominant role of the state in the tourism sector and the economy at large. Necessary economic reforms in Cuba are unlikely to come from anywhere but the Cuban military, whose tentacles reach deep into tourism (it owns almost all new Cuban hotels that have been built over the past two decades)[93] and other important sectors of the island's economy, but the opaque, top-down Cuban bureaucracy is likely to complicate the design and implementation of some crucial reforms (Bahamonde 2021). Cuban military officials who hold key positions include Prime Minister Manuel Marrero Cruz, Tourism Minister Juan Carlos García Granda, GAESA's head Ania Guillermina Lastres Morera,[94] Gaviota's president Carlos Latuff Carmenate, and others who are members of Cuba's top government bodies like the Council of State, the Council of Ministers, and the Political Bureau of the Communist Party. Besides,

Army General Raúl Castro still wields great influence on state affairs from behind the scenes at age ninety-three. It is evident that military men populate the upper reaches of power in Cuba and have become a leading entrepreneurial class with personal stakes in the country's future. In the tourism sector, Gaviota's hotel building spree and, in particular, the construction of several emblematic luxury hotels, seem a misguided strategy since the huge amount of money being used would be better spent on renovating existing facilities, upgrading infrastructure and complementary services, and reinvigorating other economic sectors, above all agriculture and energy (Mesa-Lago 2023; Rodríguez 2022). However, from another perspective, the Cuban military might be carrying out real estate investments in the best locations in Havana (Gaviota's takeover of Habaguanex's boutique hotels, stores, and restaurants in Old Havana is part of the same plan) and across Cuba to acquire wealth for itself and possibly to prepare to cash in if major unexpected political (and economic) changes take place.[95]

Fourth, Cuba has an outdated tourism model that exhibited remarkable positive results in the 1990s when the island was an emerging tourist destination in the Caribbean with ample room for growth (Cerviño and Cubillo 2005; Figueras Pérez 2002, 108), but that in recent years (even before the Covid-19 outbreak) has shown clear signs of exhaustion. The development of international tourism in Cuba in the post-Cold War era was conceived amid a profound economic crisis and the introduction of some timid yet significant free-market reforms, and from a scenario of confrontation with the United States involving US travel, trade, and financial restrictions with respect to Cuba that, to a large degree, are still in place today despite a brief easing during the Obama administration. With the urgent goal of earning badly needed foreign exchange to compensate for a sudden drop in merchandise exports, Cuba's strategy was to adopt a tourism model under the conditions of a centralized socialist economy and with limited negative social effects. Since then, Cuba's efforts to boost international tourism have centered on the rapid construction of big four- and five-star hotels, mostly all-inclusive beach resorts and, to a lesser extent, city hotels, initially through government-funded projects that included a few joint ventures with foreign investors, and since the early 2000s almost exclusively with Cuban capital and with foreign chains limiting their involvement to the administration of hotels. This kind of tourism model based on a massive building of hotel rooms brought with it the danger of immobilization of resources that occurs when hotel room capacity increases more rapidly than tourism demand (Perelló Cabrera 2015c).

Over the past three decades, when assessing tourism growth and development, Cuban authorities have focused primarily on indicators such as international visitor arrivals and the expansion of the hotel room stock, giving less emphasis to the fact that, as laid out in the economic policy guidelines adopted by the Cuban government and the Communist Party in April 2011 (and reiterated in 2016 and 2021), "the fundamental objective of the tourism sector is the direct uptake of foreign exchange, maximizing the average income per tourist."[96] Since the mid-1990s, though, overseas visitor arrivals in Cuba have grown at a faster pace than the country's total gross revenues from tourism activities, which resulted in a steady decline of gross revenues per tourist (Ayala Castro 2020). And falling short of this critical goal is not the only major shortcoming of Cuba's tourism sector. The latter suffers from low hotel occupancy rates, low repeat visits of foreign tourists, widespread inefficiencies and low hotel profitability, low quality of all kinds of services, a scarcity of material and financial resources, the lack of building maintenance at hotels, and the deterioration and obsolescence of supporting infrastructure like ports and airports, among other problems (Perelló Cabrera 2020).

In effect, there is little doubt that Cuba's tourism model must be transformed to meet the evolving preferences of international tourists and compete effectively with other Caribbean countries, as conventional sun-and-beach mass tourism, while still a significant contributor to economic growth, is steadily giving way to a multi-experience type of tourism. To be clear, it is not sun-and-beach tourism that is losing appeal. It is a model that offers only (or almost only) beach vacations that is becoming exhausted.[97] With its unique natural, cultural, social, and human assets, Cuba should embark on a new stage of tourism development that incorporates diverse offerings addressing new demands of international travelers, among them those for nature-based tourism and ecotourism, agritourism, heritage and cultural tourism, culinary tourism, wellness tourism, and community-based tourism. Taking advantage of the Caribbean region's air connectivity, these changes in Cuba's offerings should also help promote multi-destination tourism programs. Cuba indeed has made some progress in moving beyond beach tourism by building new hotels in iconic cities like Trinidad and especially Havana, where the reconversion of some architectural wonders from the colonial era into hotels and stores and the proliferation of private restaurants and businesses have expanded the tourism offer (De la Peña et al. 2020, 72). Yet, many of Cuba's greatest assets, in particular the stunning natural spots in the interior of the island, remain unexploited as over 95 percent of Cuba's total hotel rooms are in beach areas or in the cities.[98] Another serious challenge for Cuba is to meet the ever-increasing tourist demand for high-quality services that enhance the overall visitor experience. Along with the island's basic infrastructures (ports, roads, airports, energy, water, information technology) that are in great need of improvement, service standards of Cuba's four- and five-star hotels are generally sub-par compared with similar establishments in other Caribbean destinations. A big part of the problem is that Cuban companies, Gaviota above all, continue to invest far more in the construction of new hotels than in the renovation of existing ones, and foreign chains have little incentive to use their capital to upgrade the hotels they manage because they do not own them. As a result, many Cuban hotels are unable to keep up with maintenance and deteriorate quickly. Furthermore, a real shift to ultra-luxury tourism, which is a new trend in the Caribbean, is unlikely to happen in Cuba under the current conditions since that type of tourism, aside from extremely high accommodation standards, requires a perfect combination of exceptional amenities and personalized service that simply does not exist in Cuba.

Finally, and critically important, Cuba's tourism sector is inserted into a centralized, state-dominated economy with systemic flaws that curtail productivity, efficiency, and trade competitiveness (Spadoni 2023, 655). Major structural problems of the Cuban economy negatively affect the quality of domestically produced goods and services, and, despite some recent positive trends disclosed by the island's tourism authorities, stifle the interconnections or backward linkages of tourism to other economic sectors, particularly agriculture and the manufacturing industry. Citing the contribution of local agricultural producers and changes in the supply chains of various tourism companies, María del Pilar Macías, general director of operations and quality of the Cuban Ministry of Tourism, said in February 2024 that "69% of the [tourism] sector's purchases are made from the national industry."[99] Cuban media also reported that the state enterprise Muebles Ludema, in the city of Las Tunas, produced a variety of items to furnish several hotels in Havana, including luxury establishments like the Gran Hotel Manzana Kempinski and the Iberostar Tower K hotel.[100] This would mean an improvement from the pre-pandemic era in the relative contribution of domestic goods to the needs of the Cuban tourism sector but, aside from the fact that such a result occurred in the context of reduced inbound tourism, it is very unlikely that, under the current conditions, import substitution programs could successfully keep up with a major uptick in international tourism in Cuba.

As a matter of fact, official data on Cuba's agricultural and industrial outputs depict a gloomy picture that underscores the enormous challenges of boosting local sourcing in tourism. In early 2008, under the leadership of Raúl Castro, Cuba started a massive redistribution of unused state land in usufruct to private farmers and cooperatives.[101] Since then, more agricultural reforms have been implemented, among them the expansion of usufruct rights, the granting of greater freedoms to Cuban farmers to choose their crops, sell their output directly to consumers, and obtain supplies and microcredits, some limited price reforms, the revision of the system of procurement quotas, an increase in state payments to farmers, and some decentralization of distribution, but those efforts have done little to reactivate production and solve Cuba's persistent agricultural problems (Gonzalez-Corzo and Nova González 2019, 26–9). In 2022, Cuba's total non-sugar agricultural output was nearly one-third lower than its level in 2007 (Table 5.9). While certain staples like roots and tubers, plantains, and some fruits performed better than others in terms of output, the annual volumes of all staple crops in 2022 were lower than in 2007. Productions of vegetables, especially tomatoes, rice, corn, and citrus products experienced the sharpest drops during this period. There was even a decline in rationed root vegetables and tubers like potatoes, sweet potatoes, and malangas that receive priority in the supply of imported seeds, fertilizers, and plant chemicals. And the overall conditions of Cuban agriculture were precarious even before Cuba's GDP nosedived in 2020 due to the Covid-19 pandemic, stiffened US sanctions, and domestic inefficiencies. If we consider that livestock production is grossly inadequate to meet demand, it comes as no surprise that Cuba imports more than 60 percent of the food it consumes at a cost of roughly $2 billion each year.[102]

Cuba's industrial sector has serious problems as well and has yet to fully recover from its collapse in the early 1990s (Figure 5.1). Excluding the dismal performance of the once-vaunted sugar agro-industrial complex, whose production fell considerably after the disappearance of the Soviet-led Socialist Bloc and has plunged to historic lows in recent years,[103] Cuba's total manufacturing production witnessed a decline of nearly 70 percent between 1989 and 1993 as financial assistance from the Soviet Union came to an end and state investments could no longer be sustained. Notwithstanding some progress in certain subsectors, the physical output of the Cuban manufacturing industry in 2023 was still less than half its level of 1989 (only 38.6 percent if the sugar industry was included). Domestically produced equipment has almost entirely

Table 5.9 Cuba's Non-Sugar Agricultural Production, 2007–22 (thousand tons)

	Roots and Tubers	Plantains	Vegetables	Paddy Rice	Corn	Beans	Citrus	Other Fruits
2007	1,379	991	2,603	440	369	97	469	784
2008	1,393	758	2,439	436	326	97	392	739
2009	1,566	670	2,549	564	305	111	418	748
2010	1,515	735	2,141	454	325	80	345	762
2011	1,445	835	2,200	566	354	133	265	817
2012	1,452	885	2,112	642	360	127	204	965
2013	1,581	659	2,407	673	426	130	167	925
2014	1,671	836	2,499	585	429	136	97	884
2015	1,743	890	2,424	418	363	118	115	943
2016	1,843	1,016	2,385	514	405	137	120	945
2017	1,829	1,015	2,484	405	374	132	99	926
2018	1,801	961	2,454	461	346	162	71	861
2019	1,702	1,036	2,183	426	276	128	71	1,094
2020	1,269	860	1,698	267	257	66	43	863
2021	1,250	861	1,713	226	239	58	37	770
2022	1,332	910	1,434	144	167	69	27	732

Sources: ONEI 2023a, 2022a, 2021a, 2020a, 2014, 2012.

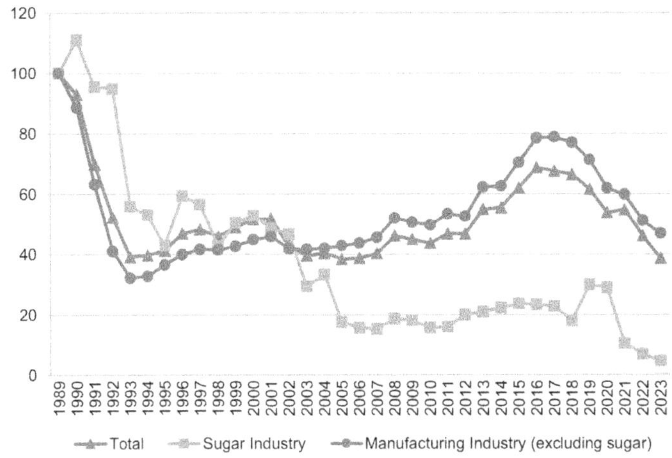

Figure 5.1 Index of industrial output, 1989–2023. Generated by the authors, based on ONEI 2024a, 2023a, 2022a, 2021a, 2020a, 2014, 2012, 2008, 2004, 1998.

disappeared today, and intermediate goods in 2023 did not reach one-sixth of their volume in 1989 (ONEI 2024a). This explains why Cuba must purchase a great deal of machinery and equipment from abroad and why intermediate products account for the lion's share of the country's merchandise imports. In another worrying sign, the physical output of consumer goods, most notably processed foods, decreased sharply in the 2018–23 period, putting further pressure on Cuba's import sector.[104] To sum up, unless structural reforms are enacted to transform Cuba's economic model into a new model that stimulates productivity and efficiency by allowing different forms of ownership to compete with each other, and market forces to drive economic activities and resource allocation (Carranza 2021), it is really hard to envision a scenario in which the demand for products and services of a thriving Cuban tourism sector (with much higher visitor volumes than the current ones) is largely met by domestic suppliers.

It is also worth underscoring that a key systemic flaw of the Cuban economy is its chronic foreign exchange shortage that has actually worsened in recent years, providing yet another reason against the costly hotel construction projects that the Cuban military is carrying out in the midst of the most challenging economic crisis on the island in decades. To demonstrate the gravity of Cuba's current financial situation, Table 5.10 collects data on Cuba's main sources of hard currency between 2018 and 2022. Cuba's total hard currency revenues in 2022 were less than two-thirds of their level in 2018, which translated into

Table 5.10 Cuba's Main Sources of Hard Currency, 2018–22 (US$ millions)

	2018	2019	2020	2021	2022
Exports of Medical Services	6,398	5,382	3,998	4,350	4,882
Remittances	3,692	3,717	2,348	1,084	2,040
Tourism	2,782	2,645	1,152	417	1,049
Exports of Goods	2,373	2,062	1,703	1,966	2,170
Telecommunications Services	723	723	807	159	128
Total	15,968	14,529	10,008	7,976	10,269

Sources: ONEI 2023a, 2022a; Morales 2023.

a loss of $5.7 billion. Nearly all major sources of foreign exchange decreased markedly during this period, especially after 2019, with remittances (which generate revenues to the Cuban government through sales in state-owned hard currency stores) and tourism receipts recording the steepest declines. Exports of goods fluctuated less than other sources, but in 2022 they amounted to less than half their level a decade earlier. After a $2.4 billion plunge in 2018–20, exports of medical services increased somewhat in 2021–2 as Covid-19 gave Cuba an opportunity to send its doctors overseas for cash, but these revenues were largely unable to compensate for the general shortfall in foreign exchange earnings. Remittances and tourism receipts grew as well in 2022 after hitting bottom the year before, yet remained far below their pre-pandemic levels. Cuban official sources also reported that revenues from the provision of telecommunications services plummeted from $807 million in 2020 to $128 million in 2022. Although it is unclear why this sudden drop occurred, it seems reasonable to assume that telecommunications revenues have fallen in recent years as voice and video calling via WhatsApp and other popular free internet-based applications have replaced expensive international calls via cell phones and landline phones. Cuba's foreign exchange revenues most likely experienced little growth, if any, in 2023 as total exports of goods and services as well as remittance inflows reportedly decreased while tourism earnings rose only by an estimated $400 million (Rodríguez 2024; Morales 2024a, 2024b; MEP 2023).

Conclusion

Fueled by arrivals from the US market and, to a lesser extent, the European and Canadian markets, international tourism in the Caribbean had one of the fastest recovery rates globally in 2022 after suffering an unprecedented shock from the Covid-19 pandemic in 2020–1. Continuing this positive trend and, again, outperforming most regions in the world, total overnight visitor arrivals to the Caribbean region finally surpassed their pre-pandemic levels in 2023, albeit by a small margin and with different performances among the destinations. Indeed, aside from the removal of Covid-19 travel barriers, improvements in air connectivity, strategic marketing campaigns, and tourism infrastructure, which played important roles in boosting tourism demand for the Caribbean, were not spread evenly across all Caribbean countries. In 2023, in terms of arrival levels, Anguilla, Aruba, Curacao, the Dominican Republic, Jamaica, Puerto Rico, St. Maarten, and the US Virgin Islands exceeded their 2019 benchmark numbers, and a few additional islands, including the Bahamas and Barbados, approached those numbers. Moreover, various destinations received a new record number of foreign tourists in a single year. The Dominican Republic, by far the biggest success story, had already surpassed the 2019 levels of stay-over visitor arrivals and tourism receipts by 2022. On the other hand, overseas visitors to Bermuda, Dominica, Trinidad and Tobago, and especially Cuba in 2023 were still far below their pre-Covid levels. Also notable, following a fifteen-month Covid-19 induced shutdown, cruise tourism in the Caribbean hit a new high in 2023 with some 31 million visits, and projections indicate that this upward trend will continue in the near-term (CTO 2024).

Unlike its main Caribbean competitors, Cuba has witnessed a frustrating dearth of overseas visitors since the end of the pandemic. The intensification of US sanctions against the island under the Trump administration that have witnessed only a modest easing under the Biden administration, Russia's war on Ukraine, reduced air connectivity and rising global airfares, and domestic shortages of power and fuel exacerbated by shrinking economic ties with Venezuela certainly contributed to the slow post-Covid recovery of international tourism in Cuba. Nevertheless, critical problems of the Cuban tourism sector, which existed well before Covid became an issue, had an adverse impact on Cuba's tourism performance. Among those problems are a weak legal framework, poor integration with multi-destination tourism programs, ill-advised business strategies, chronic capital and material shortages, insufficient maintenance and renovations of hotels, a generally low quality of complementary services, and continued reliance on an inefficient state-dominated tourism model, characterized by high operational costs and low levels of profitability, that remains overly focused on sun-and-beach mass tourism despite showing clear signs of exhaustion. And to further complicate things, Cuba's centrally planned economy has traditionally failed to ensure adequate infrastructure and supply chains for goods and services in tourism. Mired in low growth and stagnation for decades, and currently in extremely dire conditions, the Cuban economy is in urgent need of structural reforms. The island's tourism model should be transformed as well to keep up with new tourism trends and compete effectively in the Caribbean marketplace.

Notes

1 Blen Biru, "The Global Picture of Covid-19 Vaccine Approvals," Launch and Scale Speedometer, April 16, 2021, https://launchandscalefaster.org/blog/global-picture-covid-19-vaccine-approvals.

2 For historical data on the Covid-19 pandemic (deaths, cases, hospitalizations, and other key indicators) for every country in the world, see https://ourworldindata.org/coronavirus.

3 With only 8.2 percent of the world's population, Latin America and the Caribbean had 28 percent of global confirmed cases of Covid-19 and 34 percent of global deaths related to the disease by the end of September 2020. Samuel Pienknagura, Alejandro Werner, and Jorge Roldós, "Pandemic Persistence Clouds Latin America and Caribbean Recovery," IMF Blog, October 22, 2020, https://www.imf.org/en/Blogs/Articles/2020/10/22/blog-whd-reo-october-pandemic-persistence-clouds-latam-and-caribbean-recovery.

4 Between 2010 and 2019, the Caribbean's annual GDP growth averaged just around 0.5 percent (CEPAL 2023a, 2018).

5 Kejal Vyas and Collin Eaton, "A Record Oil Gusher Infuses Tiny Guyana With Wealth and Worries," *Wall Street Journal*, November 1, 2023.

6 "IDB Report Envisions Devastating Tourism Shocks to Latin America and the Caribbean," Inter-American Development Bank (IDB) News Release, July 1, 2020.

7 "St. Vincent and the Grenadines: Country Economic Review 2020," Caribbean Development Bank, June 17, 2021, https://www.caribank.org/publications-and-resources/resource-library/economic-reviews/country-economic-review-2020-st-vincent-and-grenadines.

8 Considered part of the Caribbean because of its cultural ties with the region and its long coastline on the Caribbean Sea, Belize also had its GDP above the pre-pandemic level by the end of 2021 (IMF 2024).

9 Ricardo Torres and Ricardo Gonzáles, "Cuba's Current Economic Crisis and Economic Policy Contradictions," OnCuba News, October 24, 2023.

10 "Dominican Republic Confirms First Case of Coronavirus," Reuters, March 1, 2020.

11 All statistics and calculations reported in this paragraph are from the following sources: Caribbean Public Health Agency (CARPHA) Situation Reports for Covid-19, https://carpha.org/What-We-Do/Public-Health/Novel-Coronavirus/CARPHA-Situation-Reports-for-COVID-19; Worldometer Covid-19 Statistics, https://www.worldometers.info/coronavirus/.

12 Stephanie Nolen, "W.H.O. Ends Global Health Emergency Designation for Covid," *New York Times*, May 5, 2023.

13 It should be noted that the Bahamas stopped reporting new Covid-19 cases in late February 2023.

14 For comprehensive information on the evolution of Covid-related prevention measures (outside travel-related measures), travel advisories, and reopening plans in the Caribbean by country, see the Caribbean Public Health Agency, https://carpha.org/What-We-Do/Public-Health/Novel-Coronavirus/Prevention-Measures-Travel-Advisories-and-Reopening-Plan.

15 "PM Announces Updated Protocols for Entry to Jamaica," *Jamaica Observer*, June 28, 2020.

16 "Cuba Opens Most of Country to Tourism as Enters 'New Normality,'" Reuters, October 8, 2020; "Cuba Reopens Havana Airport after 8-Month Virus Closure," Agence France-Presse, November 11, 2020.

17 "Bahamas Closes Borders to U.S. Tourists After Coronavirus Cases Spike; Other Welcome," Associated Press, July 20, 2020; Jaqueline Charles, "Bahamas Reverses Course on Ban on U.S. Travelers. All Welcome, but Must Quarantine for 14 Days," *Miami Herald*, July 25, 2020.

18 Dánica Coto, "Puerto Rico Closes Bars, Limits Beach Access as COVID-19 Cases Spike," Associated Press, July 16, 2020.

19 Jaqueline Charles, "The Caribbean Has Reopened and COVID-19 Is Spreading—But One Island Is Finding Success," *Miami Herald*, September 22, 2020.

20 Sasha Brady, "Dominican Republic Drops Mass Testing and Offers Visitors Free COVID-19 Insurance," Lonely Planet, September 4, 2020.

21 Andrea Rodríguez, "Now Largely Vaccinated, Cubans Prepare to Welcome Visitors," Associated Press, November 8, 2021.

22 Monica Pitrelli, "Vaccinated and Unvaccinated Travelers Are Going to Different Sets of Caribbean Island," CNBC, August 31, 2021. For specific information on Caribbean entry requirements (test and vaccination rules) by country as of

September 17, 2021, see https://caribbeanhotelandtourism.com/wp-content/uploads/2021/09/CHTA-Caribbean-Entry-Requirements_-September-17.pdf.

23 Justin Ocean, Allison Smith, and Marina Stanley, "Where You Can Travel Now Without a Covid Vaccine or Test," Bloomberg, September 30, 2022. For additional information by country on Caribbean entry requirements for unvaccinated travelers as of September 23, 2022, see https://caribbeanhotelandtourism.com/wp-content/uploads/2022/09/CHTA-Caribbean-Entry-Requirements_-September-23_Unvaccinated.pdf. For data by country on vaccination rates, see Caribbean Public Health Agency, "COVID-19 Vaccine Updates," https://www.carpha.org/What-We-Do/Public-Health/Novel-Coronavirus/COVID-19-Vaccine-Information. Also see Our World in Data, "Coronavirus COVID-19 Vaccinations," https://ourworldindata.org/covid-vaccinations.

24 Ximena Bustillo, "Starting Sunday, You Won't Have to Test for COVID-19 to Fly into the U.S.," NPR, June 10, 2022; "EU Countries Unlikely to Reintroduce COVID Travel Restrictions under New Recommendations," Euronews Travel, December 14, 2022; "Canada to Remove All COVID Travel Restrictions from Oct 1," Reuters, September 26, 2022.

25 Chaseedaw Giles and Carmen Heredia Rodriguez, "Tensions Rise among Residents and Travelers as US Island Territories Work to Stop COVID-19 Spread," *USA Today*, August 21, 2020.

26 Occupancy rates cited here are data from the hospitality consulting firm STR.

27 Lebawit Lily Girma, "Why the Allure of the U.S. Virgin Islands Deep into the Pandemic?" Skift, April 9, 2021.

28 STR data cited in Tourism Analytics, https://tourismanalytics.com/caribbean-statistics.html.

29 Dánica Coto, "Caribbean Sees Jump in Visitors Since Pandemic Began," Associated Press, March 7, 2023.

30 STR, "Caribbean Hotel Industry a 'Comeback Kid,'" Data Insights Blog, February 3, 2023.

31 Brian Major, "Caribbean on Target to Surpass Pre-Pandemic Visitors in 2023," Travel Pulse, March 7, 2023.

32 A codeshare agreements is a business arrangement between two or more airlines to sell seats on the same flight under their own airline designator and flight number. An interline agreement, also known as interline booking, is a commercial arrangements between individual airlines to coordinate ticketing, baggage handling, check-in, and other aspects of their service.

33 Brian Major, "Caribbean Visitor Boom Drives Expanding Flight Schedules," Travel Pulse, August 20, 2023; Brian Major, "Intra-Caribbean Air Expansion Takes Flight," Travel Pulse, February 1, 2024.

34 "US Virgin Islands, Dominican Republic Lead Caribbean Travel Boom," Caribbean Journal, September 13, 2023.

35 Eyder Peralta, "Three Things to Know about the Current Crisis in Haiti," NPR, March 6, 2024. As of March 2024, annual data on international visitor arrivals in Haiti for 2022–23 had yet to be released.

36 Jason Hanna, "Cruise Are Once Again Facing Disruption Because of Covid-19," CNN, December 27, 2021.

37 See Tourism Analytics, https://tourismanalytics.com/cruisevisitors-statistics.html.

38 For Carnival's financial data, see https://www.carnivalcorp.com/financial-information/annual-reporting. For Royal Caribbean's financial data, see https://www.rclinvestor.com/financial-info/key-statistics/.

39 Central Bank of the Bahamas, "Quarterly Economic Review," December 2023, https://www.centralbankbahamas.com/viewPDF/documents/2024-02-29-12-16-18-Quarterly-Economic-Review-December-2023.pdf; Banco Central de la Republica Dominicana, "Estadisticas Turisticas 2022," May 2023, https://cdn.bancentral.gov.do/documents/publicaciones-economicas/estadisticas-turisticas/documents/turisticas2022-12.pdf. The Dominican Republic's annual revenues from cruise tourism are not reported, but they are calculated by the authors from the average spending of cruise passengers.

40 Robert McGillivray, "The Cayman Islands Says No Cruise Ships until Next Year," October 14, 2021.

41 "Crucero Marella Discovery 2 Inicia Temporada Invernal de Turismo," Cubadebate, November 18, 2022.

42 Data of the Cuban Ministry of Tourism obtained by the authors in January 2024.

43 Nelson Acosta, "Cuba Seeks Foreign Investment, Thumbing Nose at US Sanctions," Reuters, November 6, 2023.

44 The most recent list of restricted Cuban entities and subentities as of January 8, 2021, can be found at: https://www.state.gov/cuba-restricted-list/list-of-restricted-entities-and-subentities-associated-with-cuba-effective-january-8-2021/.

45 Data of the Cuban Ministry of Tourism obtained by the authors in February and August 2020.

46 "United States Restricts Scheduled Air Service to Cuban Airports," Media Note, US Department of State, October 25, 2019; "United States Further Restricts Air Travel to Cuba," Press Statement, US Department of State, January 10, 2020; "Suspension of Private Charter Flights Between the United States and Cuba," Press Statement, US Department of State, August 13, 2020.

47 "Cuba Prohibited Accommodations List Initial Publication," Bureau of Economic and Business Affairs, September 28, 2020, https://www.state.gov/cuba-sanctions

/cuba-prohibited-accommodations-list/cuba-prohibited-accommodations-list-initial-publication/.

48 For a list of all US civil penalties and enforcement actions against companies and individuals for violations of Cuba sanctions and other sanctions programs from 2003 until today, see https://ofac.treasury.gov/civil-penalties-and-enforcement-information.

49 "Bloqueo de EEUU Obliga a Empresas de Terceros Países a Cesar Contratos de Arrendamiento con Cubana de Aviación," Cubadebate, October 23, 2019.

50 Sarah Marsh, "Trump Administration Orders Marriott to Cease Cuba Hotel Business," Reuters, June 5, 2020.

51 Kurt Eichenwald, "How Donald Trump's Company Violated the United States Embargo against Cuba," *Newsweek*, September 29, 2016; Stephanie Baker and Sharon Smyth, "Trump Considered Buying Hotels in Cuba, Iberostar Chief Says," Bloomberg, December 2, 2016; Jesse Drucker, "Did Donald Trump's Executives Violate the Cuban Embargo?" Bloomberg, July 18, 2016; Fabiola Santiago, "Trump Conned Miami's Cuban-American Supporters While Chasing Business Opportunities in Cuba," September 23, 2020.

52 See US-Cuba Trade and Economic Council, "Libertad Act and Related Lawsuits," https://www.cubatrade.org/s/Libertad-Act-Filing-Statistics-xcfw.pdf.

53 Amy Guthrie, "In a First, Parties in Helms-Burton Case Reach Settlement Over Property in Cuba," Law.com International, March 25, 2021.

54 Isabel Woodford and Anthony Esposito, "Cruise Lines to Appeal US Court's Big Award to Owner of Havana Dock," Reuters, January 3, 2023.

55 "Republic of Cuba Designation as a State Sponsor of Terrorism (SST)," *Federal Register*, Vol. 86, No. 13, January 22, 2021, 6731.

56 For additional information, see "Visa Waiver Program Improvement and Terrorist Travel Prevention Act Frequently Asked Questions," US Customs and Border Protection, https://www.cbp.gov/travel/international-visitors/visa-waiver-program/visa-waiver-program-improvement-and-terrorist-travel-prevention-act-faq. For months after the inclusion of Cuba in the US list of state sponsors of terrorism, the US Department of State apparently insisted that the ESTA ban was backdated to March 1, 2011. It later clarified that it was backdated to January 12, 2021. Simon Calder, "US U-Turns on Esta Ban for Cuba Tourists," *Independent*, February 1, 2023.

57 Lucy Davies, "Can Europeans Travel to the US If You Have Been to Cuba?" Cubania, September 8, 2023.

58 "Joe Biden Answers 10 Questions on Latin America," *Americas Quarterly*, March 4, 2020; Carmen Sesin, "Biden Slams Trump on 'Abject Failure' on Venezuela, As Well as Cuba Policies," NBC News, September 6, 2020; David Smiley, Bianco Padro

Ocasio, and Jacqueline Charles, "Joe Biden Blasts Donald Trump's Cuba Policy in Miami Visit," *Miami Herald*, October 5, 2020.

59 Jeff Mason and Steve Holland, "U.S. Issues New Cuba Sanctions, Biden Promises More to Come," Reuters, July 31, 2021.

60 "OFAC Settles with Airbnb Payments, Inc. for $91,172.29 Related to Apparent Violations of the Cuban Assets Control Regulations," US Department of the Treasury, January 3, 2022, https://ofac.treasury.gov/media/917236/download?inline.

61 "Cuban Assets Control Regulations," *Federal Register*, Vol.87, No.111, June 9, 2022, 35088-91.

62 Karen DeYoung, "Biden Administration Eases Some Economic Restrictions on Cuba," *Washington Post*, May 28, 2024.

63 Evan Dyer, "Cornered in Ukraine and Isolated by the West, the Kremlin Returns to Cuba," CBC News Canada, June 3, 2023.

64 Sarah Marsh and John Zodzi, "Cuba Punches above Weight with 'White Coat Army' during Pandemic," Reuters, September 15, 2020; Ezra Fieser and Matthew Bristow, "Cuba Is Exporting Doctors to Make Up for Lost Tourism Revenue," Bloomberg, June 15, 2020.

65 Dave Sherwood and Nelson Acosta, "Russia and Cuba Ink Deals to Revitalize Communist-run Island's Ailing Economy," Reuters, May 20, 2023; "Russia and Cuba Sign Trade and Economic Cooperation Programme Until 2030," The Russian Government (Press Service), November 1, 2023.

66 "Over 130 Russian Tourists Remain Isolated on Cuba over Suspected Covid—Consul General," Russian News Agency TASS, July 5, 2021.

67 "Cuba Relaxes COVID Requirements for Russian Tourists Arriving in Groups," Russian News Agency TASS, January 22, 2022.

68 "Russia Aeroflot Resumes Regular Flights to Cuba, Suspended since Ukraine Invasion," Reuters, July 1, 2023; Vera Bergengruen, "How Russia Is Recruiting Cubans to Fight in Ukraine," *Time*, September 18, 2023.

69 Kathryn Folliott, "Cuba Tourist Board Rolls Out Brand New Promotional Campaign," Travel Week, June 10, 2022.

70 For the latest Cuba travel advisory issued by the US Department of State, see https://travel.state.gov/content/travel/en/traveladvisories/traveladvisories/cuba-travel-advisory.html. For additional information on travel to Cuba, see the latest Cuba Country Security Report issued by the Overseas Security Advisory Council (OSAC) at: https://www.osac.gov/Content/Report/60e51af6-edd7-4996-83ff-1c8e3c4acee4.

71 "Russia Launches Its Mir Payment Cards in Cuba," Reuters, December 5, 2023.

72 "Cuba Promueve el Sector Turístico desde Rusia," Prensa Latina, March 20, 2024.

73 Joan Sebastián Vallejo, "Crece Número de Turistas en RD y Disminuyen las Habitaciones Hoteleras," *El Dinero*, December 18, 2023.

74 Ruaridh Nicoll, "Can Cuba's Tourism Adapt to Survive after Covid Devastated the Industry?" *The Guardian*, May 6, 2022.

75 "Casi 1.500 Millones de Dólares en Dos Años: La Innecesaria Inversión del Gobierno Cubano en Turismo," Diario de Cuba, May 12, 2022.

76 Vanessa Buschschlüter, "Cuba Asks UN for Help as Food Shortages Worsen," BBC News, February 29, 2024.

77 Calculations of the authors from MINTUR data obtained in May 2022 and August 2024.

78 "Cuba Presents Its Tourism Portfolio to Spanish Business People," Prensa Latina, January 17, 2022.

79 Calculations of the authors from MINTUR data obtained in May 2022 and May 2023.

80 Meliá Hotels International, "Year End Results 2023," https://www.meliahotelsinternational.com/en/shareholdersAndInvestors/PublicDocs/2023/4Q%202023/MHI%20Release%2012M2023_EN.pdf.

81 "Blue Diamond's Cayo Largo Project Launches Today with Four Hotel Openings," Travel Week, November 6, 2022.

82 James Crump, "Trump Administration Orders Marriott to Close Cuban Hotel, Chain Claims," *The Independent*, June 5, 2020.

83 For the complete list of Blue Diamond's hotels, resorts, and villas in Cuba under the company's different brands, see https://www.bluediamondresorts.com/resorts/chic. For Sunwing's agreement with Marriott in 2021, see https://www.royaltonresorts.com/press/sunwing-travel-group-announces-full-integration-of-20-blue-diamond-resorts-into-marriott-internationals-autograph-collection-and-marriott-bonvoy-loyalty-program.

84 "Cadena Hotelera Iberostar Gestionará el Nuevo Cinco Estrellas de La Habana," Cubadebate, October 20, 2022.

85 MINTUR data obtained by the authors in January 2024.

86 "OECS Wants Reduction in CTO Membership Fees," Loop News, February 14, 2023; A.J. Rao, "Territory Ended Membership in Tourism Group in 2016," *The Virgin Islands Daily News*, May 31, 2018.

87 "Embajada Dominicana en La Habana Informa Creación de Cámara de Comercio," *El Dinero*, March 12, 2023; "Air Century and Cubanacán Join Forces to Promote Multi-destination Tourism," *Dominican Today*, June 2, 2023; "Cuba Attracts Tourists in Multi-destination Packages with Cancún and the DR," *Dominican Today*, February 2, 2023.

88 Ana Ramírez, "Buscan que el Turismo Ruso Llegue a Quintana Roo Vía Cuba," *La Jornada*, January 22, 2024.

89 "The Tourism Law Needs To Be Changed Immediately in the Dominican Republic," *Dominican Today*, November 2, 2022.

90 For additional information on Jamaica's tourism-related laws, see https://www.mot.gov.jm/page/core-legislation.

91 The authors have obtained a copy of the preliminary draft of Cuba's tourism law that was elaborated in 2012.

92 Omar Everleny Perez Villanueva, "Ley de Empresas en Cuba, Necesidad Inaplazable," OnCuba, March 3, 2023.

93 Evan Dyer, "How Canadian Tourism Sustains Cuba's Army and One-party State," CBC News, August 2, 2021.

94 Ania Guillermina Lastres Morera replaced general (and Raúl Castro's former son-in-law) Luis Alberto Rodríguez López-Calleja as chief of Cuba's military-controlled conglomerate GAESA after López-Calleja died of a cardiorespiratory arrest in July 2022. "Cuban General Lopez-Calleja Dead of Heart Attack, State Media Says," Reuters, July 1, 2022; "GAESA Tiene Ahora Jefa: Ania Lastres Morera, en la Lista Oficial de Candidatos a Diputados Cubanos," Diario de Cuba, February 9, 2023.

95 Marc Frank, "Cuban Military's Tentacles Reach Deep into Economy," Reuters, June 15, 2017.

96 *Lineamientos de la Política Económica y Social del Partido y la Revolución* (as approved on April 18, 2011), guideline 255, https://www.tsp.gob.cu/sites/default/files/documentos/1%20lineamientos-politica-partido-cuba.pdf; *Actualización de los Lineamientos de la Política Económica y Social del Partido y la Revolución para el Periodo 2016-2021*, guideline 209, http://media.cubadebate.cu/wp-content/uploads/2017/07/PDF-321.pdf; *Actualización de los Lineamientos de la Política Económica y Social del Partido y la Revolución para el Periodo 2021–2026*, guideline 166, https://www.pcc.cu/sites/default/files/documentos/2024-02/actulizacion-lineamientos.pdf.

97 Michel Suarez, "El Modelo Turistico de Cuba Da Señales de Agotamiento," Diario de Cuba, December 26, 2023.

98 MINTUR data obtained by the authors in August 2024. At the end of 2023, 75.4 percent of Cuba's total hotel rooms were in beach areas and 19.9 percent were in the cities (mainly Havana).

99 Randy Alonso Falcón, Oscar Figueredo Reinaldo, and Angélica Arce Montero, "Ministro del Turismo Apuesta por la Recuperación Paulatina del Sector," Cubadebate, February 7, 2024.

100 "Empresa de Las Tunas Surtirá Mobiliario de Nuevo Hotel de 23 y K, en La Habana," Cubadebate, January 4, 2023; Liodany Arias Tamayo, "Muebles Ludema de Las Tunas por Concretar un Año de Lujo," Agencia Cubana de Noticias, November 5, 2023.

101 Marc Frank, "The State of Raul Castro's Economic Reforms in Cuba," Reuters, April 17, 2018.

102 Andrea Rodriguez, "Reforms Help Cuban Farmers, but Many Still Struggle," Associated Press, August 29, 2022; Marc Frank, "Cuba Loosens State Controls on Agricultural Production," Reuters, April 17, 2021.

103 Ed Augustin, "How Cuba's Sugar Industry Has Been Ground into Dust," Al Jazeera, May 12, 2023.

104 Cuba's production of processed foods in 2018 was 78.8 percent of its value in 1989. That share fell to 69.6 percent in 2019, 64.8 percent in 2020, 52.4 percent in 2021, 40.6 percent in 2022, and to a meager 20.4 percent in 2023 (ONEI 2024a, 2023a).

6

Rebirth Rather Than Recovery

More than four years have passed since international travel came to a standstill in the first half of 2020 as a result of border closures and strict entry rules (quarantine, testing, documentation, and later vaccine requirements) introduced by virtually all countries to limit the spread of the highly contagious novel coronavirus Covid-19. The latter precipitated a devastatingly sharp contraction of global economic activity, hitting particularly hard the tourism industry. The most tourism-dependent region in the world, the Caribbean experienced an unprecedented drop in GDP and a huge decline in foreign exchange earnings as countries saw their economic lifeblood evaporate overnight. These days, as the pandemic-related entry restrictions have been lifted everywhere, tourism is often measured on how well the sector has recovered from the Covid-19 blow. Travel to the Caribbean indeed is on the rise again, with several countries (the Dominican Republic above all) recording inbound overnight visitors in 2023 that exceeded or approached pre-pandemic levels, and some witnessing historically high levels of visitor arrivals. In contrast, Cuba, which kept its borders shuttered for a longer period than most Caribbean destinations but has been welcoming back international travelers since November 2021, is trailing behind regional competitors such as the Dominican Republic, Puerto Rico, Jamaica, the Bahamas, and Cancún in the Mexican Caribbean. Cuba also has almost completely lost the cruise business since Donald Trump banned US cruise ships from sailing to Cuba in early June 2019, ending one of the most popular forms of US-based travel to the island. Meanwhile, cruise tourism in the Caribbean is booming, with the total number of cruise passenger arrivals in the region reaching a record level in 2023 (CTO 2024).

Data from official Cuban sources indicate that overnight visitor arrivals in Cuba in 2023 were just about 63 percent of their level in 2019 (ONEI 2024b). In the first nine months of 2024, these arrivals were nearly 1.7 million (with some additional 20,000 arrivals by sea) or roughly 59 percent of the 2019 level. They were actually a bit less than their level in 2023 and well shy of the pace necessary to meet the country's goal of receiving 3.2 million international visitors in 2024.[1] Since Americans of non-Cuban origin are facing major hurdles to visiting Cuba due to US sanctions rules, and European tourist arrivals remain depressed, Havana's authorities are relying primarily on the traditional Canadian market and on the burgeoning Russian market to help boosting Cuba's ailing tourism sector. In 2023, Russian tourists were the only major group of foreign visitors to Cuba that had surpassed their level in 2019, while Canadian tourists stood at little over 80 percent of pre-pandemic levels. In the first three-quarters of 2024, compared to the same period in 2023, arrivals from Russia continued to increase while arrivals from Canada decreased slightly.[2] At any rate, Cuba's tourism sector needs far more than these two source markets to turn the tide and embark on a path of sustained growth and development. Crippled by a near-unprecedented economic crisis and widespread shortages, Cuba is counting on increased hard currency revenues generated by rising tourist numbers to help import foodstuffs, fuel, medicines, and other basic supplies.

Compared with 2017, the best year for the Cuban tourism industry in terms of inbound stopover visitor arrivals, and excluding Canadians, Russians, and US citizens of non-Cuban descent, Cuba received nearly 1.3 million fewer overnight visitors in 2023. The lost contingent of overseas visitors included 730,000 Europeans (433,000 from Western Europe),[3] about 300,000 Latin Americans, and even 160,000 Cuban residents abroad, mainly Cuban Americans who could travel to the island without restrictions (ONEI 2024a, 2023a, 2019a). Just to put things in perspective, with an average 10 percent annual growth between 2023 and 2030, which is more than double the average growth (4.9 percent) in 2015–19 when international tourism was flourishing, annual visitor arrivals in Cuba would reach some 4.7 million in 2030. This is roughly the same level as in 2017–18, which would arguably mean a lost decade for the Cuban tourism sector. In short, in the medium-term post-pandemic, Cuba's tourism policies and strategies should focus not only on tapping the Canadian and Russian markets but also on reestablishing, under new structures, strong connections with the European market, attracting more tourists from key Latin American countries such as Mexico, Argentina, Colombia, and Brazil, as well as from China and other East Asian countries,[4] and fomenting visits by Cuban permanent residents abroad who could potentially become the top segment of the Cuban tourism market. In 2023, Cuba's air connectivity index as calculated by the International Air Transport Association, which measures how well connected a country's cities are to other cities around the world critical for trade, tourism, and other economic activities, was 25 percent lower than in 2019, while the same indicator for the Dominican Republic and Jamaica was, respectively, 20 percent and 12 percent higher (IATA 2024). By early 2024, Cuba was slowly improving its air connectivity with forty-five international airlines, mostly from Latin America (19), Europe (15), Canada (4), and the United States (4), operating at Cuban airports, yet the overall air traffic was still struggling to regain its pre-pandemic levels. Most notably, aside from reduced traffic from the United States, and as acknowledged by Cuban tourism officials, Cuba's air connections to Europe showed little growth compared to 2019 as weekly flights to Cuban airports from European cities were less than half their number in 2019.[5]

Regarding the future of the Cuban tourism industry, a fundamental aspect is that Cuba should not limit its efforts to spearheading a recovery in international tourist arrivals from the Covid-induced shock. This would surely provide help but also bring back all the problems that have plagued the island's tourism sector for decades, above all the inability to boost efficiency and profitability levels, the comparatively low quality of hotel services, the profound instability of supply chains, chronic financial troubles, and the difficulties to adapt to ever-changing tourism trends in an increasingly globalized world. Major adjustments should be introduced to effectively address, or at least minimize, these problems. In other words, the post-Covid path for the Cuban tourism industry should lead to a rebirth of the industry rather than a simple recovery. The same reasoning applies to the overall Cuban economy because tourism by itself does not develop any country. In fact, with respect to the functioning of the economy, Cuba has tested different models of management over the past sixty years, and none of them has led the country toward a pathway of sustainable development despite some significant achievements in the areas of healthcare and education.

Here are the principal issues confronting the Cuban tourism industry and the strategic actions that Cuba should take:

- Implement commercialization strategies to boost sun-and-beach tourism from the traditional Canadian and European markets given that over 70 percent of Cuba's total hotel rooms, mainly in four- and five-star all-inclusive resorts administered by renowned foreign chains, are in beach areas.[6]
- Diversify source markets and identify new segments. At present, only four groups of travelers, such as Canadians, Cuban residents abroad, Russians,

and US citizens of non-Cuban descent, account for approximately two-thirds of all international arrivals in Cuba. Diversification is all the more important because competing destinations in the Caribbean can rely on two important groups of tourists like US overnight and cruise tourists who cannot travel to Cuba due to restrictions imposed by their government.[7]

- Foment multi-destination tourism programs with the Dominican Republic, Jamaica, and Mexico (Cancún), especially now that those countries are recovering strongly with tourists to spare. In addition to pursuing negotiations with travel agencies of those countries and with regional airlines, a useful move would be to put in place a system of special visas that avoids stamps on passports of tourists who could travel without problems to Cuba and other countries during the same trip.
- Specialize and focus on different types of tourism products besides sun-and-beach tourism, improve tourism marketing campaigns, and increase service quality.
- Integrate more effectively into regional tourism organizations and associations.
- Reduce the pace of construction of new hotels and accelerate investments in hotel renovations as well as non-hotel facilities and infrastructure (airports, roads, railways, and the like). Significant infrastructure investments to enhance internet connectivity and incorporate the new generation of information technology should be pursued as well to improve competitiveness.[8] Inadequate infrastructure, coupled with the relatively poor quality of goods and services, makes it difficult for Cuba to seek high-end tourism.
- Increase the role of foreign investment in the tourism sector (hotels and other projects) through joint ventures, management contracts, and eventually 100 percent foreign-owned enterprises. Invite prospective overseas investors to propose projects to be incorporated into Cuba's portfolio of investment opportunities published annually by the Cuban Ministry of Foreign Trade and Investment.
- Grant greater decision-making autonomy to Cuban hotel corporations and other entities of the tourism sector and free them from excessive state control and auditing.
- Expand and improve training programs for tourism workers (Salinas, Eros, and Mundet 2019, 41).

- Stimulate and stabilize productive chains with economies of scale from the local markets to improve the supply and quality of domestic goods for the tourism sector, redirecting resources to financially support local manufacturing activities and producers of agricultural cooperatives through low interest loans (Pulido Catasús 2023).

Finally, as noted before, reforming the Cuban tourism sector will do little for Cuba's development without reforming the country's economic system with structural changes aimed, among other things, at increasing the multiplier effects of tourism on the overall economy. First of all, Cuba must transform its heavily centralized agricultural model by dismantling the vast and highly inefficient state network responsible for providing supplies to farmers, purchasing most farm output, setting prices, and distributing products (Nova González 2022, 565–6). Moreover, stronger efforts should be made to stimulate economic sectors and production activities with substantial linkages with tourism, particularly, besides agriculture, the local manufacturing industry that remains largely obsolete and decapitalized. A true deregulation of state enterprises is necessary as well. The strict regulatory framework under central planning curtails the ability of Cuban state companies to compete, innovate, assume risks, and express their full production potential (Díaz Fernández 2015, 154–5). There should also be an expansion of the private sector (and of non-agricultural cooperatives) with more opportunities for private businesses to supply goods and services to hotels and other tourism facilities. Additional critical measures must include the granting of greater freedoms to private enterprises to independently carry out export and import activities and access wholesale markets, a real change of mentality in creating the conditions to attract foreign investment, and a more extensive use of market mechanisms and incentives to govern the economy.

To conclude, there is a bumpy road ahead for the Cuban tourism industry whose current conditions are in stark contrast with those of competing Caribbean destinations that are receiving record numbers of international tourists in the post-pandemic era. Because of its ability to generate foreign exchange and contribute to economic growth, tourism is key to Cuba's plans to find a way out of the harshest economic crisis in three decades. However, a significant improvement in Cuba's economic situation depends on the willingness of Cuban authorities to introduce major changes to the island's outdated and increasingly inadequate tourism model, and to the inefficient economic system of centralized planning and state dominance over the market and private enterprises. Since mid-September 2022, Cuba has launched the tourism marketing campaign *Cuba Única* (Unique) in numerous countries to emphasize the unique nature of the island's attractions and people, and the main differences between Cuba and competing Caribbean destinations.[9] While natural beauty and the warmth and friendliness of its people are undeniable features of Cuba, the inability of the Cuban government to address the serious problems of the tourism sector and of the country's overall economy is making Cuba every day more unique, but, unfortunately, for all the wrong reasons.

Notes

1 Calculations of the authors from data of the Cuban Ministry of Tourism (MINTUR) obtained in October 2023 and October 2024; "Cuba Experimenta un Aumento en la Llegada de Turistas," Excelencias, March 28, 2024.

2 Data of the Cuban Ministry of Tourism (MINTUR) obtained by the authors in October 2024.

3 Ruaridh Nicoll, "Europeans Have Abandoned Cuba, and It's All America's Fault," *The Telegraph*, November 6, 2023.

4 "Interview: Cuba Gears Up for More Chinese Tourist Arrivals in Upcoming Peak Season, Tourism Insider Says, Xinhua, October 7, 2023.

5 "Cuba Ya Recibe a 45 Aerolíneas Pese a Quejas por Calidad del Producto," Reportur, April 15, 2024; "Destacan Prioridades del Turismo Cubano para el 2024," Agencia Cubana de Noticias, February 6, 2024.

6 Roberto F. Campos, "Turismo Cubano, Encrucijada y Opiniones," Prensa Latina, June 9, 2020.

7 Edilberto Carmona Camayo and José Luis Perelló, "Cuba en Datos: El Turismo Internacional en Cuba Cierra el 2023 con 2.4 Millones de Turistas," Cubadebate, January 19, 2024.

8 Yisel Martínez García, "Cuba, un Destino Turístico Mas Conectado," *Granma*, August 20, 2019.

9 Samantha Mayling, "New Cuban Campaign Highlights 'Unique' Tourism Offerings," Travel Weekly, September 14, 2022.

Bibliography

Acevedo, Sebastian, and Joyce Wong. 2017. "Cuba Awakening: Potential Risks and Opportunities." In *Unleashing Growth and Strengthening Resilience in the Caribbean*, edited by Trevor Serge Coleridge Alleyne, Inci Ötker, Uma Ramakrishnan, and Krishna Srinivasan, 63–77. Washington, D.C.: International Monetary Fund.

Acevedo, Sebastian, Nicole Laframboise, and Joyce Wong. 2017a. "Caribbean Tourism in the Global Marketplace: Trends, Drivers, and Challenges." In *Unleashing Growth and Strengthening Resilience in the Caribbean*, edited by Trevor Serge Coleridge Alleyne, Inci Ötker, Uma Ramakrishnan, and Krishna Srinivasan, 39–62. Washington D.C.: International Monetary Fund.

Acevedo, Sebastian, Trevor Alleyne, and Rafael Romeu. 2017b. "Revisiting the Potential Impact to the Rest of the Caribbean from Opening US-Cuba Tourism." Working Paper WP/17/100, Washington D.C.: International Monetary Fund

Airbnb. 2017. *Airbnb and Cuba: Two Years of Connecting People and Generating Economic Opportunity for Individuals and Families*. San Francisco: Airbnb.

Alvarez, A., and D. Amat. 1996. "El Mercado Emisor Turístico Estadounidense Hacia el Caribe." Unpublished diploma work, University of Havana.

Ambrosie, Linda M. 2015. *Sun and Sea Tourism: Fantasy and Finance of the All-Inclusive Industry*. Newcastle: Cambridge Scholars Publishing

Arteaga Garavito, Maricruz, Diether Beuermann, Laura Giles Álvarez, Victor Gauto, Jeetendra Khadan, Henry Mooney, and Lodewijk Smets. 2020. *COVID-19: The Caribbean Crisis: Results from an Online Socioeconomic Survey*. Washington, D.C.: Inter-American Development Bank.

Ayala Castro, Héctor. 2020. "Comportamiento de Ingresos por Turismo Internacional en Cuba 2014–2018." *Economía y Desarrollo* 163(1), http://scielo.sld.cu/pdf/eyd/v163n1/0252-8584-eyd-163-01-e13.pdf.

Ayala Castro, Héctor. 2014. "Evolución de Indicadores Turísticos de Cuba 2011–2013." *Retos Turísticos* 13(2), http://retos.mes.edu.cu/index.php/retojs/article/view/62/56.

Babun, Jr., Teo A. 2007. "Reforming Cuba's Port Operations during a Transition." In *Cuba in Transition* 17, 50–8. Washington D.C.: Association for the Study of the Cuban Economy.

Bahamonde, Tamarys L. 2021. "Cuban Bureaucracy and Its Impact on the Economy and Society." *Horizonte Cubano*, Columbia Law School, November 11, https://horizontecubano.law.columbia.edu/news/cuban-bureaucracy-and-its-impact.

Baklanoff, Eric N. 2009. "A Bountiful Legacy: U.S. Investment and Economic Diversification in Cuba During the 1950s." In *Cuba in Transition* 19, 325–35. Washington D.C.: Association for the Study of the Cuban Economy.

Behsudi, Adam. 2020. "Wish You Were Here: Tourism-Dependent Economies Are Among Those Harmed the Most by the Pandemic." *Finance and Development*, International Monetary Fund, https://www.imf.org/en/Publications/fandd/issues/2020/12/impact-of-the-pandemic-on-tourism-behsudi.

Bellinger, III, John B., Thomas A. Shannon Jr., John P. Barker, Baruch Weiss, and Sean A. Mirski. 2021. "Two Years of Title III: Helms-Burton Lawsuits Continue to Face Legal Obstacles." *Arnold & Porter*, May 10, https://www.arnoldporter.com/en/perspectives/advisories/2021/05/two-years-of-title-iii-helmsburton-lawsuits.

Betancourt, Rafael. 2020. "Turismo Internacional y el Sector Privado en Cuba." In *Turismo de Pequeña Escala: Una Mirada Desde Cuba y Costa Rica*, edited by Rafael Betancourt and Ellen Rugh, 32–44. Washington, DC: Center for Responsible Travel.

Blume, Kenneth J. 2012. *Historical Dictionary of the U.S. Maritime Industry*. Lanham, MD: Scarecrow Press.

BNC (Banco Nacional de Cuba). 1988. *Informe Económico*. Havana: BNC.

BREA (Business Research and Economic Advisors). 2018. *Economic Contribution of Cruise Tourism to the Destination Economies: A Survey-based Analysis of the Impacts of Passenger, Crew and Cruise Line Spending*. Phillipsburg, NJ: BREA.

Brenner, Philip. 1988. *From Confrontation to Negotiation: U.S. Relations with Cuba*. Boulder, CO: Westview Press.

Bye, Vegard. 2020. *Cuba, From Fidel to Raúl and Beyond*. London: Palgrave Macmillan.

Cabezas, Amalia L. 2009. *Economies of Desire: Sex and Tourism in Cuba and the Dominican Republic*. Philadelphia: Temple University Press.

Cabezas, Amalia L. 2006. "The Eroticization of Labor in Cuba's All-Inclusive Resorts: Performing Race, Class and Gender in the New Tourist Economy." *Social Identities* 12(5): 507–21.

Cannonier, Colin, and Monica Galloway Burke. 2019. "The Economic Growth Impact of Tourism in Small Island Developing States—Evidence from the Caribbean." *Tourism Economics* 25(1): 85–108.

Capitanía General. 1800. *Calendario Manual, y Guía de Forasteros en la Isla de Cuba, Para el Año de 1800*. Havana: Capitanía General.

Cárdenas O'Farrill, Andrés. 2021. *Cooperation Networks and Economic Development: Cuba's High-Tech Potential*. New York: Routledge.

Carranza, Julio. 2024. "La Economía Cubana y un Nuevo Anuncio." *Horizonte Cubano*, Columbia Law School, February 13, https://horizontecubano.law.columbia.edu/news/la-economia-cubana-y-un-nuevo-anuncio.

Carranza, Julio. 2021. "Cuba: La Reforma Económica y la Percepción del Tiempo." *Horizonte Cubano*, Columbia Law School, May 24, https://horizontecubano.law.columbia.edu/news/cuba-la-reforma-economica-y-la-percepcion-del-tiempo.

Carrazana Amador, Tania Caridad, and Pedro Luis Herrera Cárdenas. 2022. "Las Relaciones Internacionales Canadá-Cuba y el Turismo Canadiense." *Economía y Desarrollo* 166(2), http://scielo.sld.cu/pdf/eyd/v166n2/0252-8584-eyd-166-02-11.pdf

CEE (Comité Estatal de Estadísticas). 1991. *Anuario Estadístico de Cuba 1989*. Havana: CEE.

CEE (Comité Estatal de Estadísticas). 1987. *Anuario Estadístico de Cuba 1986*. Havana: CEE.

CEPAL (Comisión Económica para América Latina y el Caribe). 2023a. *Economic Survey of Latin America and the Caribbean, 2023*. Santiago: CEPAL.

CEPAL (Comisión Económica para América Latina y el Caribe). 2023b. *Preliminary Overview of the Economies of Latin America and the Caribbean, 2023*. Santiago: CEPAL.

CEPAL (Comisión Económica para América Latina y el Caribe). 2022. *Preliminary Overview of the Economies of Latin America and the Caribbean, 2022*. Santiago: CEPAL.

CEPAL (Comisión Económica para América Latina y el Caribe). 2019. *Foreign Direct Investment in Latin America and the Caribbean, 2019*. Santiago: CEPAL.

CEPAL (Comisión Económica para América Latina y el Caribe). 2018. *Economic Survey of Latin America and the Caribbean, 2018*. Santiago: CEPAL.

CEPAL (Comisión Económica para América Latina y el Caribe). 2010. *Estudio Económico de América Latina y el Caribe 2009–2010*. Santiago: CEPAL.

CEPAL (Comisión Económica para América Latina y el Caribe). 2009. *Cuba: Evolución Económica Durante 2008 y Perspectivas para 2009*. Mexico City: CEPAL.

CEPAL (Comisión Económica para América Latina y el Caribe). 2008a. *Foreign Direct Investment in Latin America and the Caribbean, 2008*. Santiago: CEPAL.

CEPAL (Comisión Económica para América Latina y el Caribe). 2008b. *Tourism Life Cycle, Tourism Competitiveness and Upgrading Strategies in the Caribbean*. Port of Spain: CEPAL

CEPAL (Comisión Económica para América Latina y el Caribe). 2006. *Cuba: Evolución Económica Durante 2005 y Perspectivas para 2006*. Mexico City: CEPAL.

CEPAL (Comisión Económica para América Latina y el Caribe). 2005. *Issues and Challenges in Caribbean Cruise Ship Tourism*. Port of Spain: CEPAL

CEPAL (Comisión Económica para América Latina y el Caribe). 2004. *Política Social y Reformas Estructurales: Cuba a Principios del Siglo XXI*. Mexico City: CEPAL

CEPAL (Comisión Económica para América Latina y el Caribe). 2003a. *Caribbean Tourism, Trends, Policies, and Impact, 1985-2002*. Port of Spain: CEPAL.

CEPAL (Comisión Económica para América Latina y el Caribe). 2003b. *A Review of Caribbean Tourism in the 1990s and at the Beginning of the New Century*. Port of Spain: CEPAL

CEPAL (Comisión Económica para América Latina y el Caribe). 2000. *La Economía Cubana: Reformas Estructurales y Desempeño en los Noventa*. Mexico City: CEPAL.

CEPAL (Comisión Económica para América Latina y el Caribe). 1995. *La Inversión Extranjera en Cuba: Aspectos Recientes.* Santiago: CEPAL.

Cerviño, Julio, and José Antonio Fraiz. 2014. "How Much Have Cuban Hotel Chains Learned From Foreign Partners? A Customer Satisfaction Comparison Between Cuban and Foreign-Managed Hotels." *Presentation, Twenty-Fourth Annual Meeting of the Association for the Study of the Cuban Economy* (ASCE), Miami, August 1.

Cerviño, Julio, and José María Cubillo. 2005. "Hotel and Tourism Development in Cuba: Opportunities, Management Challenges, and Future Trends." *Cornell Hotel and Restaurant Administration Quarterly* 46(2): 223–46.

Cintra Roque, Leopoldo. 2017. "La Inversión Extranjera Directa (IED) en Cuba y Sus Implicaciones en el Desarrollo del Turismo." Thesis, Department of Tourism, University of Havana.

CHTA (Caribbean Hotel & Tourism Association). 2021a. *Activity Report: October 2020 – September 2021.* Coral Gables: CHTA.

CHTA (Caribbean Hotel & Tourism Association). 2021b. *Caribbean Tourism Performance Scores High Marks Despite the Pandemic.* Coral Gables: CHTA, September 8.

CHTA (Caribbean Hotel & Tourism Association). 2015. *Cuba: The Great Disruption for the Good of the Caribbean.* Coral Gables: CHTA, June 18.

Clarke, June E., Cynthia Mayo, and Cecil Cornwall. 2015. "A Case Study Analysis on the Competitiveness of Small Lodging Operations in Jamaica." *Journal of Hospitality and Tourism Cases* 5(3): 71–82.

Cleverdon, Robert. 1979. *The Economic and Social Impact of International Tourism on Developing Countries.* London: Economist Intelligence Unit.

CLIA (Cruise Lines International Association). 2023. *State of the Cruise Industry 2023.* Washington, D.C.: CLIA.

CLIA (Cruise Lines International Association). 2022. *State of the Cruise Industry Outlook 2022.* Washington, D.C.: CLIA.

CLIA (Cruise Lines International Association). 2019. *State of the Cruise Industry Outlook 2020.* Washington, D.C.: CLIA.

Colantonio, Andrea, and Robert B. Potter. 2006a. *Urban Tourism and Development in the Socialist State: Havana during the "Special Period"*. Burlington, VT: Ashgate.

Colantonio, Andrea, and Robert B. Potter. 2006b. "The Rise of Urban Tourism in Havana since 1989." *Geography* 91(1): 23–33.

Crahan, Margaret E., and Ariel C. Armony. 2007. "Rethinking Civil Society and Religion in Cuba." In *Debating Cuban Exceptionalism*, edited by Bert Hoffmann and Laurence Whitehead, 139–64. New York: Palgrave Macmillan.

Crespo, Nicolás, and Santos Negrón Díaz. 1997. "Cuban Tourism in 2007: Economic Impact." In *Cuba in Transition* 7, 150–61. Washington D.C.: Association for the Study of the Cuban Economy.

CTO (Caribbean Tourism Organization). 2024. "Caribbean Tourism Experiences Strong Growth in 2023, Recovery to Continue into 2024." March 15, https://www

.onecaribbean.org/caribbean-tourism-experiences-strong-growth-in-2023-recovery-to-continue-into-2024/.
CTO (Caribbean Tourism Organization). 2023. *State of Tourism Industry 2022.* Barbados: CTO.
CTO (Caribbean Tourism Organization). 2021. *Caribbean Tourism Performance Report 2020.* Barbados: CTO.
CTO (Caribbean Tourism Organization). 2020. *Annual Statistical Report 2019.* Barbados: CTO.
CTO (Caribbean Tourism Organization). 2018. *Annual Statistical Report 2017.* Barbados: CTO.
CTO (Caribbean Tourism Organization). 2016. *Annual Statistical Report 2015.* Barbados: CTO.
CTO (Caribbean Tourism Organization). 2015. *Annual Statistical Report 2014.* Barbados: CTO.
CTO (Caribbean Tourism Organization). 2013. "Key Policy Issues in Caribbean Tourism and CTO's Statistical Inventory," https://unstats.un.org/unsd/tradeserv/workshops/Dominica%202013/Presentations/3%20-%20CTO%20-%20Policy%20Questions.pdf.
CTO (Caribbean Tourism Organization). 2011. "State of Industry 2010." Online presentation, February 27, http://www.onecaribbean.org/content/files/CaribbeanTourismStateIndustryFeb272011.pdf.
Da Costa, Michael, and Cyrus Rustomjee. 2022. "IMF Engagement with Small Developing States—Caribbean Case Studies." Independent Evaluation Office (IEO) Background Paper BP/22-01/03, International Monetary Fund, March 30.
Davies, Brian, and Jean Mangan. 1992. "Family Expenditure on Hotels and Holidays." *Annals of Tourism Research* 19(4): 691–9.
Davis, Janet M. 2016. *The Gospel of Kindness: Animal Welfare and the Making of Modern America.* New York: Oxford University Press.
De la Fuente, Alejandro. 2001. *A Nation for All: Race, Inequality, and Politics in Twentieth-Century Cuba.* Chapel Hill, NC: University of North Carolina Press.
De la Peña, Mario Raúl, David Martín-Barroso, Jacobo Núñez, Juan A. Núñez-Serrano, Jaime Turrión, and Francisco J. Velázquez. 2020. "Cuban Tourism Competitiveness: Anything beyond Sun, Beaches, and Son Music?" In *Cooperation between the EU and Cuba for Economic and Productive Reforms: The Challenges of Economic Reform in Cuba*, edited by José Antonio Alonso, 57–74. Barcelona: Barcelona Centre for International Affairs (CIDOB).
De las Cuevas Toraya, Juan. 2001. *500 Años de Construcciones en Cuba.* Havana: Chavín.
Deere, Carmen Diana, and Frederick S. Royce. 2019. "Provisioning Cuba's Private Restaurants: Linking Tourism and Agriculture." *Tourism Planning & Development* 16(6): 675–95.

Díaz-Briquets, Sergio, and Jorge Pérez-López. 2006. *Corruption in Cuba: Castro and Beyond*. Austin: University of Texas Press.

Díaz Fernández, Ileana. 2015. "Autonomía de las Empresas Estatales: Transformaciones Actuales." In *Cuba en un Nuevo Escenario: Perspectivas de Desarrollo Económico*, edited by Ricardo Torres Pérez, Omar Everleny Pérez Villanueva, Fulvio Castellacci, and Claes Brundenius, 144–58. Havana: Publicaciones Acuario.

Djiofack, Calvin, Rei Odawara, Maros Ivanic, and Julie Lohi. 2015. "The Impact of Easing the US Travel Restrictions against Cuba on Tourism in the OECS." Macroeconomics and Fiscal Management Practice Notes 4, World Bank.

D'Meza Pérez, Gustavo, Martha Zaldívar Puig, and Ramón Martín Fernández. 2017. "Costo de los Contratos de Administración Extranjera para la Industria Hotelera Cubana." *Economía y Desarrollo* 158(2): 141–50.

Dupeyras, Alain, and Neil MacCallum. 2013. "Indicators for Measuring Competitiveness in Tourism: A Guidance Document." OECD Tourism Papers 2013/02, OECD Publishing.

Durán Cárdenas, Alejandro. 2000. "Turismo y Economía Cubana: Un Análisis al Final del Milenio." *Economía y Desarrollo* 127(2): 30–59.

Duval, David Timothy. 2004. "Trends and Circumstances in Caribbean Tourism." In *Tourism in the Caribbean: Trends, Development, Prospects*, edited by David Timothy Duval, 3–22. New York: Routledge.

Eckstein, Susan, and Lorena Barberia. 2001. "Cuban-American Cuba Visits: Public Policy, Private Practices." Cambridge, MA: Center for International Studies, MIT.

EIU (Economist Intelligence Unit). 2023. *Business Environment and the Long Term Forecast: Cuba*. London: EIU.

EIU (Economist Intelligence Unit). 2020. *Country Report: Cuba*. London: EIU.

EIU (Economist Intelligence Unit). 2016. *Country Report: Cuba*. London: EIU.

Elliott, Sheryl Marie, and Lisa Delpy Neirotti. 2008. "Challenges of Tourism in a Dynamic Island Destination." *Tourism Geographies* 10(3): 375–402.

English, T. J. 2008. *How the Mob Owned Cuba and Then Lost It to the Revolution*. New York: William Morrow.

Espino, María Dolores. 2013. "Diaspora Tourism: Performance and Impact of Nonresident Nationals on Cuba's Tourism Sector." In *Cuba in Transition* 23, 416–23. Washington D.C.: Association for the Study of the Cuban Economy.

Espino, María Dolores. 2010. "The Cuban Tourism Sector: A Note on Performance in the First Decade of the 21st Century." In *Cuba in Transition* 20, 364–9. Washington D.C.: Association for the Study of the Cuban Economy.

Espino, María Dolores. 2008. "International Tourism in Cuba: An Update." In *Cuba in Transition* 18, 130–7. Washington D.C.: Association for the Study of the Cuban Economy.

Espino, María Dolores. 2000. "Cuban Tourism During the Special Period." In *Cuba in Transition* 10, 360–73. Washington D.C.: Association for the Study of the Cuban Economy.

Espino, María Dolores. 1993. "Tourism in Cuba: A Development Strategy for the 1990s?" *Cuban Studies* 23, 49–69.

Feinberg, Richard E. 2016. *Open for Business: Building the New Cuban Economy*. Washington, D.C.: Brookings Institution Press.

Feinberg, Richard E., and Richard S. Newfarmer. 2016. *Tourism In Cuba: Riding the Wave Toward Sustainable Prosperity*. Washington D.C.: Brookings Institution.

Feinsilver, Julie M. 2008. "Cuba's Medical Diplomacy." In *A Changing Cuba in a Changing World,* edited by Mauricio A. Font, 273–85. New York: Bildner Center for Western Hemisphere Studies.

Figueras Pérez, Miguel Alejandro. 2020. "Foreign Participation in the Development of Tourism in Cuba," *Horizonte Cubano*, Columbia Law School, February 27, https://horizontecubano.law.columbia.edu/news/foreign-participation-development-tourism-cuba.

Figueras Pérez, Miguel Alejandro. 2019. "Impactos del Turismo en la Economía: Avances y Retroceso." In *La Letra de Temas: ¿Cómo Viene el 2019?* January 24, http://www.temas.cult.cu/catalejo/impactos-del-turismo-en-la-econom-avances-y-retroceso

Figueras Pérez, Miguel Alejandro. 2013. "Turismo: Pasado, Presente, y Sus Impactos en la Economía y Sociedad." In *Cuba: La Ruta Necesaria del Cambio Económico*, edited by Omar Everleny Pérez Villanueva and Ricardo Torres Pérez, 204–55. Havana: Editorial de Ciencias Sociales.

Figueras Pérez, Miguel Alejandro. 2002. "El Turismo Internacional y la Formación de Clusters Productivos en la Economía Cubana." In *Cuba: Reflexiones Sobre Su Economía,* edited by Omar Everleny Pérez Villanueva, 99–118. Havana: Universidad de La Habana.

Figueras Pérez, Miguel Alejandro, and David J. Pajón Espina. 2021. "Pensar el Turismo en Cuba en una Etapa Poscovid." In *Miradas a la Economía Cubana: Elementos Claves para la Sostenibilidad*, edited by Ricardo Torres Pérez and Dayma Echevarría León, 115–27. Havana: Centro de Estudios de la Economía Cubana.

Figueras Pérez, Miguel Alejandro, and Yanedy Pérez Cárdenas. 2015. "Competitividad del Destino Turístico a Cuba: Impacto Económico." *Economía y Desarrollo* 153: 177–89.

Fitzgerald, Marguerite. 2017. *Taking the Long View on Cuba's Tourism Opportunity*. Boston: Boston Consulting Group.

Fonseca, Brian, and John Polga-Hecimovich. 2020. *Two Nations, One Revolution: The Evolution of Contemporary Cuba-Venezuela Relations*. Washington, D.C.: Wilson Center.

Foreign Policy Association. 1935. *Problems of the New Cuba*. New York: Foreign Policy Association.

ForwardKeys. 2023a. *ForwardKeys' Destination Insights: Global Travel Trends 2023*. Valencia, Spain: ForwardKeys.

ForwardKeys. 2023b. *Caribbean Travel Trends Report*. Valencia: ForwardKeys.

ForwardKeys. 2022. *The 2022 Most Visited Destinations Report*. Valencia: ForwardKeys.

Fox, Peter R., and Natalia Delgado. 2023. "Four Years Since the Triggering of Title III of Helms-Burton: The Plaintiffs Have Not Done Well." *Horizonte Cubano*, Columbia Law School, November 15, https://horizontecubano.law.columbia.edu/news/four-years-triggering-title-iii-helms-burton-plaintiffs-have-not-done-well.

Franco Rodríguez, Mercedes del Carmen, Lourdes Cisneros Mustelier, and Maité Echarri Chávez. 2021. "El Trabajo por Cuenta Propia en el Turismo: Las Lecciones por Aprender." Universidad de La Habana (UH), Electronic Publication n. 292, May 6.

Freeman, J. Brian. 2011. *La Carretera Panamericana, el Circuito del Caribe y el Turismo Norteamericano en México y Cuba, 1898–1961*. Mexico City: Universidad Nacional Autónoma de México (UNAM).

Freyre, Pedro A., Martin Domb, and Augusto E. Maxwell. 2022. "Cuba: Helms-Burton Act Trends and Development." *Akerman LLP*, January 11, https://www.akerman.com/en/perspectives/cuba-helms-burton-act-trends-and-developments.html.

García Jiménez, Alfredo. 2010. *Turismo y Su Incidencia en el Desarrollo Económico en Cuba*. Havana: Instituto Nacional de Investigaciones Económicas.

García Jiménez, Alfredo. 1999. "La Economía del Turismo en Cuba." *Estudios y Perspectivas en Turismo* 8(3–4): 213–31.

García Jiménez, Alfredo, Pilar Caballero Figueroa, Alfonso Nichar Gladys, and Maricela Esperón Zaldivar. 2006. *Turismo: Desempeño y Futuro*. Havana: Instituto Nacional de Investigaciones Económicas.

García Molina, Jesús M. 2005a. *La Economía Cubana desde el Siglo XVI al XX: Del Colonialismo al Socialismo con Mercado*. Mexico City: Comisión Económica para América Latina y el Caribe (CEPAL).

García Molina, Jesús M. 2005b. *La Reforma Monetaria en Cuba*. Mexico City: CEPAL.

Gomes de Menezes, Antonio, Ana Moniz, and José Cabral Vieira. 2008. "The Determinants of Length of Stay of Tourists in the Azores." *Tourism Economics* 14(1): 205–22.

Gómez García, Olga, Henry Mooney, David Rosenblatt, and María Alejandra Zegarra. 2021. "Imagining a Post-COVID Tourism Recovery: Regional Overview." *Caribbean Quarterly Bulletin* 10(1): 1–25, Inter-American Development Bank (IDB), May.

Gonzalez-Corzo, Mario A., and Armando Nova González. 2019. "Cuban Agriculture after a Decade of Reforms." *The Journal of Business and Economic Studies* 23(2): 24–41.

González Moncada, Verónica. 2020. *Impact of COVID-19 on Transport and Logistics Connectivity in the Caribbean*. Santiago: Economic Commission for Latin America and the Caribbean.

González Núñez, Gerardo. 2022. *Turismo en el Caribe: Pasado, Presente y Futuro*. Lajas, Puerto Rico: Editorial Akelarre.

González Núñez, Gerardo, y Roberto Orro Fernández. 2011. "El Turismo Cubano: ¿Cuantos Norteamericanos Puede Cuba Recibir? In *Cuba in Transition* 21, 415–22. Washington D.C.: Association for the Study of the Cuban Economy.

González-Ripoll Navarro, M. Dolores. 1999. *Cuba, la Isla de los Ensayos. Cultura y Sociedad (1790–1815)*. Madrid: Consejo Superior de Investigación Científica.

Goodman, Robert. 1995. *The Luck Business: The Devastating Consequences and Broken Promises of America's Gambling Explosion*. New York: Free Press.

Gössling, Stefan, Daniel Scott, and C. Michael Hall. 2018. "Global Trends on Length of Stay: Implications for Destination Management and Climate Change." *Journal of Sustainable Tourism* 26(12): 2087–101.

Gott, Richard. 2005. *Cuba: A New History*. New Haven, CT: Yale University Press.

Grandoit, Janouska. 2005. "Tourism as a Development Tool in the Caribbean and the Environmental By-Products: The Stresses on Small Island Resources and Viable Remedies." *Journal of Development and Social Transformation* 2(2): 89–97.

Griswold, Wesley S. 1954. "New Road to Turn History's Vivid Places." *Popular Science* 165(3): 146–8.

Gustavsen, John Andrew. 2009. "Tension Under the Sun: Tourism and Identity in Cuba, 1945–2007." PhD dissertation, University of Miami.

Gutiérrez-Castillo, Orlando, and Nélida Gancedo Gaspar. 2002. "Tourism Development for the Cuban Economy." *ReVista: Harvard Review of Latin America*, Winter, https://revista.drclas.harvard.edu/tourism-development-for-the-cuban-economy/.

Gutiérrez Guerra, Ivis, and Orlando Gutiérrez Castillo. 2011. *Estrategia de Comercialización de Productos Turísticos para el Segmento de Mercado de Cubanos Residentes en el Exterior*. Havana: Centro de Estudios de la Economía Cubana.

Hagedorn, Katherine J. 2001. *Divine Utterances: The Performance of Afro-Cuban Santeria*. Washington, D.C.: Smithsonian Institution Press.

Henken, Ted A. 2018. "'The Revenge of the Jealous Bureaucrat': A Critical Analysis of Cuba's New Rules for *Cuentapropistas*." In *Cuba in Transition* 28, 231–8. Washington D.C.: Association for the Study of the Cuban Economy.

Henthorne, Tony L. 2018. *Tourism in Cuba: Casinos, Castros, and Challenges*. Bingley: Emerald Publishing.

Hernández-Catá, Ernesto. 2019. "Cuba's Swaps of Petroleum for Doctors and the Impact of Cutbacks in Venezuelan Oil Deliveries." In *Cuba in Transition* 29, 22–34. Washington, DC: Association for the Study of the Cuban Economy.

Hernández-Catá, Ernesto. 2001. "The Fall and Recovery of the Cuban Economy in the 1990s: Mirage or Reality?" Working Paper 01/48. Washington, D.C.: IMF.

Honey, Martha. 2018a. "Repaso del Turismo de Cruceros." In *Por el Mar de las Antillas . . . 50 Años de Turismo de Cruceros en el Caribe*, edited by Martha Honey,

José Luis Perelló Cabrera, Jannelle Wilkins, and Rafael Betancourt, 34–41. Havana: Ediciones Temas.

Honey, Martha. 2018b. "Historia y Crecimiento del Turismo de Cruceros." In *Por el Mar de las Antillas . . . 50 Años de Turismo de Cruceros en el Caribe*, edited by Martha Honey, José Luis Perelló Cabrera, Jannelle Wilkins, and Rafael Betancourt, 42–59. Havana: Ediciones Temas.

Honey, Martha. 2018c. "El Modelo Economico. Los Impactos del Turismo de Cruceros" In *Por el Mar de las Antillas . . . 50 Años de Turismo de Cruceros en el Caribe*, edited by Martha Honey, José Luis Perelló Cabrera, Jannelle Wilkins, and Rafael Betancourt, 122–54. Havana: Ediciones Temas.

Honey, Martha. 1999. *Ecotourism and Sustainable Development: Who Owns the Paradise?"* Washington, D.C.: Island Press.

Honey, Martha, and Rafael Betancourt. 2020. "La 'Montaña Rusa' de los Viajes de Estadounidenses a Cuba." In *Turismo de Pequeña Escala: Una Mirada Desde Cuba y Costa Rica*, edited by Rafael Betancourt and Ellen Rugh, 2–12. Washington, D.C.: Center for Responsible Travel.

IATA (International Air Transport Association). 2024. *Quarterly Air Transport Chartbook: IATA Sustainability & Economics Q4 2023*. Montreal: IATA.

IATA (International Air Transport Association). 2023. *Quarterly Air Transport Chartbook: IATA Economics Q4 2022*. Montreal: IATA.

IBRD (International Bank for Reconstruction and Development). 1951. *Report on Cuba: Findings and Recommendations of an Economic and Technical Mission.* Washington D.C.: IBRD.

IDB (Inter-American Development Bank). 2023. *Global and Regional Economies at a Crossroads*. Washington, D.C.: IDB.

IDB (Inter-American Development Bank). 2020. *Assessment of the Effects and Impacts of Hurricane Dorian in the Bahamas*. Washington, D.C.: IDB.

IEO (Independent Evaluation Office). 2022. *IMF Engagement with Small Developing States*. Washington, D.C.: International Monetary Fund.

IMF (International Monetary Fund). 2024. *World Economic Outlook. Steady but Slow: Resilience amid Divergence*. Washington, D.C.: IMF.

Issa, John J., and Chandana Jayawardena. 2003. "The 'All-Inclusive' Concept in the Caribbean." *International Journal of Contemporary Hospitality Management* 15(3): 167–71.

Jatar-Hausmann, Ana Julia. 1999. *The Cuban Way: Capitalism, Communism, and Confrontation*. West Hartford, CT.: Kumarian Press.

Jayawardena, Chandana. 2003. "Revolution to Revolution: Why Is Tourism Booming in Cuba." *International Journal of Contemporary Hospitality Management* 15(1): 52–58.

Jiménez, Marguerite Rose. 2008. "The Political Economy of Leisure." In *A Contemporary Cuba Reader: Reinventing the Revolution*, edited by Philip Brenner, Marguerite Rose

Jiménez, John M. Kirk, and William M. Leogrande, 146–55. Lanham, MD: Rowman and Littlefield.

Jones, Kirby. 2008. "Doing Business in Cuba and Lessons Learned." Bethesda, MD: Alamar.

JTB (Jamaica Tourist Board). 2020. *Annual Travel Statistics 2019*. Kingston: JTB.

Kapcia, Antoni. 2005. *Havana: The Making of Cuban Culture*. Oxford: Berg.

Kirk, Emily. 2011. "Operation Miracle: A New Vision of Public Health?" *International Journal of Cuban Studies* 3(4): 366–81.

Kozak, Kim E. 2016. "Everyday Encounters: Tourists in Casas Particulares." *World Leisure Journal* 58(2): 98–108.

Laframboise, Nicole. 2015. "Yanquis in Havana." *Finance & Development* 52(3): 31–3.

Le Riverend, Julio. 1967. *Historia Económica de Cuba*. Havana: Instituto Cubano del Libro.

Leal Spengler, Eusebio. 1988. *La Habana, Ciudad Antigua*. Havana: Editorial Letras Cubanas.

LeoGrande, William M. 2023. "The Cuban Military: An Economic Powerhouse." In *Contemporary Cuba: The Post-Castro Era*, edited by Hope Bastian, Philip Brenner, John M. Kirk, and William M. LeoGrande, 176–84. Lanham, MD: Rowman and Littlefield.

LeoGrande, William M. 2021: "Not a Top Priority: Why Joe Biden Embraced Donald Trump's Cuba Policy." American University, Center for Latin American and Latino Studies, https://www.american.edu/centers/latin-american-latino-studies/upload/leogrande_us-cuban-relations.pdf.

LeoGrande, William M. 2017. "The Trouble With Cuba's New Economy." *Americas Quarterly*, October 11.

LeoGrande, William M., and Richard S. Newfarmer. 2017. "The Real Economic Impact of Trump's Sanctions on Cuba Travel." Washington D.C.: Brookings Institution.

Lobo Montalvo, María Luisa. 2000. *Havana: History and Architecture of a Romantic City*. New York: Monacelli Press.

Luis, Luis R. 2020. "Cuba's Uncertain Revenues from Medical Exports." *Association for the Study of the Cuban Economy (ASCE) Blog*, July 22, https://www.ascecuba.org/cubas-uncertain-revenues-medical-exports/.

Mabrian. 2023. *Perspectivas de Temporada Otoño-Invierno para Canarias y Caribe*. Menorca: Mabrian.

Maliska, Nicholas. 2009. *Prospects for a Resurgent Cuban Tourism Market*. Washington, D.C.: Council on Hemispheric Affairs.

Maloney, Patrick, Lyudmyla Kompaniyets, Hussain Yusuf, Luis Bonilla, Carmen Figueroa, and Macarena Garcia. 2023. "The Effects of Policy Changes and Human Mobility on the COVID-19 Epidemic in the Dominican Republic, 2020–2021." *Preventive Medicine Reports* 36: 102459.

Maribona, Armando. 1959. *Turismo en Cuba*. Havana: Editorial Lex.

Martín Fernández, Consuelo, Antonio Aja Díaz, Ángela Casaña Mata, and Magali Martín Quijano. 2007. *La Emigración de Cuba desde Fines del Siglo XX y Principios del XXI: Lecturas y Reflexiones Mirando a la Ciudad de la Habana.* Havana: Centro de Estudios de Migraciones Internacionales.

McGuigan, Michael P. 2012. "Fulgencio Batista's Economic Policies, 1952–1958." PhD dissertation, University of Miami.

McIntyre, Meredith A. 2017. "Tax Incentives: To Use or Not To Use?" In *Unleashing Growth and Strengthening Resilience in the Caribbean*, edited by Trevor Serge Coleridge Alleyne, Inci Ötker, Uma Ramakrishnan, and Krishna Srinivasan, 101–22. Washington D.C.: International Monetary Fund.

MEP (Ministerio de Economía y Planificación). 2023. *Objetivos y Metas de la Economía para el Año 2024.* Havana: MEP.

MEP (Ministerio de Economía y Planificación). 2008. *Informe sobre los Resultados Económicos del 2008 y los Lineamientos del Plan Económico y Social para el 2009.* Havana: MEP.

Merrill, Dennis. 2009. *Negotiating Paradise: U.S. Tourism and Empire in Twentieth-Century Latin America.* Chapel Hill: University of North Carolina Press.

Mesa-Lago, Carmelo. 2023. "Cuba's Economy in Times of Crisis: 2020-2022 and Prospects for 2023." Occasional Paper, Florida International University, January.

Mesa-Lago, Carmelo. 2019. "The Cuban Economy After Six Decades of Socialism: Changes, Continuities and the Worsening Crisis." In *Cuba in Transition* 29, 1–21. Washington D.C.: Association for the Study of the Cuban Economy.

Mesa-Lago, Carmelo. 1993. "The Economic Effects on Cuba of the Downfall of Socialism in the USSR and Eastern Europe." In *Cuba After the Cold War*, edited by Carmelo Mesa-Lago, 133–96. Pittsburgh: University of Pittsburgh Press.

Mesa-Lago, Carmelo, and Jan Svejnar. 2020. "The Cuban Economic Crisis: Its Causes and Possible Policies for a Transition." Miami, FIU Vaclac Havel Program for Human Rights & Diplomacy, October.

Mesa-Lago, Carmelo, and Jorge Pérez-López. 2013. *Cuba Under Raúl Castro: Assessing the Reforms.* Boulder: Lynne Rienner.

Mesa-Lago, Carmelo, and Pavel Vidal Alejandro. 2019. "El Impacto en la Economía Cubana de la Crisis Venezolana y de las Políticas de Donald Trump." Working Paper 9/2019. Madrid: Real Instituto Elcano.

Mesa-Lago, Carmelo, and Pavel Vidal Alejandro. 2010. "The Impact of the Global Crisis on Cuba's Economy and Social Welfare." *Journal of Latin American Studies* 42(4): 689–717.

Meyer, Carrie A. 2020. "Tourism and Economic Growth in the Dominican Republic: 1985–2018." Working Paper No. 20-09, Department of Economics, George Mason University.

MICONS (Ministerio de la Construcción). 1980. *La Arquitectura de Hoteles en la Revolución Cubana.* Havana: MICONS.

Miller, Mark M., and Tony L. Henthorne. 1997. *Investment in the New Cuban Tourist Industry: A Guide to Entrepreneurial Opportunities*. Westport, CT: Quorum Books.
MINCEX (Ministerio del Comercio Exterior y la Inversión Extranjera). 2021. *Cuba: Cartera de Oportunidades de Inversión Extranjera 2021–2022*. Havana: MINCEX.
MINTUR (Ministerio de Turismo). 2018. *Oportunidades de Negocios para la Inversión Extranjera en el Sector del Turismo: Cuba*. Havana: MINTUR.
MINTUR (Ministerio de Turismo). 2017. *Oportunidades de Negocios para la Inversión Extranjera en el Sector del Turismo: Cuba*. Havana: MINTUR.
MINTUR (Ministerio de Turismo). 2016. *Oportunidades de Negocios para la Inversión Extranjera en el Sector del Turismo: Cuba*. Havana: MINTUR.
MINTUR (Ministerio de Turismo). 2015a. *Oportunidades de Negocios para la Inversión Extranjera en el Sector del Turismo: Cuba*. Havana: MINTUR.
MINTUR (Ministerio de Turismo). 2015b. *Actualización Sobre el Perfeccionamiento del Ministerio de Turismo*. Havana: MINTUR.
Monreal, Pedro. 2020. "Recuperando el Turismo en Medio de la Mayor Crisis Económica Global Desde 1929: ¿Un Viaje Sin 'Mapas'?" *El Estado Como Tal (Blog)*, May 23.
Mooney, Henry, and Maria Alejandra Zegarra. 2020. "Extreme Outlier: The Pandemic's Unprecedented Shock to Tourism in Latin America and the Caribbean." Policy Brief No. IDB-PB–339, Inter-American Development Bank (IDB), June.
Morales, Emilio. 2024a. "El Sistema Cubano Colapsó: A 71 Años del Asalto al Cuartel Moncada." *Cuba Siglo 21*, N. 23, July, https://cubasiglo21.com/el-sistema-cubano-colapso-a-71-anos-del-asalto-al-cuartel-moncada/.
Morales, Emilio. 2024b. "Cuba: Crece la Emigración, pero Se Hunden las Remesas." *Cuba Siglo 21*, N. 15, January, https://cubasiglo21.com/cuba-crece-la-emigracion-pero-se-hunden-las-remesas/
Morales, Emilio. 2023. "Remesas: GAESA Socava la Prosperidad de los Cubanos." *Cuba Siglo 21*, N.11, September, https://cubasiglo21.com/remesas-gaesa-socava-la-prosperidad-de-los-cubanos/.
Morales, Emilio. 2020. "COVID-19 Hits the Remittance Market Hard in Latin America." *The Havana Consulting Group*, March 31, http://www.thehavanaconsultinggroup.com/en/Articles/Article/80.
Morales, Emilio. 2019. "Remittances, an Investment Route for Cubans?" *The Havana Consulting Group*, September 27, http://www.thehavanaconsultinggroup.com/en/Articles/Article/69.
Morales, Emilio. 2016. "Cuba: The Fastest Growing Remittances Market in Latin America." *Havana Consulting Group*, June 23, http://www.thehavanaconsultinggroup.com/en-US/Articles/Article/20.
Morales, María Eugenia, Sócrates Barinas, Sandy Ramírez, and Elia Martinez. 2021. "Transforming the Tourism Sector in the Dominican Republic: From Mass-Market to a Resilient System," UNDP (United Nations Development Programme) Strategic

Innovation, March 16, https://medium.com/@undp.innovation/transforming-the-tourism-sector-in-the-dominican-republic-from-mass-market-to-a-resilient-system-faf072234c11.

Morley, Morris. 1984. "The United States and the Global Economic Blockade of Cuba: A Study in Political Pressures on America's Allies." *Canadian Journal of Political Science* 17(1): 25–48.

Morris, Emily. 2024. "Economic Crisis in Cuba: Government Missteps and Tightening US Sanctions Are To Blame." *The Conversation*, January 24, https://theconversation.com/economic-crisis-in-cuba-government-missteps-and-tightening-us-sanctions-are-to-blame-220985.

Morris, Emily. 2008. "Cuba's New Relationship with Foreign Capital: Economic Policy-Making Since 1990." *Journal of Latin American Studies* 40(4): 769–92.

Murphy, M. M., S. M. Jeyaseelan, C. Howitt, N. Greaves, H. Harewood, K. R. Quimby, N. Sobers, R. C. Landis, K. D. Rocke, and I. R. Hambleton. 2020. "COVID-19 Containment in the Caribbean: The Experience of Small Island Developing States." *Research in Globalization* 2, 100019.

Neiberg, John E., Shachak Pe'eri, Susan L. Slocum, Matthew Rice, Maction Komwa, and Donglian Sun. 2021. "Planning and Preparation for Cruising Infrastructure: Cuba as a Case Study." *Sustainability* 13(5), https://www.mdpi.com/2071-1050/13/5/2951.

Nova González, Armando. 2022. "Nuevo Modelo de Gestión Económico-Productiva y Sus Implicaciones." In *Agricultura en Cuba: Entre Retos y Transformaciones*, edited by Armando Nova González, 564–72. Havana: Editorial Caminos.

OAS (Organization of American States). 1963. *The Pan American Highway System*. Washington D.C.: OAS.

OECD (Organization for Economic Co-operation and Development). 2020a. *Production Transformation Policy Review of the Dominican Republic: Preserving Growth, Achieving Resilience*. Paris: OECD.

OECD (Organization for Economic Co-operation and Development). 2020b. *COVID-19 in Latin America and the Caribbean: An Overview of Government Responses to the Crisis*. Paris: OECD.

ONEI (Oficina Nacional de Estadística e Información). 2024a. *Anuario Estadístico de Cuba 2023*. Havana: ONEI

ONEI (Oficina Nacional de Estadística e Información). 2024b. *Turismo: Indicadores Seleccionados Enero-Diciembre 2023*. Havana: ONEI.

ONEI (Oficina Nacional de Estadística e Información). 2023a. *Anuario Estadístico de Cuba 2022*. Havana: ONEI.

ONEI (Oficina Nacional de Estadística e Información). 2023b. *Turismo Nacional e Internacional: Indicadores Seleccionados Enero-Diciembre 2022*. Havana: ONEI.

ONEI (Oficina Nacional de Estadística e Información). 2022a. *Anuario Estadístico de Cuba 2021*. Havana: ONEI.

ONEI (Oficina Nacional de Estadística e Información). 2022b. *Turismo Nacional e Internacional: Indicadores Seleccionados Enero-Diciembre 2021*. Havana: ONEI.
ONEI (Oficina Nacional de Estadística e Información). 2021a. *Anuario Estadístico de Cuba 2020*. Havana: ONEI.
ONEI (Oficina Nacional de Estadística e Información). 2021b. *Turismo: Llegadas de Visitantes Internacionales Enero-Octubre 2020*. Havana: ONEI.
ONEI (Oficina Nacional de Estadística e Información). 2020a. *Anuario Estadístico de Cuba 2019*. Havana: ONEI.
ONEI (Oficina Nacional de Estadística e Información). 2020b. *Turismo Internacional: Indicadores Seleccionados Enero-Diciembre de 2019*. Havana: ONEI.
ONEI (Oficina Nacional de Estadística e Información). 2020c. *Turismo Nacional: Indicadores Seleccionados Enero-Diciembre de 2019*. Havana: ONEI.
ONEI (Oficina Nacional de Estadística e Información). 2020d. *Turismo: Llegadas de Visitantes Internacionales Enero-Febrero 2020*. Havana: ONEI.
ONEI (Oficina Nacional de Estadística e Información). 2019a. *Turismo Internacional: Indicadores Seleccionados Enero-Diciembre de 2018*. Havana: ONEI.
ONEI (Oficina Nacional de Estadística e Información). 2019b. *Anuario Estadístico de Cuba 2018*. Havana: ONEI.
ONEI (Oficina Nacional de Estadística e Información). 2018a. *Anuario Estadístico de Cuba 2017*. Havana: ONEI.
ONEI (Oficina Nacional de Estadística e Información). 2018b. *Turismo Internacional: Indicadores Seleccionados Enero-Diciembre de 2018*. Havana: ONEI.
ONEI (Oficina Nacional de Estadística e Información). 2017. *Turismo Internacional: Indicadores Seleccionados Enero-Diciembre de 2018*. Havana: ONEI.
ONEI (Oficina Nacional de Estadística e Información). 2016a. *Anuario Estadístico de Cuba 2015*. Havana: ONEI.
ONEI (Oficina Nacional de Estadística e Información). 2016b. *Turismo Internacional: Indicadores Seleccionados Enero-Diciembre 2015*. Havana: ONEI.
ONEI (Oficina Nacional de Estadística e Información). 2015a. *Anuario Estadístico de Cuba 2014*. Havana: ONEI.
ONEI (Oficina Nacional de Estadística e Información). 2015b. *Turismo Internacional: Indicadores Seleccionados Enero-Diciembre de 2014*. Havana: ONEI.
ONEI (Oficina Nacional de Estadística e Información). 2015c. *Turismo Nacional: Indicadores Seleccionados Enero-Diciembre de 2014*. Havana: ONEI.
ONEI (Oficina Nacional de Estadística e Información). 2015d. *Anuario Estadístico de La Habana 2014*. Havana: ONEI.
ONEI (Oficina Nacional de Estadística e Información). 2014. *Anuario Estadístico de La Habana 2013*. Havana: ONEI.
ONEI (Oficina Nacional de Estadística e Información). 2013. *Turismo Internacional: Indicadores Seleccionados Enero-Diciembre de 2012*. Havana: ONEI

ONEI (Oficina Nacional de Estadística e Información). 2012. *Anuario Estadístico de La Habana 2011*. Havana: ONEI.

ONEI (Oficina Nacional de Estadística e Información). 2011a. *Anuario Estadístico de Cuba 2010*. Havana: ONEI.

ONEI (Oficina Nacional de Estadística e Información). 2011b. *Turismo: Principales Indicadores Diciembre 2010*. Havana: ONEI.

ONEI (Oficina Nacional de Estadística e Información). 2009a. *Anuario Estadístico de Cuba 2008*. Havana: ONEI.

ONEI (Oficina Nacional de Estadística e Información). 2009b. *Anuario Estadístico de Ciudad de La Habana 2008*. Havana: ONEI.

ONEI (Oficina Nacional de Estadística e Información). 2009c. *Turismo: Principales Indicadores Diciembre 2008*. Havana: ONEI.

ONEI (Oficina Nacional de Estadística e Información). 2008. *Principales Indicadores del Turismo Diciembre 2007*. Havana: ONEI.

ONEI (Oficina Nacional de Estadística e Información). 2006. *Anuario Estadístico de Cuba 2005*. Havana: ONEI.

ONEI (Oficina Nacional de Estadística e Información). 2005. *Anuario Estadístico de Cuba 2004*. Havana: ONEI.

ONEI (Oficina Nacional de Estadística e Información). 2004. *Anuario Estadístico de Cuba 2003*. Havana: ONEI.

ONEI (Oficina Nacional de Estadística e Información). 2003. *Anuario Estadístico de Cuba 2002*. Havana: ONEI.

ONEI (Oficina Nacional de Estadística e Información). 2000. *Anuario Estadístico de Cuba 1999*. Havana: ONEI.

ONEI (Oficina Nacional de Estadística e Información). 1998. *Anuario Estadístico de Cuba 1996*. Havana: ONEI.

Parra, Juan Carlos, Samuel Pienknagura, and Tahseen Sayed. 2018. *Dominican Republic: Systematic Country Diagnostic*. Washington, D.C.: World Bank.

Perelló Cabrera, José Luis. 2020. "El Turismo en Cuba Ante un Contexto Global Adverso: Los Mercados Emergentes." *Horizonte Cubano*, Columbia Law School, March 23, https://horizontecubano.law.columbia.edu/news/el-turismo-en-cuba-ante-un-contexto-global-adverso-los-mercados-emergentes.

Perelló Cabrera, José Luis. 2019. "El Turismo en Cuba: Cambios y Tendencias." *Horizonte Cubano*, Columbia Law School, March 9, https://horizontecubano.law.columbia.edu/news/el-turismo-en-cuba-cambios-y-tendencias.

Perelló Cabrera, José Luis. 2017. "Turismo de Cruceros en el Caribe." Working Paper DT-02-08/17, Center for Responsible Travel (CREST), February.

Perelló Cabrera, José Luis. 2016. "Desarrollo del Turismo en Cuba: Tendencias, Políticas e Impactos Futuros." In *Cuba in Transition* 26, 229–33. Washington D.C.: Association for the Study of the Cuban Economy.

Perelló Cabrera, José Luis. 2015a. "El Sector Privado y el Turismo en Cuba Antes un Escenario de Relaciones con Estados Unidos." In *Cuba in Transition* 25, 51–8. Washington D.C.: Association for the Study of the Cuban Economy.

Perelló Cabrera, José Luis. 2015b. "Actualización del Turismo en Cuba: Desempeño del Turismo en 2014." Working Paper, Tourism Studies Department, University of Havana.

Perelló Cabrera, José Luis. 2015c. "Turismo y Desarrollo Territorial en el Modelo Económico y Social de Cuba." *TURyDES* 8(18), online publication, https://dialnet.unirioja.es/descarga/articulo/8068905.pdf.

Perelló Cabrera, José Luis. 2011a. "Antecedentes Históricos de la Sunshine Circle Route (Ruta del Golfo)." Working Paper, Tourism Studies Department, University of Havana.

Perelló Cabrera, José Luis. 2011b. "Dinámicas de las Migraciones y los Flujos Turísticos en el Espacio Caribeño." *Revista Ciencia y Tecnología* 2: 20–30.

Perelló Cabrera, José Luis, and Rafael Betancourt. 2018. "Turismo de Cruceros en Cuba." In *Por el Mar de las Antillas . . . 50 Años de Turismo de Cruceros en el Caribe*, edited by Martha Honey, José Luis Perelló Cabrera, Jannelle Wilkins, and Rafael Betancourt, 60–121. Havana: Ediciones Temas.

Pérez, Lorenzo L. 2009. "The Impact of the Global Financial and Economic Crisis on Cuba." In *Cuba in Transition* 19, 116–23. Washington, D.C.: Association for the Study of the Cuban Economy.

Pérez Jr., Louis A. 2008. *On Becoming Cuban: Identity, Nationality, and Culture*. Chapel Hill: University of North Carolina Press.

Pérez Jr., Louis A. 2001. *Winds of Change: Hurricanes and the Transformation of Nineteenth-Century Cuba*. Chapel Hill: University of North Carolina Press.

Pérez Jr., Louis A. 1986. *Cuba Under the Platt Amendment: 1902–1934*. Pittsburgh: University of Pittsburgh Press.

Pérez-López, Jorge F. 2017. "Cuba's Never-Ending External Sector Crisis. In *Cuba in Transition* 27, 30–44. Washington D.C.: Association for the Study of the Cuban Economy.

Pérez-López, Jorge F. 2011a. "Cuba's External Sector and the VI Party Congress. In *Cuba in Transition* 21, 437–50. Washington D.C.: Association for the Study of the Cuban Economy.

Pérez-López, Jorge F. 2011b. "The Global Financial Crisis and Cuba's External Sector." In *The Cuban Economy: Recent Trends,* edited by José Raúl Perales, 31–49. Washington, DC: Woodrow Wilson Center for International Scholars.

Pérez-López, Jorge F. 2004. "Foreign Investment in Cuba: An Inventory." In *Cuba in Transition* 14, 103–15. Washington D.C.: Association for the Study of the Cuban Economy.

Pérez-López, Jorge F. 1999. "Foreign Investment in Cuba in the Second Half of the 1990s." Paper presented at the international symposium "The Cuban Economy:

Problems, Policies, Perspectives," Carleton University, Ottawa, Canada, September 28–30.

Pérez-López, Jorge F. 1995. *Cuba's Second Economy: From Behind the Scenes to Center Stage*. New Brunswick: Transaction Publishers.

Pérez-López, Jorge F., and Sergio Díaz-Bríquets. 2011. "The Diaspora and Cuba's Tourism Sector." In *Cuba in Transition* 21, 314–25. Washington D.C.: Association for the Study of the Cuban Economy.

Pérez-Stable, Marifeli. 2011. *The United States and Cuba: Intimate Enemies*. New York: Routledge.

Pérez-Stable, Marifeli. 2007. "Looking Forward: Democracy in Cuba?" In *Looking Forward: Comparative Perspectives on Cuba's Transition,* edited by Marifeli Pérez-Stable, 17–46. Notre Dame: University of Notre Dame Press.

Pérez-Stable, Marifeli. 1999. *The Cuban Revolution: Origins, Course, and Legacy*. New York: Oxford University Press.

Pérez Villanueva, Omar Everleny. 2013. "Análisis de la Evolución Reciente de la Economía Cubana." In *Miradas a la Economía Cubana: Entre la Eficiencia Económica y la Equidad Social,* edited by Omar Everleny Pérez Villanueva and Ricardo Torres Pérez, 17–27. Havana: Editorial Caminos.

Pérez Villanueva, Omar Everleny. 2012. "Foreign Direct Investment in China, Vietnam, and Cuba: Pertinent Experiences for Cuba." In *Cuban Economic and Social Development: Policy Reforms and Challenges in the 21st Century,* edited by Jorge I. Domínguez, Omar Everleny Pérez Villanueva, Mayra Espina Prieto, and Lorena Barberia, 193–225. Cambridge: Harvard University Press.

Pérez Villanueva, Omar Everleny. 2004. "The Role of Foreign Direct Investment in Economic Development: The Cuban Experience." In *The Cuban Economy at the Start of the Twenty-First Century,* edited by Jorge I. Domínguez, Omar Everleny Pérez Villanueva, and Lorena Barberia, 161–97. Cambridge: Harvard University Press.

Pérez Villanueva, Omar Everleny. 1999. *La Inversión Extranjera Directa en Cuba: Peculiaridades*. Havana: Centro de Estudios de la Economía Cubana.

Pérez Villanueva, Omar Everleny, and José Luis Perelló Cabrera. 2020. "El Papel de los Visitantes de Estados Unidos en la Economía Cubana: Historia y Realidad." In *90 Millas: Relaciones Económicas Cuba-Estados Unidos, 1898–2020*, edited by Antonio Santamaría García and José Manuel Azcona Pastor, 351–69. Madrid, Spain: Dykinson.

Peters, Philip. 2012. *Cuba's Entrepreneurs: Foundation of a New Private Sector*. Arlington, VA: Lexington Institute.

Peters, Philip. 2002. *International Tourism: The New Engine of the Cuban Economy.* Arlington, VA: Lexington Institute.

Pierre-Charles, Gérard. 1978. *Génesis de la Revolución Cubana*. Mexico D.F.: Siglo Veintiuno Editores.

Planas, Juan M. 1943. "La Carretera Central, Ramal Principal de la Carretera Panamericana." *Revista de la Sociedad Cubana de Ingenieros* 38–9: 303–18.

Pons Pérez, Saira. 2013. "Hacia una Nueva Fiscalidad en Cuba." In *Miradas a la Economía Cubana: Entre la Eficiencia Económica y la Equidad Social*, edited by Omar Everleny Pérez Villanueva and Ricardo Torres Pérez, 43–55. Havana: Editorial Caminos.

Pulido Catasús, Jesús. 2023. "Reflexiones Acerca de la Situación Económica Actual de Nuestro País." *Cuba y la Economía (Blog)*, January 2.

Quintana, Rogelio, Manuel Figuerola, Mariano Chirivella, Damarys Lima, Miguel Alejandro Figueras, and Alfredo García. 2004. *Efectos y Futuro del Turismo en la Economía Cubana*. Havana: Instituto Nacional de Investigaciones Económicas.

Ram, Justin, Damien Reeves, and Ronald James. 2018. "Air Transport Competitiveness and Connectivity in the Caribbean." CDB Working Paper NO: CDB/WP/18/02, Caribbean Development Bank.

Randall, Ruby. 2006. "Eastern Caribbean Tourism: Developments and Outlook." In *The Caribbean: From Vulnerability to Sustained Growth*, edited by David Robinson, Paul Cashin, and Ratna Sahay, 285–306. Washington D.C.: International Monetary Fund.

Reimondo Barrios, Yanet Maria, and Clóvis Reis. 2019. "Authentic Cuba: Analysis of Tourism Advertising, 2002–2016". *Brazilian Journal of Tourism Research* 13(1): 125–40.

Ritter, Archibald R. M. 1988. "Cuba's Convertible Currency Debt Problem." *CEPAL Review* 36: 117–40.

Ritter, Archibald R. M., and Ted A. Henken. 2015. *Entrepreneurial Cuba: The Changing Policy Landscape*. Boulder, CO: First Forum Press.

Robyn, Dorothy, James D. Reitzes, and Bryan Church. 2002. *The Impact on the U.S. Economy of Lifting Restrictions on Travel to Cuba*. Washington, D.C.: Center for International Policy.

Rodríguez, José Luis. 2024. "La Economía Cubana en 2023 y Perspectivas para 2024." *Cubadebate*, February 17, http://www.cubadebate.cu/opinion/2024/02/17/la-economia-cubana-en-2023-y-perspectivas-para-2024-i/.

Rodríguez, José Luis. 2022. "Cuba: Factores de la Compleja Coyuntura Económica en el Primer Semestre del 2022." *Cubadebate*, August 26, http://www.cubadebate.cu/opinion/2022/08/26/cuba-factores-de-la-compleja-coyuntura-economica-en-el-primer-semestre-del-2022-i/.

Rodríguez, José Luis. 1993. "Cuba en la Economía Internacional: Nuevos Mercados y Desafíos de los Años Noventa." *Estudios Internacionales* 26(103): 415–53.

Rodríguez Domínguez, Mercedes. 2009. *Turismo y Republica*. Havana: Ediciones Balcón.

Rodríguez García, Victor Alejandro. 2009. "Estructura de Capital y Formas de Participación Extranjera en Cuba: Un Modelo para Medir la Rentabilidad País." Unpublished master's thesis, University of Havana.

Rojas, Rafael. 2020. "The New Cuban Executive Branch: Constitutional Changes in the Power Structure." Briefings on Cuba, Florida International University (FIU), November.

Romero, Carlos A. 2010. "La Cooperación 'Sur-Sur' entre Venezuela y Cuba." In *Cooperación Sur-Sur: Un Desafío al Sistema de la Ayuda*, 127–35. Medellín, Colombia: Asociación Latinoamericana de Organizaciones de Promoción.

Romeu, Rafael. 2008. "Vacation Over: Implications for the Caribbean of Opening U.S.-Cuba Tourism." Working Paper WP/08/162, Washington D.C.: International Monetary Fund.

Ross, James E., and María Antonia Fernández Mayo. 2002. "Overview of Cuba's Dollar Food Market: An Exploration of the Purchasing and Distribution System." In *Cuba in Transition* 12, 276–84. Washington D.C.: Association for the Study of the Cuban Economy

Rutty, Michelle, and Robert B. Richardson. 2019. "Tourism Research in Cuba: Gaps in Knowledge and Challenges for Sustainable Tourism." *Sustainability* 11(12): 1–7.

Sáenz Rovner, Eduardo. 2008. *The Cuban Connection: Drug Trafficking, Smuggling, and Gambling in Cuba from the 1920s to the Revolution*. Chapel Hill: University of North Carolina Press.

Salinas, Eduardo, Salinas Eros, and Lluis Mundet. 2019. "El Turismo en Cuba: Desarrollo, Retos y Perspectivas." *Rosa dos Ventos – Turismo y Hospitalidade*, 11(1): 23–49.

Salinas, Eros. 1998. "Turismo en Cuba: Desarrollo, Retos y Perspectivas." *Estudios y Perspectivas en Turismo* 7(2): 151–64.

Salinas, Eros, Lluis Mundet, and Eduardo Salinas. 2018. "Historical Evolution and Spatial Development of Tourism in Cuba, 1919–2017: What Is Next?" *Tourism Planning & Development* 15(3): 216–38.

Sánchez Baena, Juan José. 2009. *El Terror de los Tiranos: La Imprenta en la Centuria que Cambió Cuba (1763–1868)*. Castellón de la Plana: Jaume I University.

Sánchez Egozcue, Jorge Mario. 2011. *Crecimiento Económico y Sector Externo en Cuba*. Havana: Centro de Estudios de la Economía Cubana.

Sánchez Egozcue, Jorge Mario, and Juan Triana Cordoví. 2008. "Un Panorama Actual de la Economía Cubana, las Transformaciones en Curso y sus Retos Perspectivos." Working Paper 31/2008, Real Instituto Elcano, Madrid.

Santamarina, Juan Carlos. 2011. "The Cuba Company and Eastern Cuba's Economic Development, 1900–1959." *Essays in Economic and Business History* 29: 75–90.

Santamarina, Juan Carlos. 2000. "The Cuba Company and the Expansion of American Business in Cuba, 1898–1915." *The Business History Review* 74(1): 41–83.

Santamarina, Victor. 1948. "La Industria Turística Cubana." Paper presented at Conferencia para el Progreso de la Economía Nacional, Havana, Cuba, November 15–20.

Santamarina, Victor. 1944. *El Turismo, Industria Nacional*. Havana: Patronato del Balneario de san Diego de los Baños.

Sarduy González, Mariuska, Saira Pons Pérez, and Maday Traba Muñoz. 2015. "¿Por qué Evaden Impuestos los Trabajadores por Cuenta Propia?" In *Miradas a la Economía Cubana: Análisis del Sector No Estatal,* edited by Omar Everleny Pérez Villanueva and Ricardo Torres Pérez, 95–102. Havana: Editorial Caminos.

Scarpaci, Joseph L., Roberto Segre, and Mario Coyula. 2002. *Havana: Two Faces of the Antillean Metropolis.* Chapel Hill: University of North Carolina Press.

Schwalb, Alvaro, Eleonora Armyra, Melissa Méndez-Aranda, and César Ugarte-Gil. 2022. "COVID-19 in Latin America and the Caribbean: Two Years of the Pandemic." *Journal of Internal Medicine* 292(3): 409–27. https://doi.org/10.1111/joim.13499.

Schwartz, Rosalie. 1997. *Pleasure Island: Tourism and Temptation in Cuba.* Lincoln: University of Nebraska Press.

SECTUR (Secretaría de Turismo). 2015. *Resultados de la Actividad Turística: México, Año 2014.* Mexico City: SECTUR.

Sharpley, Richard, and Martin Knight. 2009. "Tourism and the State in Cuba: From the Past to the Future." *International Journal of Tourism Research* 11(3): 241–54.

Sierra, Alexander. 2017. "Desarrollo del Turismo en Cuba," Presentation at Conference "Iniciativa de Gestión del Conocimiento para la Sostenibilidad del Desarrollo en Cuba." Havana, Hotel Comodoro, February13.

Silveira, Yahilina, Ricardo Silveira Martins, and Graciela Castellanos. 2010. "Impacto Social y Económico de la Industria Turística Cubana." *TURyDes* 3(8), online publication, http://www.eumed.net/rev/turydes/08/index.htm.

Simoni, Valerio. 2016. *Tourism and Informal Encounters in Cuba.* New York: Berghahn Books.

Skwiot, Christine. 2010. *The Purpose of Paradise: U.S. Tourism and Empire in Cuba and Hawai'i.* Philadelphia: University of Pennsylvania Press.

Smith, Wayne S. 1988. *The Closest of Enemies: A Personal and Diplomatic History of the Castro Years.* New York: W. W. Norton.

Spadoni, Paolo. 2023. "The External Sector of Cuba's Economy: Performance and Challenges." *FIU Law Review* 17(3): 635–76.

Spadoni, Paolo. 2021. "Cuba's Struggling External Sector: Internal Challenges and Outside Factors." In *Cuban International Relations at 60: Reflections on Global Connections*, edited by Mervyn J. Bain and Chris Walker, 223–59. Lanham, MD: Lexington Books.

Spadoni, Paolo. 2017a. "Cuba's External Sector: Conditions, Challenges, and Opportunities." *Florida Journal of International Law* 29(1): 99–133.

Spadoni, Paolo. 2017b. "U.S.-Cuba Business Relations under the Obama Administration and Prospects under the Trump Administration." In *Cuba in Transition* 17, 235–52. Washington D.C.: Association for the Study of the Cuban Economy.

Spadoni, Paolo. 2014. *Cuba's Socialist Economy Today: Navigating Challenges and Change.* Boulder, CO: Lynne Rienner.

Spadoni, Paolo. 2012. "Cuban Economic Policies, 1990–2010: Achievements and Shortcomings." In *The Oxford Handbook of Latin American Political Economy*, edited by Javier Santiso and Jeff Dayton-Johnson, 168–90. New York: Oxford University Press.

Spadoni, Paolo. 2010. *Failed Sanctions: Why the U.S. Embargo Against Cuba Could Never Work*. Gainesville, FL: University Press of Florida.

Spadoni, Paolo. 2001. "The Impact of the Helms-Burton Legislation on Foreign Investment in Cuba." In *Cuba in Transition* 11, 18–36. Washington D.C.: Association for the Study of the Cuban Economy.

Spadoni, Paolo, and Julia Sagebien. 2013. "Will They Still Love Us Tomorrow? Canada-Cuba Business Relations and the End of the US Embargo." *Thunderbird International Business Review* 55(1): 77–93.

Statistics Canada. 2010. *Canada Year Book 2010*. Ottawa: Statistics Canada.

STR. 2019. "Hotel Performance in the Caribbean Region." June, https://caribbeanhotelandtourism.com/wp-content/uploads/2019/07/STR-Caribbean-Hotel-Performance-Report-June-2019.pdf

Sullivan, Mark P. 2022. *Biden Administration's Cuba Policy Changes*. Washington, D.C.: Congressional Research Service.

Sullivan, Mark P. 2012. *Cuba: U.S. Restrictions on Travel and Remittances*. Washington, D.C.: Congressional Research Service.

Sullivan, Mark P. 2009. *Cuba: U.S. Restrictions on Travel and Remittances*. Washington, D.C.: Congressional Research Service.

Terry, Philip T. 1929. *Terry's Guide to Cuba*. Boston: Houghton Mifflin.

THCG (The Havana Consulting Group). 2017. *THCG Business Report*. Miami: THCG, February (No. 1).

Thomas, Desmond. 2015. "The Caribbean Tourism Industry in the 21st Century: An Assessment." KGLACC Working Paper No.3/2015, Kimberly Green Latin American and Caribbean Center, Florida International University.

Thomas, John H., Miranda Kitterlin-Lynch, and Daymaris Lorenzo Del Valle. 2015. "Cuba's Future Hospitality and Tourism Business: Opportunities and Obstacles." *Cornell Hospitality Report* 15(1): 4–16.

Torres-Cuevas, Eduardo, and Oscar Loyola Vega. 2001. *Historia de Cuba 1492–1998: Formación y Liberación de la Nación*. Havana: Editorial Pueblo y Educación.

Torres Pérez, Ricardo. 2015. "Exportaciones y Desarrollo Económico en Cuba: Entre Reformas y el 17D." In *Cuba en un Nuevo Escenario: Perspectivas de Desarrollo Económico*, edited by Ricardo Torres, Omar Everleny Pérez, Fulvio Castellacci, and Claes Brundenius, 161–77. Havana: Publicaciones Acuario.

Travieso-Diaz, Matias F. 1997. *The Laws and Legal System of a Free-Market Cuba: A Prospectus for Business*. Westport, CT: Quorum Books.

Travieso-Diaz, Matias F., and Charles P. Trumbull IV. 2002. "Foreign Investment in Cuba: Prospects and Perils." In *Cuba in Transition* 12, 179–97. Washington, DC: Association for the Study of the Cuban Economy.

Trelles, Carlos M. 1907. *Ensayo de Bibliografía Cubana de los Siglos XVII y XVIII*. Matanzas: El Escritorio.

Triana Cordoví, Juan. 2019. "Menos Turistas Vs Más Ingresos por Turista." OnCuba (oncubanews.com), August 26.

Trumbull, Charles. 2001. "Prostitution and Sex Tourism in Cuba." In *Cuba in Transition* 11, 356–71. Washington D.C.: Association for the Study of the Cuban Economy

Umakanthan, Srikanth, Anuradha Chauhan, Madan Mohan Gupta, Pradeep Kumar Sahu, Maryann M Bukelo, and Vijay Kumar Chattu. 2021. "COVID-19 Pandemic Containment in the Caribbean Region: A Review of Case-Management and Public Health Strategies." *AIMS Public Health* 8(4): 665–81.

UNCTAD (United Nations Conference on Trade and Development). 2017. *The Role of the Services Economy and Trade in Structural Transformation and Inclusive Development*. Geneva: UNCTAD.

UNDP (United Nations Development Programme). 2017. *Regional Overview: Impact of Hurricanes Irma and Maria*. New York: UNDP.

United Nations. 2010a. *Tourism Satellite Account: Recommended Methodological Framework 2008*. Luxembourg: United Nations.

United Nations. 2010b. *International Recommendations for Tourism Statistics 2008*. New York: United Nations.

United Nations. 2006. "Necessity of Ending the Economic, Commercial, and Financial Embargo Imposed by the United States of America against Cuba." Report of the Secretary-General, A/61/132.

UNWTO (United Nations World Tourism Organization). 2024. *World Tourism Barometer*. Madrid: UNWTO, Volume 22, Issue 2, May.

UNWTO (United Nations World Tourism Organization). 2023a. *World Tourism Barometer*. Madrid: UNWTO, Volume 21, Issue 4, November.

UNWTO (United Nations World Tourism Organization). 2023b. *International Tourism Highlights: The Impact of COVID-19 on Tourism (2020–2022)*. Madrid: UNWTO.

UNWTO (United Nations World Tourism Organization). 2023c. *Compendium of Tourism Statistics, Data 2017–2021, 2023 Edition*. Madrid: UNWTO.

UNWTO (United Nations World Tourism Organization). 2022a. *Tourism Doing Business, Invirtiendo en República Dominicana*. Madrid: UNWTO.

UNWTO (United Nations World Tourism Organization). 2022b. *Yearbook of Tourism Statistics, Data 2016-2020, 2022 Edition*. Madrid: UNWTO.

UNWTO (United Nations World Tourism Organization). 2022c. *World Tourism Barometer*. Madrid: UNWTO.

UNWTO (United Nations World Tourism Organization). 2021a. *World Tourism Barometer*. Madrid: UNWTO.

UNWTO (United Nations World Tourism Organization). 2021b. *International Tourism Highlights, 2020 Edition*. Madrid: UNWTO.

UNWTO (United Nations World Tourism Organization). 2020. *Yearbook of Tourism Statistics, Data 2014–2018, 2020 Edition*. Madrid: UNWTO.
UNWTO (United Nations World Tourism Organization). 2019. *International Tourism Highlights, 2019 Edition*. Madrid: UNWTO.
UNWTO (United Nations World Tourism Organization). 2018. *Tourism Highlights, 2018 Edition*. Madrid: UNWTO.
UNWTO (United Nations World Tourism Organization). 2017. *Tourism Highlights, 2017 Edition*. Madrid: UNWTO.
UNWTO (United Nations World Tourism Organization). 2016. *Tourism Highlights, 2016 Edition*. Madrid: UNWTO.
UNWTO (United Nations World Tourism Organization). 2015. *Tourism Highlights, 2015 Edition*. Madrid: UNWTO.
UNWTO (United Nations World Tourism Organization). 2011. *Yearbook of Tourism Statistics*. Madrid: UNWTO.
UNWTO (United Nations World Tourism Organization). 2010. *Yearbook of Tourism Statistics*. Madrid: UNWTO.
UNWTO (United Nations World Tourism Organization). 2009. *Yearbook of Tourism Statistics*. Madrid: UNWTO.
UNWTO (United Nations World Tourism Organization). 2006. *Yearbook of Tourism Statistics*. Madrid: UNWTO.
UNWTO (United Nations World Tourism Organization). 2004. *Yearbook of Tourism Statistics*. Madrid: UNWTO.
UNWTO (United Nations World Tourism Organization). 2001. *Yearbook of Tourism Statistics*. Madrid: UNWTO.
UNWTO (United Nations World Tourism Organization). 2000. *Tourism Highlights, 2000 Edition*. Madrid: UNWTO.
UNWTO (United Nations World Tourism Organization). 1997. *Tourism Market Trends World, 1985–1996*. Madrid: UNWTO.
Uriarte, Miren. 2008. "Social Impact of the Economic Measures." In *A Contemporary Cuba Reader: Reinventing the Revolution,* edited by Philip Brenner, Marguerite Rose Jimenez, John M. Kirk, and William M. Leogrande, 285–91. Lanham, MD: Rowman & Littlefield.
U.S. Civil Aeronautics Board. 1945. *Economic Decisions of the Civil Aeronautics Board,* Vol. 3, December 1942 to June 1944. Washington, D.C.: Government Publishing Office (GPO).
USCTEC (U.S.-Cuba Trade and Economic Council). 2018. "New Cuba Sailings Continue To Move Cuba Above US$2 Billion in Marketplace Value to U.S. Companies." March 5, https://www.cubatrade.org/blog/2018/3/5/new-cuba-sailings-continue-to-move-cuba-above-us2-billion-in-marketplace-value-to-us-companies.
USCTEC (U.S.-Cuba Trade and Economic Council). 2017. "US Cruise Lines Continue to Add Capacity to Cuba, Revenue from Cuba, and Revenue to Cuba." November 4,

https://www.cubatrade.org/blog/2017/11/4/us-cruise-lines-continue-to-add-capacity-to-cuba-revenue-from-cuba-and-revenue-to-cuba.

USCTEC (U.S.-Cuba Trade and Economic Council). 2009. "Annual Report—2008 U.S. Export Statistics for Cuba." *Economic Eye on Cuba*, February.

USCTEC (U.S.-Cuba Trade and Economic Council). 2003. "2003 Commercial Highlights." http://www.cubatrade.org/.

USCTEC (U.S.-Cuba Trade and Economic Council). 1998. "1998 Commercial Highlights." http://www.cubatrade.org/.

USITC (United States International Trade Commission). 2007. *U.S. Agricultural Sales to Cuba: Certain Economic Effects of U.S. Restrictions*. Washington, D.C.: USITC.

USITC (United States International Trade Commission). 2001. *The Economic Impact of U.S. Sanctions with Respect to Cuba*. Washington, D.C.: USITC.

Valle Ojeda, Amir. 2006. *Jineteras*. Bogotá: Editorial Planeta.

Vidal Alejandro, Pavel. 2012. "Desafíos Monetarios y Financieros." In *Miradas a la Economía Cubana: El Proceso de Actualización*, edited by Pavel Vidal Alejandro and Omar Everleny Pérez Villanueva, 99–111. Havana: Editorial Caminos.

Vidal Alejandro, Pavel. 2010. *Cuban Economic Policy under the Raúl Castro Government*. Chiba, Japan: Institute of Developing Economies, Japan External Trade Organization.

Vidal Alejandro, Pavel. 2009. *El Salario Real y la Productividad en Cuba: Actualización*. Havana: Centro de Estudios de la Economía Cubana.

Vidal Alejandro, Pavel, and Omar Everleny Pérez Villanueva. 2013. *La Reforma Monetaria en Cuba Hasta el 2016: Entre Gradualidad y "Big Bang."* Washington, D.C.: Brookings Institution.

Vidal Alejandro, Pavel, and Omar Everleny Pérez Villanueva. 2012. "Apertura al Cuentapropismo y la Microempresa, una Pieza Clave del Ajuste Estructural." In *Miradas a la Economía Cubana: El Proceso de Actualización*, edited by Pavel Vidal Alejandro and Omar Everleny Pérez Villanueva, 43–52. Havana: Editorial Caminos.

Villalba Garrido, Evaristo. 1993. *Cuba y el Turismo*. Havana: Editorial de Ciencias Sociales.

Ward, Evan R. 2007. "A Means of Last Resort: The European Transformation of the Cuban Hotel Industry and the American Response, 1987–2004." In *The Business of Tourism: Place, Faith, and History*, edited by Philip Scranton and Janet F. Davidson, 213–38. Philadelphia: University of Pennsylvania Press.

Werner, Alejandro, and Trevor Alleyne. 2015. "The Caribbean at a Crossroads: Improved Prospects, Higher Risks." Powerpoint presentation, *The 2015 High-Level Caribbean Forum*, St. Kitts, September 3.

Wylie, Lana L. 2021. "Cuba's Response to COVID-19: Lessons for the Future." *Journal of Tourism Futures* 7(3): 356–63.

Woodard, Blair D. 2018. "Shifting Sands: Cuban Tourism, Politics, and the Transformation of Varadero, 1945–2016." *The Latin Americanist* 62(2): 186–212.

Wright, Irene Aloha. 1910. "New Hotels in Cuba." *Bulletin of the International Bureau of the American Republics* 30(1): 57–65.

Wright, Robert. 2009. "'Northern Ice': Jean Chretien and the Failure of Constructive Engagement in Cuba." In *Our Place in the Sun: Canada and Cuba in the Castro Era*, edited by Robert Wright and Lana Wylie, 195–222. Toronto: University of Toronto Press.

WTTC (World Travel & Tourism Council). 2023a. *World Economic Impact Report 2023.* London: WTCC.

WTTC (World Travel & Tourism Council). 2023b. *Travel & Tourism Economic Impact 2023: Caribbean.* London: WTCC.

WTTC (World Travel & Tourism Council). 2022a. *Cuba 2022 Annual Research: Key Highlights.* London: WTCC.

WTTC (World Travel & Tourism Council). 2022b. *Travel and Tourism in the Caribbean: Prospects for Growth.* London: WTTC.

WTTC (World Travel & Tourism Council). 2022c. *Travel & Tourism Economic Impact 2022: Global Trends.* London: WTTC.

WTTC (World Travel & Tourism Council). 2021. *Travel & Tourism Economic Impact 2021: Global Economic Impact and Trends 2021.* London: WTTC.

WTTC (World Travel & Tourism Council). 2019. *Cuba 2019 Annual Research: Key Highlights.* London: WTCC.

WTTC (World Travel & Tourism Council). 2018. *Travel & Tourism Economic Impact 2018: Cuba.* London: WTCC

WTTC (World Travel & Tourism Council). 2015. *Travel & Tourism Economic Impact 2015: Cuba.* London: WTTC.

WTTC (World Travel & Tourism Council). 2004. *The Caribbean: The Impact of Travel & Tourism on Jobs and the Economy.* London: WTTC.

About the Authors

Paolo Spadoni is an associate professor in the Department of Social Sciences at Augusta University. Dr. Spadoni has carried out extensive research on the Cuban economy and the economic impact of US sanctions against Cuba. He has published widely in scholarly and academic venues and is the author of *Cuba's Socialist Economy Today: Navigating Challenges and Change* (Lynne Rienner, 2014), and of *Failed Sanctions: Why the US Embargo against Cuba Could Never Work* (University Press of Florida, 2010).

José Luis Perelló Cabrera is a retired professor at the Faculty of Tourism of the University of Havana and an advisor to the Development Directorate of the Ministry of Tourism of Cuba. Dr. Perelló Cabrera has taught numerous postgraduate courses in Cuba and abroad and has participated as a visiting professor at universities in the United States and Central and South America. He has published several books and various articles in national and foreign specialized journals.

Index

Note: 'n' refers to Footnote number

www.ingramcontent.com/pod-product-compliance
Ingram Content Group UK Ltd.
Pitfield, Milton Keynes, MK11 3LW, UK
UKHW020220290926
15470UKWH00001B/303

9 781498 590433